Binding and Dissociation Kinetics for Different Biosensor Applications Using Fractals

Binding and Dissociation Kinetics for Different Biosensor Applications Using Fractals

Ajit Sadana

Chemical Engineering Department
Composite Structures and Nanoengineering Research
University of Mississippi
University, MS, USA

ELSEVIER

Amsterdam • Boston • Heidelberg • London • New York • Oxford
Paris • San Diego • San Francisco • Singapore • Sydney • Tokyo

Elsevier
Radarweg 29, PO Box 211, 1000 AE Amsterdam, The Netherlands
The Boulevard, Langford Lane, Kidlington, Oxford OX5 1GB, UK

First edition 2006

Library of Congress Cataloging-in-Publication Data
A catalog record for this book is available from the Library of Congress

British Library Cataloguing in Publication Data
A catalogue record for this book is available from the British Library

ISBN-13: 978-0-444-52784-4
ISBN-10: 0-444-52784-2

Printed and bound in The Netherlands

06 07 08 09 10 10 9 8 7 6 5 4 3 2 1

This book is dedicated to my daughters Neeti and Richa

Foreword

Binding and Dissociation Kinetics for Different Biosensor Applications Using Fractals is certainly a book to revolutionize the way engineers design products, and physicians and patients view medicine. From an engineering perspective, the idea that a biosensor, essentially a microprocessor, can be created at the level of nanotechnology to be applied for day-to-day use is truly challenging. The concept of using these microprocessors to rapidly and accurately detect and quantify disease is remarkable.

Biosensors involve recognition and transduction elements allowing biological events to be converted to chemical, electrical, magnetic, mechanical, optical, and even thermal signals. These signals are then recognized by the desired elements such as enzymes, antibodies, DNA, etc. and are amplified, processed, and displayed. This book allows those of us new to this concept of biosensors to learn the basics. Those who are familiar with this sophisticated technology can find applications to further advance this evolving field. Unique to this book is the final chapter involving market size and economics. It is difficult to predict the way investors will view new and exciting technology, and this book presents information in an easy and factual manner that all can appreciate.

In conclusion, it is my privilege and honor that Dr. Ajit Sadana, the author of this book and my father, asked me to write this foreword. I am humbled in this experience as I stop to take the time to appreciate all that my Dad and his colleagues throughout the world spend their lives doing. They seek valuable pieces of information that patiently changes the way that my colleagues and I practice quality medicine. I take great pride in recommending this book to all those who want to understand and contribute to the relationship between research and medicine.

Neeti Sadana, MD
University of Miami

Foreword

Preface

The last few years have seen an almost exponential increase in the applications for biosensors in diverse areas. The initial biosensor was developed for the detection of glucose for the effective management of diabetes. This is incidentally still the major market for biosensors; however, other markets of biosensor applications especially in the medical area and for the detection of harmful pathogens are gradually coming into the forefront. This book is the fourth in the series on analyte–receptor kinetics on biosensor kinetics by the same author, and in some sense reflects the expanding applications of biosensors in traditional and non-traditional areas of applications due to their simplicity of use.

Fractal mathematics have been used to model the binding and dissociation kinetics for the different analyte–receptor reactions occurring on the different types of biosensor surfaces. Fractals, which exhibit self-similarity, have been used to model the heterogeneity that exists on the biosensor surface. The fractal analysis permits a convenient means to provide a lumped parameter analysis of the diffusion-limited reactions occurring on these structured biosensor surfaces. It is anticipated that the fractal analysis presented will provide novel physical insights into these types of reactions occurring on biosensor surfaces, and also assist in improving the biosensor performance parameters such as sensitivity, selectivity, response time, stability, etc.

Chapter 1 provides a definition of biosensors, their current and future applications, and the economics of biosensors. Chapter 2 outlines the basic fractal theory used to model the binding and dissociation kinetics used. Simple single-fractal analysis models are initially introduced. This is followed by the presentation of more complex models wherein dual- and triple-fractal analysis may be used, when a simple model does not provide an adequate fit.

The detection of harmful pathogens and toxins is an extremely important area of investigation considering the present-day geopolitical events and environment. Large sums of money and other resources are being spent by governmental agencies worldwide to help combat this problem. The current trend indicates that this will continue to hold in the foreseeable future. Chapter 3 analyzes the kinetics of detection of toxins and pathogens on biosensor surfaces. Heart ailments impact not only a single individual but also his or her family, besides being very expensive especially when extended hospital stays are involved. Chapter 4 analyzes the detection of heart-related compounds by biosensors. The intent over here is that if one is able to predict the occurrence of a heart attack by detecting a precursor(s), then hopefully one would be able to manage this ailment better.

No introduction to cancer is necessary. Common knowledge suggests that the detection of cancer at an early stage leads to a better prognosis. Hopefully, biosensors can play an important role here. Chapter 5 analyzes the kinetics of binding and dissociation of cancer markers on biosensor surfaces. Cancer may also be considered as an auto-immune disease. Auto-immune diseases are slow and insidious, and often their etiology is not known. Chapter 6 analyzes the binding and dissociation kinetics of auto-immune disease markers for diseases such as diabetes, systemic lupus erythematosus (SLE), arthritis, etc. Once again, the intent being that if one can catch these diseases early, then one can put the individual on an appropriate protocol, and help manage the disease better.

Prion proteins are involved in Alzheimer's disease. This is an extremely debilitating disease that generally manifests itself in older populations. Currently, there is significant interest in the treatment of this disease, wherein proteins misfold, aggregate, and then deposit in different locations in the brain. An early detection by biosensors of the aggregate forms during their early stages of formation would significantly assist in the management of this disease. Chapter 7 analyzes the binding and dissociation kinetics of prion proteins on biosensor surfaces. Environmental contaminants need no introduction. Chapter 8 analyzes the binding interactions of different environmental contaminants present in the air and in streams (liquid phases) on biosensor surfaces.

Biosensors are being increasingly used in drug discovery and design. Biosensors may be used to correctly identify a potential drug from a list of probable candidates. Chapter 9 analyzes the binding and dissociation kinetics of proteins involved in drug design on biosensor surfaces. Sol gels have unique features that may be used to advantage for biosensor applications, though there are problems that need to be addressed. Chapter 10 analyzes the binding and dissociation kinetics of analytes on sol gels used as biosensors.

Novel types of biosensors are continuously appearing in the literature. Chapter 11 analyzes only some of these types of biosensor applications that have very recently appeared in the literature, and seem to exhibit potential for further applications. The examples presented may only be considered as a sample and not all inclusive by any means. Chapter 12 analyzes the binding and dissociation kinetics of a wide variety of compounds on biosensor surfaces.

Chapter 13 analyzes the markets for biosensors presented from an academic perspective. The pitfalls and what it takes to set up a biosensor industry is presented. This chapter may be considered as the 'capstone' chapter. This type of information is critical, but it is difficult to obtain in the open literature. Understandably, this type of 'financial information' is carefully, almost fiercely, guarded by industries or individuals. However, change is afoot, as may be noted by the session on this topic at Biosensors 2006 held in Toronto, Canada in May 2006. This should be very beneficial for the biosensor industry as a whole.

The author wishes to express his appreciation to Dr. Kai-Fong Lee, Dean, School of Engineering at the University of Mississippi, for his continued support and encouragement for research that facilitates the writing of treatises like this.

A lot of credit goes to my students who have patiently helped me develop over the years the fractal theory for biosensor applications. My present student, Mr. Atul M. Doke has helped me considerably in getting this book ready for Elsevier.

Contents

– 1 –

Introduction

1.1 DEFINITION OF BIOSENSORS

Biosensors are small devices that use biological reactions to detect target analytes (Taylor and Schultz, 1996). An elementary definition of a biosensor is that it is an analytical device that converts a biological recognition event to an electrical signal that may be measured. There is an increasing tendency to make these devices smaller and smaller with the advent of nanotechnology and nanobiotechnology. Biosensors have been used in health care, drug design, environmental monitoring (in air, soil, and water), detection of biological, chemical, and toxic agents (for example, bacteria, virus, or tissue components), biotechnology, aviation, physics, oceanography, and the protection of civilian and engineering infrastructures. Recently, there has been an explosion of sorts of areas where they may be used.

An important and promising area of application is Point of Care Testing (POCT) for cancer diagnostics (National Cancer Institute, 2005). This is a diagnostic testing at the bedside where the feedback vis-à-vis patient care is almost immediate. The early detection by biosensors of early biomarkers for cancer is presently a hot area of investigation. Lieber and his colleagues at Harvard University in Boston, Massachusetts have used nanowires to detect cancer markers (Analytical Current, 2005). These authors with their label-free detection device based on a silicon-nanowire field effect were able to detect cancer markers at the femtomolar concentrations with high selectivity. Lieber and his group attached monoclonal antibodies (mAbs) for the cancer markers to the nanowires. Their device exhibited a conductance change when the cancer marker bound to the monoclonal antibody was attached to the nanowire. Recently, the NIST (National Institute of Standards and Technology) located in Gaithersburg, Maryland and the National Cancer Institute (NCI) located in Bethesda, Maryland exchanged ideas and discussed technologies for the early detection of cancer (Smith, 2005). Ideally, a cancer marker should clearly distinguish between a diseased state and a healthy state. Besides, it should exhibit a high sensitivity and selectivity. Biosensors are ideally suited to help detect these biomarkers for cancer and other insidious diseases. In fact, one of the goals of this book is to promote the use of biosensors for the early detection of autoimmune diseases such as cancer, systemic lupus erythematosus, and arthritis to name a few. In fact, the current line of thinking in the medical field is that most, if not all, diseases may be considered to be of an autoimmune nature.

Ghosh *et al.* (2005) have recently described a biosensor incorporating cell barrier architectures for the detection of biomarkers during the early stages of cancer. Gao *et al.* (2005)

have fabricated a microarray sensor using a temperature-responsive elastin fusion protein for the simultaneous detection of multiple tumor markers.

Tonneson and Withrow (2006) describe an interesting application of a biosensor, wherein they define a biosensor as "a neural interface technology that is capable of detecting nerve and muscle activity." These biosensors may be electrodes wherein they sit on the skin above a muscle or a nerve to be sampled. For example, they may be used to detect eye movement by placing these biosensors on the forehead or under the eyes. These authors indicate that just as the brain uses electrical signals to control the different functions in the body, these biosensors may use the electrical signals by way of appropriate software to control electronic devices external to the body. These authors emphasize that even a paralyzed individual exhibits electrical activity that may be detected.

Biosensors involve a biological (recognition) element and a transduction element. Figure 1.1 shows the main components of a biosensor in their simplest form (Chaplin, 2004). Some of these components may be combined with each other. In essence, the receptor (the biocatalyst in this case) converts the substrate to a product (step a). In step b, the transducer converts the reaction (product) to an electrical signal. This electrical signal may be amplified (step c), processed (step d), and displayed (step e).

The biological or recognition element may be an antibody, an enzyme, DNA, RNA, a whole cell, or a whole organ or system. The transduction element wherein the biological event or signal is converted to a measurable signal may include any one of the following forms: chemical, electrical, magnetic, mechanical, optical, or thermal. There is a choice of recognition elements such as enzymes, antibodies, aptamers, DNA strands, etc., and of the transduction elements (electrode, fiber optic, thermal, chemical, magnetic, etc.). One needs to combine these in an optimum manner to suit one's application so as to obtain, ideally a simple, rapid, and label-free application. Biosensor performance parameters may be improved significantly by providing (a) the proper interface between the biological and the transduction element and (b) by manipulating the structure of the interface, for example, the degree of heterogeneity on the biosensor surface. This heterogeneity on the biosensor surface may be due to the orientation of the receptors on the biosensor surface, or to the inherent nature of the biosensor surface itself. Other factors may also lead to changes in the degree of heterogeneity on the biosensor surface. The influence of this heterogeneity on biosensor surface is one of the main themes of this entire book.

Carter (2006) emphasizes that biosensors combine the sensitivity and specificity of biological systems with the computational abilities of microprocessors. The intention of this

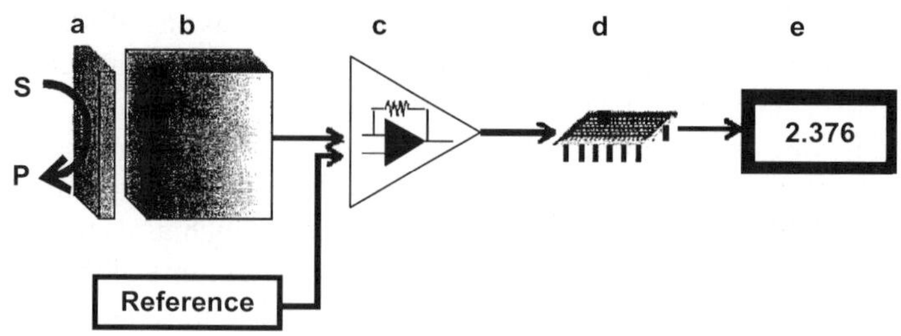

Figure 1.1 The main components of a biosensor (Chaplin, 2004).

book is to analyze the binding and dissociation kinetics of analyte–receptor reactions occurring on biosensor surfaces, and to help improve the biosensor performance parameters such as sensitivity and specificity mentioned above.

Wang (2000) indicates that there are two types of biosensors: bioaffinity devices and biocatalytic devices. In the bioaffinity devices, the analyte in solution binds selectively to a receptor immobilized on the biosensor surface. In the biocatalytic devices, an enzyme immobilized on the biosensor surface catalyzes the target substance. Wang (2000) has reviewed DNA sensors and emphasizes that the future of DNA sensors is bright, though these devices need to be small, fast, and easy to use. This author indicates that DNA sensors permit one to obtain sequence-specific information quickly, in a simple manner, and more cheaply when compared to traditional hybridization assays.

By necessity, a group working on biosensors either in a university setting or at an industrial or government organization needs to be of a multidisciplinary nature with individuals of different backgrounds that include: chemistry, physics, biology, optoelectronics, robotics, engineering, etc. The emphasis on making these future biosensors cost effective, miniaturized, and wireless is bound to include people with manufacturing backgrounds, electrical engineering, and nanotechnology also in the group to name just a few. These types of individuals may already be present in the group and may be making contributions right from the start to the biosensor group. If commercialization of the biosensor is the eventual goal, which should be in most cases, then this is apparently an appropriate approach, so that corrections or modifications may be made at an early stage in the development process to arrive at a competitive product, rather than at a later stage when it may be either more difficult or may not be economically feasible. Commercialization of biosensors is an important area, and the last chapter of the book analyzes a few of the concepts and facets and pitfalls involved. Finally, as the applications of biosensors grow, as expected, individuals with widely different backgrounds will be required in different biosensor groups worldwide.

Health care is an important and a growing area of biosensor application. Besides, the biosensor used should demonstrate speed in the sampling procedure, be sensitive, be of low cost, and exhibit ease of handling. Different companies have come up with biosensors that demonstrate these properties to varying degrees of satisfaction. Often, when one improves one biosensor property (or efficiency parameter), then another presumably exhibits some deterioration in its property. Thus, there is always, as expected, a 'balancing' act. An example of a company is now perhaps appropriate that develops (or has developed) a biosensor that demonstrates the above-mentioned parameters to a satisfactory degree in its biosensor.

IMEC (2005), an industrial company that does research on and makes biosensors for use in the health-related fields, indicates that its use of planar silicon technology and novel microelectronics permits its biosensors to acquire the expected speed, use of small analyte samples for detection, attain the required sensitivity levels, are easy to handle, and above all are available at low cost. The company emphasizes that its sensors are capable of measuring ions, metabolites, and medicines in biological fluids and in cell cultures effectively.

There are similarities in the detection of WMDs (weapons of mass destruction) used (or to be potentially used) by terrorists and in the detection of infectious diseases. In both cases, these harmful organisms (bacteria, viruses, etc.) are to be rapidly and reliably detected in dilute complex solutions (such as blood, urine, in the body, etc.) or in the atmosphere. For the detection of WMDs time is a very important factor (perhaps of the order of minutes), and this can be critical in a prognosis, since, in general, a large amount of people have to be moved and the WMD has to be quickly contained by emergency personnel. There is no leeway for any sort of error.

Robbins (2003) indicates the possibility of implantable biosensors that may serve as onboard doctors. These biosensors would have the possibility of detecting diseases early, and by using a feedback-type mechanism they would deliver drugs straight into the bloodstream, as for example, in measuring glucose levels and delivering insulin. This author mentions that environmental monitoring sensors (for WMD detection) may take 12–24 h to detect the harmful compound such as anthrax, with the better ones detecting this compound in the order of 20–30 min. Sandia National Laboratories in Alburquerque, New Mexico is attempting to develop a biosensor that may detect biotoxins in 5–10 min. If successful, this biosensor can then be employed, for example, in subway transit systems.

For the detection of infectious diseases, time is important, though not absolutely critical (perhaps of the order of minutes to hours). The medical staff do have a little luxury of time before corrective and remedial action needs to take place. Besides, the medical staff has the advantage of exploring different possibilities, and may even have had some previous experience in the treatment of a particular infectious disease or a similar type of disease. In both cases, for the detection of WMDs and infectious diseases false positives are largely unacceptable, thus reliability and sensitivity of biosensors come to the forefront.

In the United States, another aspect of WMD detection has very recently been mentioned by Prosnitz (2005). In a science and law policy forum section this author emphasizes that the development of detection devices (such as biosensors) for WMDs should not be developed in a legal vacuum. One needs to carefully balance (national) security with civil liberties. This author mentions the following criteria for the development of WMD detection devices such as biosensors that should be met: (a) sensor discretion, (b) sensor performance should be well documented, (c) sensor deployment should be effectively demonstrated, and (d) readily available biosensors. These criteria, which this author indicates will help carefully balance the civil liberties with impending and realistic threats.

1.2 CURRENT AND FUTURE APPLICATIONS

Traditionally, biosensor applications have found increasing applications in the biomedical areas, with the initial development of biosensors being the detection and quantitative determination of glucose in blood for the effective management of diabetes. This still remains the major market for biosensors. However, over the years the application of biosensors has rapidly expanded to other areas. Some of the biosensor applications that have recently appeared in the literature are briefly mentioned and described below.

1.2.1 Nanotechnology and nanobiotechnology applications for sensors

One of the recent areas of applications for nanotechnology and nanobiotechnology include biosensors. Lee *et al.* (2005) have recently described nanoparticle assemblies with molecular springs for biosensor applications. Detection limits of biosensors is an important efficiency parameter. Hong *et al.* (2005) have very recently indicated that metallic nanoparticles may be used as a fluorescence enhancer for lowering the detection limit of a fluorophore-mediated immunosensor.

The detection of hydrogen by biosensors is an important area of investigation. Srinivasan *et al.* (2005) have recently developed a hydrogen sensor by coating a surface acoustic wave (SAW) sensor with a palladium nanoparticle-coated tobacco mosaic virus sensing layer. The detection of harmful biologicals (for example, WMDs), as indicated earlier, is becoming of increasing importance. Doty and Fernig (2005) have recently developed a biosensor using protein-like silver (Ag) nanoparticles. Lee (2005) has recently reviewed nanotechnology applications to diagnostic devices.

Barone *et al.* (2005) have recently developed a single-walled carbon nanotube optical sensor for the *in vivo* fluorescence detection of glucose. These authors compared the optical properties of commonly used organic and nanoparticle fluorescent probes with regard to quantum yield and human-tissue penetration. Both types of sensors exhibited almost a linear response to blood glucose levels. Furthermore, they noted that their optical biosensor is more stable to membrane fouling since it transduces glucose concentration and not the flux.

1.2.2 Surface plasmon resonance (SPR) biosensor

The SPR biosensor continues to dominate the biosensor market. However, the software that comes along with (Biacore, 1999) does not take into the presence of external diffusional limitations and the degree of heterogeneity that exists on the biosensor surface. This may lead to erroneous results. Zeng *et al.* (2005) have recently examined methods to overcome these mass transfer limitations on the SPR biosensor surface. Nevertheless, the SPR biosensor along with its software has been used to determine the binding and the dissociation kinetics of (a) anti-human interleukin monoclonal antibody (SCGH 55700) and (b) human interleukin-5 interactions (DiGiacomo *et al.*, 2004). Other examples of analyte–receptor binding and dissociation kinetics determination by the SPR biosensor are readily available in the literature. Zubritsky (2005) indicates that Saerens *et al.* (2005) have developed an SPR-based assay for the determination of human prostrate-specific antigen (hPSA) that can detect clinically relevant concentrations less than 1 ng/ml in about 15 min. These authors used a single-domain antigen-binding fragment (VUH).

Wear *et al.* (2005) have recently used an SPR biosensor to screen and rank several small molecule ligands in a competitive binding assay with cyclosporin A (CsA). They describe a simple protocol to generate an SPR biosensor surface of recombinant human hexahistidine cyclophin A (His-CypA). Kinetic rate constants for the interaction between His-CypA and CsA were obtained. These authors emphasize that for the measuring of the binding rate constants the surface immobilized molecules need to be stably attached, besides being highly active. They further indicate that cyclophins are a potential drug target for diseases such as HIV and malaria (Towers *et al.*, 2003; Gavigan *et al.*, 2003).

1.2.3 Acoustic wave sensor

Velasquez *et al.* (2005) have developed an acoustic wave sensor to determine the bacterial count in municipal wastewater. Nonspecific binding leads to erroneous measurements in biosensors. Cular *et al.* (2005) have used SAW to remove nonspecific binding on microsensors. The detection of contrabands and drugs of abuse is a constant problem for governmental drug enforcement agencies. Staples and Viswanathan (2005) have developed an ultrafast SAW/Gas chromatograph (GC) vapor analyzer to detect contrabands and drugs of abuse.

1.2.4 Electrochemical sensors

Lee (1997) indicates that the hepatitis B virus (HBV) causes acute and chronic hepatitis, and is the ninth leading cause of death in the world (Dusheiko, 1991; Mast *et al.*, 1999). Electrochemical monitoring of DNA hybridization has being investigated (Wang *et al.*, 1996, 2001; Erdem and Ozsoz, 2002; Kobayashi *et al.*, 2001). Ariksoysal *et al.* (2005) indicate that genosensors may provide a low-cost and hand-held biosensor that may assist in the detection of a few inherited diseases (Ozsoz *et al.*, 2003; Wang and Kawde, 2001). Ariksoysal *et al.* (2005) have developed a label-free electrochemical hybridization genosensor for the detection of HBV genotype on the development of lamivudine [(-)-β-L-2′,3′-dideoxythiacytidine, 3TC] resistance. Lamivudine is one of the two main drugs used to treat chronic hepatitis B. Fattovich *et al.* (1998) indicate that long-term treatment is required. However, the cure rate is low and lamivudine resistance develops over the long term.

Lutz *et al.* (2005) indicate that hydrogen (H_2) is a possible alternate to the currently used fossil-based transportation fuels. The use of H_2 needs to be safe as far as production, storage, and usage is concerned since it has a low explosive limit in air. Thus, the need for reliable sensors. These authors have developed a hydrogen sensor using enzyme-catalyzed electrochemical detection. These authors used the soluble hydrogenase of aerobic β-proteobacterium *Ralstonia eutropha* strain H16 for the electrochemical detection of H_2. Clark *et al.* (2005) have recently developed an electrochemical glucose sensor for diabetes management.

Karasinski *et al.* (2005) indicate that for medical diagnosis, environmental studies, homeland security, and in the agricultural and food industry it is important to correctly identify and differentiate bacterial contamination. These authors have recently developed an electrochemical biosensor with pattern recognition techniques that is able to detect and classify bacteria at subspecies and strain level. Their multi-array technique (96-well-type electrodes array) used a DOX (dissolved oxygen sensor)-PCA (principal component analysis) technique to analyze and classify *Corynebacterium glutamicum*, *Micrococcus luteus*, *Staphylococcus epidermidis*, *Yersinia ruckeri*, *Escherichia adcarboxylata*, *Comamonas acidovorans*, *Alcaligenes odorans*, *Bacillus globigii*, and three strains of *Escherichia coli* (K12, SM10, and ATCC 25922). The 'principle' that they used is that under identical experimental conditions the different bacteria would consume oxygen at different rates. Furthermore, these different types of bacteria would be affected in different ways by selected antibiotics.

1.2.5 Quantum dots

Hahn *et al.* (2005) have recently used semiconductor quantum dots (QDs) as a fluorescence assay to detect single cells of the pathogen, *Escherichia coli* 0157:H7 serotype.

These authors indicate that their QD system is more than two orders of magnitude more sensitive than the common dyes used. They further state that their semiconductor QDs are more stable than their organic dye counterparts. This permits the real-time analysis of certain longer time biological processes (Hoshino *et al.*, 2004; Chen and Gerion, 2004). Hahn *et al.* (2005) indicate that QDs fluoresce brighter than organic dyes. Thus, they are more sensitive labels for fluorescence-based assays.

1.2.6 Hydrogels

Heo and Crooks (2005) have recently used an array of hydrogel-entrapped enzymes to detect glucose (single analyte) and glucose and galactose (multiple analytes). These authors emphasize that the enzymes are large enough so that they do not escape or leave the sol–gel matrix, and the analytes are small enough that they enter the hydrogel, interact with the entrapped enzymes, and the detectable products may leave the hydrogel matrix. The authors indicate that these hydrogels have been used to entrap proteins (Arenkov *et al.*, 2000; Guschin *et al.*, 1997), cells (Koh *et al.*, 2002, 2003), and DNA (Zangmeister and Tarlov, 2003, 2004). Heo and Crooks (2005) indicate that when enzymes are entrapped in hydrogels to be used as biosensors, they should (a) keep the enzyme active within the hydrogen, (b) nonspecific adsorption should be minimized (this is a perennial problem in most, if not all biosensors), (c) the number of probes (receptors) should be optimized, and (d) most importantly, the mass transport of the analyte to and the product away from the probe or receptor should be minimized. As expected, this external mass transport limitation will be a problem in most, if not all cases, of hydrogels used for biosensor purposes, and needs to be carefully addressed.

Dukova *et al.* (2005) have recently developed hydrogel-glycan microarrays wherein the microchip consisting of hydrogel drops is separated by a hydrophobic surface. These authors used these hydrogel-glycan microarrays for the quantitative assay of (a) antibodies against blood group antigens and (b) for the assay of lectins with fluorescent detection. These authors indicate that these three-dimensional hydrogel-based microchips comprise hydrogel drops, which are in nanoliters and are separated from each other by a hydrophobic glass or a plastic surface. They further indicate that some of the advantages of these three-dimensional microchips over the traditional two-dimensional microchips are increased immobilization capacity and increased stability.

1.2.7 Immunosensors

Shen *et al.* (2005) indicate that most affinity-based sensors that use antibodies as recognition elements may be termed as immunosensors. These authors indicate that immunosensors may use absorbance, fluorescence, amperometric or radiochemical transduction mechanisms. They developed a piezoimmunosensor by binding to a recombinant single-chain fragment variable (scFv) antibody (designated A10B). Shen *et al.* (2005) used this piezoimmunosensor to analyze the binding kinetics of antigen/scFv antibody binding. These authors indicate that the recombinant scFv are small heterodimers. These heterodimers comprise the antibody variable heavy (V_H) and light (V_L) pairs. Finally,

Shen *et al.* (2005) indicate that their recombinant scFv-based immunosensors were more sensitive and displayed less nonspecific binding than their intact monoclonal IgG and Fab fragment-based counterpart immunosensors. Tang *et al.* (2005) have developed a multi-analyte, fiber-optic immunosensor for rapid anticoagulant deficiency diagnosis.

Pyun *et al.* (2005) have developed a new immobilization method for immunoaffinity sensors using thiolated proteins. These authors indicate that their method provides a high surface density as well as stability for the receptor layer on the surface. In their technique, their molecular recognition protein is first thiolated by covalent conjugation of mercaptopropionic acid. Thereafter, the thiolated protein is attached to the gold surface by the transducer. The molecular recognition layer is often called the immunoaffinity layer (Luppa *et al.*, 2001; Ekins, 1999; Warsinnke *et al.*, 2000). Pyun *et al.* (2005) used horseradish peroxidase (HRP) and its antibody as a model system.

1.2.8 Aptamers

Jayasena (1999) indicates that aptamers are oligonucleotide sequences that are a class of molecules that rival antibodies in both therapeutic and diagnostic applications. They are distinctly different from antibodies but they do mimic their properties. They could serve as an important source of molecular recognition compounds. Weeks and Merino (2005) have recently developed a simple method to convert aptamers to biosensors. They indicate that their approach of creating 2′-ribose-derivatized aptamers provides a method to create reagentless biosensors for small molecules. These authors developed fluorescent adducts of three aptamers that recognized adenosine monophosphate (AMP), argininamide, and tyrosinamide. Upon binding to their respective ligands, these authors noted a greater than 40% increase in the fluorescence intensity.

1.2.9 Quartz crystal microbalance

Carmon *et al.* (2005) have recently used a quartz crystal microbalance (QCM) to detect estrogenic substances using a genetically engineered construct of the hormone-binding domain of the α-estrogen receptor. These authors used a single exposed cysteine to immobilize and form a uniformly oriented α-estrogen receptor on the quartz crystal surface. These authors emphasize that xenoestrogens are a cause for public health concerns since they affect reproductive capacities. These endocrine disrupting chemicals (EDCs) will affect hormone synthesis, storage, and metabolism. Mueller and Korach (2001) have emphasized the importance of detection devices to screen and to detect these types of chemicals. These EDCs may be found in water or in the air due to the usage of chemicals such as detergents, surfactants, plastics, and pesticides (Crews *et al.*, 2000).

1.2.10 Transient grating method

Khan *et al.* (2005) very recently indicate that although different methods are available that detect protein binding (Schubert *et al.*, 2003; Chen and Gerion, 2004; Northup, 2004; Casper *et al.*, 2004), including the highly sensitive SPR method, there are a few shortcomings: (a) the target protein should be fixed to a surface (this may change its conformation) and (b) it may take tens of minutes for the accumulation of proteins on the surface in order to detect them. This precludes the study of chemical intermediates. These chemical species may play a significant role in transduction processes (Khan *et al.*, 2005). In order to alleviate these types of shortcomings, these authors developed the transient grating method, which is based on diffusion measurement in the time domain. These authors have used a photosensitive and photoactive protein to demonstrate their technique on protein–protein interaction, protein–DNA interaction, and protein–small molecule binding. In essence, the principle of their technique is "if protein association occurs, then molecular size becomes larger, and the diffusion coefficient becomes smaller."

The next section briefly presents biosensor economics, which is often neglected in the open literature. More often than not, one has to pay for expensive reports (or books) that describe this aspect of biosensors.

1.3 BIOSENSOR ECONOMICS

In this section, we briefly provide the economic markets (numbers) for biosensors. Numbers are taken from literature sources as well as from internet sources. These numbers are difficult to get, and those taken from internet sources should be treated with caution. Market projections, by definition, are the only projections, and should be treated as such. More detailed economic numbers are provided in the last chapter (Chapter 13).

Yurish *et al.* (2005) have recently analyzed the trends in the world sensors and micromechanical systems (MEMS) markets. Our interest lies only in biosensor markets, so our comments will be restricted to them only. These authors emphasize the drawbacks that are hindering the development of biosensors: significant up front investment and prolonged research and development timelines. These two factors make it difficult to justify the high expense involved for the relatively 'small' market for biosensors. These authors estimate that the global market for fiber-optic biosensors is \$304.3 million for the year 2006, and it is expected to increase at a rate of 4.1% till the year 2011 (Business Communications Company, Report G-116N, Fiber Optic Sensors, 2005). For the years 2006, 2007, 2008, 2009, 2010, and 2011, the estimated worldwide market for biosensors is in \$ million (U.S.) 304.3, 316.8, 329.8, 343.3, 357.4, and 372, respectively. This is an expensive report and costs \$3950.

In a report on nanosensors by the same company (Business Communications Company, Report GB-310 Nanosensors, 2005), the estimated market for nanosensors for the years 2006 through 2009 is expected to grow at 25.5%. For the years 2006, 2007, 2008, and 2009, the estimated market for nanosensors is estimated to be in \$ million (U.S.) 299.25, 375.56, 471.33, and 591.52, respectively. This is also an expensive report and costs \$3950. The report, however, does provide (suggest) factors that will influence the long-term development of the biosensor market, besides providing an estimate of the market share and industry structure.

The SPR biosensor is an important player in the biosensor market, especially the one manufactured by Biacore. Mukhopadhyay (2005) has recently reviewed the SPR instruments available in the market. This author indicates that the SPR instruments have diversified and an increasing number of vendors have entered the SPR spectroscopy and imaging market. The SPR technology is being used by other companies, besides Biacore to develop instruments. This author emphasizes that the commercially available SPR instruments (used mainly in biological applications) are used for spectroscopy and for imaging. They may be categorized into three classes based on the cost per instrument (from Mukhopadhyay, 2005):

(a) less than \$10,000 — NanoSPR, Biosuplar-2, and Biosuplar-3
(b) between \$11,000 and \$99,000 — Autolab ESPIRIT, SPRimager II, Multiskop, SPT100 Module
(c) \$100,000 and higher — Biacore T100.

All numbers are in U.S. \$.

More details about the prices may be found either from Mukhopadhyay's (2005) article or directly from the manufacturers. Nonspecific interactions are a problem with biosensors. Mukhopadhyay (2005) emphasizes that researchers are always manipulating the SPR (or other biosensor) surface to minimize these nonspecific interactions, and to minimize the adsorption or binding of errant molecules. Nevertheless, miscellaneous materials do compete with the required analyte (target) in solution to bind to the receptors on the SPR (or other biosensor) surface.

As indicated above, more details about biosensor economics and the trends in biosensor markets are presented in the last chapter (Chapter 13). The above was just presented to provide a brief glimpse into this area of biosensors, which is often neglected by academics and is very closely guarded, as expected, by industrial and manufacturing sources.

1.4 OVERVIEW

The material to be presented in Chapters 2 through 13 is now briefly presented. The theory behind the single-, dual-, and triple-fractal analysis is presented in Chapter 2. Expressions for both the binding and the dissociation phases are also presented in this chapter. Chapter 3 analyzes the binding and the dissociation of toxins and pathogens on biosensor surfaces. The detection of ricin and yessotoxin (its effects on human health are presently unknown) by biosensors is analyzed in this chapter.

In Chapter 4, we analyze the binding and dissociation kinetics of heart-related compounds. Some of the compounds analyzed include the low-density lipoproteins (LDLs), high-density lipoproteins (HDLs), digoxin (digitalis), and lecithin cholesterol acyltransferase (LCAT). LCAT catalyzes the transfer of *sn*-2 fatty acid of phosphatidyl-choline to the 3-OH group of cholesterol converting cholesterol to cholesterol ester. This is a key step in reverse cholesterol transport, wherein the free cholesterol is returned to the liver for catabolism (Fielding and Fielding, 1995).

The detection of cancer biomarkers is a vigorous area of investigation. Chapter 5 analyzes the binding and dissociation of the vascular endothelial growth factor (VEGF) to its receptor (VEGF-receptor; sFlt-1) immobilized on a biosensor surface (Von Tiedemann and Bilitewski, 2002). The VEGF receptor plays a significant role in angiogenesis (which is the formation of new capillaries from pre-existing blood cells). In Chapter 6, we analyze the binding and dissociation of autoimmune disease markers on biosensor surfaces. Autoantibodies such as IgG-anti-F$(ab')_2$ are involved in the natural pathogenesis of certain autoimmune diseases (Welschoff *et al.*, 1997). These IgG-anti-F$(ab')_2$ autoantibodies belong to the large family of anti-IgG antibodies, generally termed as rheumatoid factors (RFs).

Chapter 7 analyzes the binding and dissociation of prion proteins on biosensor surfaces. Prions are 'misfolded' proteins that lead to intractable diseases such as Alzheimer's disease. These prion proteins have generated, and will presumably continue to generate a considerable amount of interest in the research community. The analysis of environmental pollutants is an important area of investigation. Chapter 8 presents the binding and dissociation kinetics of different environmental contaminants such as toluene, xylene, different aromatic and nonaromatic compounds, and lead (Pb) to biosensor surfaces.

Drug–lipid interactions are an important and crucial step in drug discovery. The drug discovery market for biosensor usage occupies a significant share of the total market for biosensors. Chapter 9 presents the binding and dissociation kinetics of proteins (such as kalata B1 and kalata B2) involved in drug design on biosensor surfaces (Kamimori *et al.*, 2005). Sol–gel derived materials are suitable for biosensor applications since they may be synthesized under mild conditions, they are chemically flexible, and they adhere to a variety of substrates. Chapter 10 analyzes the binding and dissociation kinetics of analytes in solution to sol–gel surfaces used as biosensors.

Novel types of biosensors are continuously being developed for different applications. Chapter 11 analyzes the binding and dissociation kinetics of analytes on novel biosensors such as the diffraction-based sandwich immunoassay (Goh *et al.*, 2003), the SPIT-FRI (solid-phase immobilized tripod for fluorescent renewable immunoassay) (Volland *et al.*, 2005), and the double-wavelength technique for SPR measurements (Zybin *et al.*, 2005). Chapter 12 analyzes the binding and dissociation kinetics of different compounds on biosensor surfaces. Some of the compounds analyzed include adenosine triphosphate (ATP) (Liaudet *et al.*, 2005), human myeloid protein 14 (MRP 14) (Grote *et al.*, 2005), and alkaline phosphatase (Serra *et al.*, 2005).

In Chapter 13, we try to address the market size and economics for biosensors. The importance of this aspect is being recognized even by the academic community as is evidenced by a session of this topic at the Biosensors 2006 conference held recently in Toronto, Canada from May 10–12, 2006. The different aspects that are involved in the market size for biosensors are presented. Such type of information is not available free of charge in the open literature. On the contrary, one has to pay, in general, thousands of dollars (U.S.) to obtain some type of market forecast for the next 5–10 years for the different types of biosensor applications. As expected, different types of biosensor companies will guard this type of information very carefully. This chapter provides a 'necessary'; balance to the whole book, even though it is the last one. In some sense, one may consider it as the 'capstone' chapter. Every attempt has been made to provide accurate information considering that quite a large fraction of the sources are from the internet, and there may be a reliability factor here.

Predictions of market size should be viewed with caution, with particular attention being paid to who is making these predictions. An appropriate caveat, and this is fair to indicate up front, that besides some consultancy experience, the author has not been involved in a biosensor start-up company. Nevertheless, the information presented together should provide the readers with an initial estimate and rationale of market size and economics of biosensors for their specific application. If one may be allowed to draw an analogy between the solution of a nonlinear problem by an iterative solution technique, which requires an initial solution or estimate, then similarly the information presented should hopefully provide the means for an initial estimate of the market size for the different (or one's specific) biosensor applications.

REFERENCES

Analytical Currents. (2005). Detecting cancer markers with nanowires. *Analytical Chemistry*, (November 1), 406A, originally published in the *Journal of the American Chemical Society*, **127,** 12766–12767.

Arenkov, P., Kukhtin, A., Gemmel, A., Voloshchuk, S., Chupeeva, V., and Mirzabekov, A. (2000). Protein microchips: use for microassay and enzyme reactions. *Analytical Biochemistry*, **278**, 123–131.

Ariksoysal, D.O., Karadeniz, H., and Ozsoz, M. (2005). Label-free electrochemical hybridization genosensor for the detection of hepatitis B virus genotype on the development of lamivudine resistance. *Analytical Chemistry*, **77**, 4908–4917.

Barone, P.W., Parker, R.S., and Strano, M.S. (2005). *In vivo* fluorescence detection of glucose using a single-walled carbon nanotube optical sensor: design, fluorophore properties, advantages, and disadvantages. *Analytical Chemistry*, **77**, 7556–7562.

Business Communications Company, Inc. (2005). Report G-11N Fiber Optic Sensors, Norwalk, CT, USA, http://www.bccresearch.com/instrum/

Business Communications Company, Inc. (2005), Report G-310 Nanosensors, Norwalk, CT, USA, http://www.bccresearch.com/instrum/

Carmon, K.S., Balthus, R.E., and Luck, L.A. (2005). A biosensor for estrogenic substances using the quartz crystal microbalance. *Analytical Biochemistry*, **345**, 277–283.

Carter, S.A. (2006). Biosensors, http://physics.ucsc.edu/~sacarter/biosensors.shtml

Casper, D., Bukhtiyarova, M., and Springman, E.B. (2004). A Biacore biosensor for detailed kinetic binding analysis of small molecule inhibitors of p38 α nitrogen-activated protein kinase. *Analytical Biochemistry*, **325**, 126–136.

Chaplin, M. (2004). What are Biosensors? http://www.lsbu.ac.uk/biologuy/enztech/biosensors.html

Chen, F., and Gerion, D. (2004). Fluorescent CdSe/Zns nanocrystal-peptide conjugates for long-term nontoxic imaging and nuclear targeting in living cells. *Nano Letters*, **4**, 1827–1832.

Clark, B.L., Mugwera, A., and Pishko, M.V. (2005). Electrochemical glucose sensor for diabetes management. Papers 526e and 595d, American Chemical Institute of Chemical Engineers Meeting, Cincinnati, Ohio, October 30–November 4.

Crews, D., Willingham, E., and Skipper, J.K. (2000). Endocrine disruptors: present issues, future directions. *Quarterly Reviews in Biology*, **75**, 243–260.

Cular, S., Branch, D.W., Meyer, G.D., Craighead, H., and Bhethanabotla, V.R. (2005). Removal of nonspecific binding on microsensors using surface plasmon acoustic waves. Paper 526c, American Institute of Chemical Engineers Annual Meeting, Cincinnati, Ohio, October 30–November 4.

DiGiacomo, R.A., Xie, L., Cullen, C., and Indelicato, S.R. (2005). Development and validation of a kinetic assay for analysis of anti-human interleukin-5 monoclonal antibody (SCH 55700) and human interleukin-5 interactions using surface plasmon resonance. *Analytical Biochemistry*, **327**, 165–175.

Doty, R.C., and Fernig, D.G. (2005). Protein-like nanoparticles for biological sensing applications. Paper 74f, American Institute of Chemical Engineers Annual Meeting, Cincinnati, Ohio, October 30–November 4.

Dukova, V.I., Dementieva, E.I., Zubtsov, D.A., Galanina, O.E., Bovin, N.V., and Rubina, A.Y. (2005). Hydrogel glycan microarrays. *Analytical Biochemistry*, **347**, 94–105.

Dusheiko, G. (1991). In: *Oxford Textbook of Clinical Hepatology* (ed. N. McIntyre), Oxford University Press, New York, p. 571.

Ekins, R. (1999). Immunoassay and other ligand assays: from isotopes to luminescence. *Journal of Clinical and Ligand Assay*, **22**, 61–77.

Erdem, A., and Ozsoz, M. (2002). Electrochemical DNA biosensors based on drug–DNA interactions. *Electroanalysis*, **14**(14), 965.

Fattovich, G., Brollo, L., Aliberti, A., Pontisso, P., Giustina, G., and Realdi, G. (1998). Long-term follow-up of anti-HBe-positive chronic active hepatitis B. *Hepatology*, **8**, 1651–1664.

Fielding, C.J., and Fielding, P.E. (1995). Molecular physiology of reverse cholesterol transport. *Journal of Lipid Research*, **36**, 211.

Gao, D., McBean, N., Schultz, J.S., Mucchandani, A., and Chen, W. (2005). Fabrication of microarray sensors using a temperature-responsive elastin fusion protein for simultaneous detection of multiple tumor markers. Paper 423d, American Institute of Chemical Engineers Annual Meeting, Cincinnati, Ohio, October 30–November 4.

Gavigan, C.S., Kiely, S.P., Hirtzlin, J., and Bell, A. (2003). Cyclosporin-binding proteins of *Plasmodium falciparum*. *International Journal of Parasitology*, **33**, 987–996.

Ghosh, G., Anderson, K., Bachas, L., and May, K.M.L. (2005). Biosensors incorporating cell barrier architectures for detecting biomarkers indicative of early stages of cancer. Paper 423c, American Institute of Chemical Engineers Annual Meeting, Cincinnati, Ohio, October 30–November 4.

Goh, J.B., Tam, P.L., Loo, R.W., and Goh, M.C. (2003). A quantitative diffraction-based sandwich immunoassay. *Analytical Biochemistry*, **313**, 262–266.

Grote, G., Dankbar, N., Gefdig, E., and Koening, S. (2005). Surface plasmon resonance/mass spectrometry interface. *Analytical Chemistry*, **77**, 1157–1162.

Guschin, D., Yershov, G., Zaslavsky, A., Gemmell, A., Schick, V., Proudnikov, D., Arenkov, P., and Mirzabekov, A. (1997). *Analytical Biochemistry*, **250**, 203–211.

Hahn, M.A., Tabb, J.S., and Krauss, T.D. (2005). Detection of single bacterial pathogens with semiconductor quantum dots. *Analytical Chemistry*, **77**, 4861–4869.

Heo, J., and Crooks, R.M. (2005). Microfluidic biosensor based on an array of hydrogel-entrapped enzymes. *Analytical Chemistry*, **77**, 6843–6851.

Hong, B., Ren, Y., Tang, L., and Kang, K.A. (2005). Metallic nanoparticle as fluorescence enhancer for lowering the detection limit of fluorophore immunosensor. Paper 74b, American Institute of Chemical Engineers Annual Meeting, Cincinnati, Ohio, October 30–November 4.

Hoshino, A., Hanaki, K.I., Suzuki, K., and Yamamoto, K. (2004). Applications of T-lymphoma labeled with fluorescence quantum dots to cell tracing markers in body. *Biochemical and Biophysical Research Communication*, **314**, 46–53.

IMEC (2005). Biosensors, http://www.imec.be/ovinter/static_research/biosensors.shtml

Jayasena, S.D. (1999). Aptamers: an emerging class of molecules that rival antibodies in diagnostics. *Clinical Chemistry*, **45**, 1628–1650.

Kamimori, H., Hall, K., Craik, D.J., and Aguilar, M.I. (2005). Studies on the membrane interactions of the cyclotides kalata B1 and kalata B6 on model membrane systems by surface plasmon resonance. *Analytical Biochemistry*, **337**, 149–153.

Karasinski, J., Andreescu, S., Sadik, O.A., Lavine, B., and Vora, M.N. (2005). Multiarray sensors with pattern recognition for the detection, classification, and differentiation of bacteria at subspecies and strain levels. *Analytical Chemistry*, **77**, 7941–7949.

Khan, J.S., Imamoto, Y., Yamazaki, Y., Tokunaga, F., and Terazima, M. (2005). A biosensor in the time domain based on the diffusion coefficient measurement: intermolecular interaction on an intermediate of photoactive yellow protein. *Analytical Chemistry*, **77**, 6625–6629.

Kobayashi, M., Mizukami, T., Morita, Y., Murakami, Y., Yokoyama, K., and Tamiya, E. (2001). *Electrochemistry*, **69**, 1013.

Koh, W.G., Itle, L.J., and Pishko, M.V. (2003). Molding of hydrogel microstructures to create multiphenotype cell microarrays. *Analytical Chemistry*, **75**(21), 5783–5789.

Koh, W.G., Revzin, A., and Pishko, M.V. (2002). Poly(ethylene glycol) hydrogel microstructures encapsulating living cells. *Langmuir*, **18**, 2459–2462.

Lee, G.U. (2005). Application of nanotechnology to diagnostic devices. Paper 305c, American Institute of Chemical Engineers Annual Meeting, Cincinnati, Ohio, October 30–November 4.

Lee, J., Lilly, G.D., Govorov, A.O., and Kotov, N.A. (2005). Nanoparticle assemblies with molecular springs. Paper 74a, American Institute of Chemical Engineers Annual Meeting, Cincinnati, Ohio, October 30–November 4.

Lee, W.M. (1997). Medical progress: hepatitis B virus infection. *New England Journal of Medicine*, **337**, 1733.

Liaudet, E., Sonja, H., Droniou, M., and Dale, N. (2005). Microelectrode biosensor for real-time measurement of ATP in biological tissue. *Analytical Chemistry*, **77**, 3267–3273.

Luppa, P.B., Sokoll, L.J., and Chan, D.W. (2001). Immunosensors: principles and applications to clinical chemistry. *Clinica Chimica Acta*, **314**, 1–26.

Lutz, B.J., Fan, Z.H., Burgdorf, T., and Friedrich, B. (2005). Hydrogen screening by enzyme-catalyzed electrochemical detection. *Analytical Chemistry*, **77**, 4969–4975.

Mast, E.E., Alter, M.J., and Margolis, H.S. (1999). Strategies to prevent and control hepatitis B and C virus infections: a global perspective. *Vaccine*, **17**, 1730.

Mueller, S.O., and Korach, K.S. (2001). Mechanisms of estrogen receptor-mediated agonistic and antagonistic effect. In: *The Handbook of Environmental Chemistry* (ed. O. Hutzinger), Vol. 3, Springer, LLC, New York, NY, p. 1001.

Mukhopadhyay, R. (2005). Surface plasmon resonance instruments diversify. *Analytical Chemistry*, **77**(August 1), 313A–317A.

National Cancer Institute. (2005). Moving biosensors to point-of-care cancer diagnostics. Conference Meeting Chair, Dr. Steven Soper, and Co-Chair, Dr. James W. Jacobson, June 8–9, Rockville, Maryland, 20852, US.

Northup, J. (2004). *Methods in Molecular Biology*, **261**, 93–112.

Ozsoz, M., Erdem, A., Kerman, K., Ozkan, D., Tugrul, B., Topcuoglu, N., Erken, H., and Taylan, M. (2003). Electrochemical genosensor based on colloidal gold particles for the detection of factor V leiden mutation using disposable pencil graphite electrodes. *Analytical Chemistry*, **75**, 2181–2187.

Prosnitz, D. (2005). WMD sensors – search and seizure, *Science*, **310**(11 November), 978.

Pyun, J.C., Kim, S.D., and Chung, J.W. (2005). New immobilization technique for immunoaffinity biosensors by using thiolated proteins. *Analytical Biochemistry*, **347**, 227–233.

Robbins, A. (2003). Health Care: Biosensors, PCmag.com, July 1, http://www.pcmag.com/article2/0,4149,1137664,00.asp

Saerens, D., Fredrix, F., Reekmans, G., Conrath, K.J., Brys, L., Huang, L., Bosman, E., Maes, G., Borghs, G., and Muyldermans, S. (2005). Engineering camel single-domain antibodies and immobilization chemistry for human prostate specific antigen sensing. *Analytical Chemistry*, **77**(23), 7547–7555.

Schubert, F., Zettl, H., Haffner, W., Krauss, G., and Krausch, G. (2003). Engineering camel single-domain antibodies and immobilization chemistry for human prostate-specific antigen sensing. *Biochemistry*, **42**(34), 10288–10294.

Serra, B., Maralcs, M.D., Reviejo, A.J., Hall, E.A.H., and Pingarron, J.M. (2005). Rapid and highly sensitive electrochemical determination of alkaline phosphatase. *Analytical Biochemistry*, **336**, 289–294.

Shen, J.H., Weinmann, S.W., and Schoenfisch, M.H. (2005). Sol-gel derived amperometric nitric oxide microsensor. *Analytical Chemistry*, **77**, 3494–3501.

Smith, N. (2005). NIST–NCI workshop on standards, methods, assays, reagents, and technologies: NIST and early detection of cancer. *Analytical Chemistry*, **77**(November 1), 413A.

Srinivasan, K., Cular, S., Bhethanabotla, V.R., Lee, S.Y., Harris, M.T., and Culver, J.N. (2005). Palladium nanoparticle coated tobacco mosaic virus sensing layer based surface acoustic wave hydrogen sensors. Paper 74d, American Institute of Chemical Engineers Annual Meeting, Cincinnati, Ohio, October 30–November 4.

Staples, E.J., and Viswanathan, S. (2005). Detection of contrabands and drugs of abuse using an ultra-fast SAW/GC vapor analyzer. Paper 563a, American Institute of Chemical Engineers Annual Meeting, Cincinnati, Ohio, October 30–November 4.

Tang, L., Hong, B., Ren, Y., and Kang, K.A. (2005). Multi-analyte, fiber-optic immunobiosensor with automatic control system for rapid anticoagulant deficiency diagnosis. Paper 423f, American Institute of Chemical Engineers Annual Meeting, Cincinnati, Ohio, October 30–November 4.

Taylor, R.F., and Schultz, J.S. (1996). *Handbook of Chemical and Biological Sensors*, Institute of Physics Publishing, Bristol, UK.

Tonneson, C., and Withrow, G. (2006). Biosensors, http://www.hitl.washington.edu/scivw/EVE/I.D.1.c.Biosensors.html

Towers, G.J., Hatziioannou, T., Cowan, S., Goff, S.P., Luban, J., and Bieniasz, P.D. (2003). Cyclophin A modulates the sensitivity of HIV-1 to host restriction factors. *Nature Medicine*, **9**, 1138–1143.

Velasquez, A.H., Cular, S., Bhetanabotla, V.R., Showalter, S.K., and Ceronsek, R.W. (2005). Municipal water bacteria content monitoring through acoustic wave sensors. Paper 481d, American Institute of Chemical Engineers Annual Meeting, Cincinnati, Ohio, October 30–November 4.

Volland, H., Neuberger, L.M., Schultz, E., Grassi, J., Perraut, F., and Creminon, C. (2005). Solid-phase immobilized tripod for fluorescent renewable immunoassay. A concept for continuous monitoring of an immunoassay including a regeneration of the solid phase. *Analytical Chemistry*, **77**, 1896–1905.

Von Tiedemann, B., and Bilitewski, U. (2002). Characterization of the vascular endothelial growth factor–receptor interaction and determination of the recombinant protein by an optical receptor sensor. *Biosensors & Bioelectronics*, **17**, 983–991.

Wang, J. (2000). Survey and summary: from DNA biosensors to gene chips. *Nucleic Acids Research*, **28**(10), 3011–3016.

Wang, J., and Kawde, A.N. (2001). Pencil-based renewable biosensor for label-free electrochemical detection of DNA hybridization. *Analytica Chimica Acta*, **431**(2), 219.

Wang, J., Kawde, A.N., Erdem, A., and Salazar, M. (2001). Magnetic bead-based label-free electrochemical detection of DNA hybridization. *Analyst*, **126**(11), 2020.

Wang, J., Paleek, E., Nielsen, P., Rivas, G., Cai, X., Shirashi, H., Dontha, H., Lo, D., and Farias, P.A. (1996). Peptide nucleic acid probes for sequence-specific DNA biosensors. *Journal of the American Chemical* Society, **118**(33), 7667–7670.

Warsinnke, A., Benkert, A., and Scheller, F.W. (2000). Electrochemical immunoassays. *Journal of Analytical Chemistry*, **366**, 622–634.

Wear, M.A., Patterson, A., Malone, K., Dunsmore, C., Turner, N.J., and Walkinshaw, M.D. (2005). A surface plasmon resonance-based assay for small molecule inhibitors of human cyclophilin A, *Analytical Biochemistry*, **345,** 214–226.

Weeks, K., and Merino, E. (2005). Facile conversion of aptamers into sensors using a 2(-ribose-linked fluorophore. *Journal of the American Chemical Society*, **127**, 12766–12767.

Welschoff, M., Terness, P., Kirpiyanov, S.M., Stanescu, D., Boeitling, F., Dorsan, H., Dubel, S., Little, M., and Opelz, G. (1997). The antigen–antibody domain of a human IgG-anti-F(ab′)$_2$ autoantibody. *Proceedings of the National Academy of Sciences USA*, **94**, 1902–1907.

Yurish, S.Y., Kirianaki, N.V., and Myshkin, I.L. (2005). World sensors and MEMS markets: analysis and trends. *Sensors & Transducers Magazine (S&T e-Digest)*, **62**(12), December, 456–461.

Zangmeister, R.A., and Tarlov, M.J. (2003). UV graft polymerization of polyacrylamide hydrogel plugs in microfluidic channels. *Langmuir*, **19**, 6901–6904.

Zangmeister, R.A., and Tarlov, M.J. (2004). DNA displacement assay integrated into microfluidic channels. *Analytical Chemistry*, **76**, 3655–3659.

Zeng, L., Li, X., Chishti, A., and Takoudis, C.G. (2005). Overcoming mass transport limitations in plasmon resonance biosensor. Paper 342e, American Institute of Chemical Engineers Annual Meeting, Cincinnati, Ohio, October 30–November 4.

Zubritsky, E. (2005). Camel antibodies for PSA assays. *Analytical Chemistry*, **77**(December 1), 451A.

Zybin, A., Grunwald, C., Mirsky, V.M., Kuhlmann, J.,Wolfbeis, O.S., and Niemax, K. (2005). Double-wavelength technique for surface plasmon resonance measurements: basic concept and applications for single sensors and two-dimensional arrays. *Analytical Chemistry*, **77**, 2393–2399.

– 2 –

Modeling and Theory

2.1 INTRODUCTION

In a biosensor-based assay the molecule to be detected (analyte) is present in solution and an appropriate receptor is immobilized on a solid surface. The interaction between the analyte and the receptor on the solid biosensor surface is detected either by a change in the refractive index (in SPR (surface plasmon resonance) instruments) or by changes in the fluorometric intensity, ultraviolet light intensity, etc. The SPR biosensor protocol analyzes the binding (and dissociation where applicable) kinetic curves using classical saturation models involving analyte–receptor binding with 1:1, 1:2, etc. ratios, generally under diffusion-free conditions and assuming that the receptors are homogeneously distributed over the sensor surface. Computer programs and software that come with the equipment provide values of the binding (and the dissociation) rate coefficients. Though a careful analysis and experimental protocol may eliminate or minimize the influence of diffusional limitations; realistically speaking, it is more appropriate to include a heterogeneous distribution on the sensing surface. Heterogeneity on the sensing surface and in the biosensor systems itself may be due to other reasons, such as for example, non-specific binding, inherent irregularities on the sensing surface, mixture of receptors on the surface, and mixture of analytes in solution which includes the analyte of interest.

Two factors need to be addressed while analyzing the analyte–receptor binding and dissociation kinetics. The system by its design is heterogeneous. For example, and as indicated above, the receptors immobilized on the biosensor surface may exhibit some heterogeneity, i.e surface roughness. No matter how careful one is in immobilizing the receptors on the biosensor surface, there will be some degree of heterogeneity on the surface.

Henke *et al.* (2002) have used the atomic force microscopy (AFM) technique to determine the effects of the cleaning of fused silica and glass on surface roughness. This is for biosensor use. Note that prior to the immobilization of receptors on the surface, the surface needs to be cleaned to remove contaminants and to create surface attachment sites, for hydroxyl groups.

For the analyte–receptor binding (and dissociation) to take place by the diffusion process the analyte must come within the 'proximity' of the active site on the receptor. Mass transport limitations may be minimized or eliminated if the system is either properly designed or properly operated or both. In most cases, however, both diffusional effects and heterogeneity aspects will be present in biosensor systems, and their influence on binding and dissociation kinetics needs to be determined. Ideally, one would like to determine the influence of each of these separately on the binding and dissociation kinetics. In the theoretical analysis presented below (the Havlin analysis (1989), the effects of diffusion and heterogeneity are presented coupled together. One possible way of accounting for the presence of diffusional limitations and the heterogeneity that exists on the surface is by using fractals. Ideally, and as indicated above, one would prefer to decouple the influence of diffusion and heterogeneity. Presumably, an approach other than fractal analysis is required to decouple these two effects.

A characteristic feature of fractals is self-similarity at different levels of scale. Fractals exhibit dilatational symmetry. Fractals are disordered systems, and the disorder is described by non-integral dimensions (Pfeifer *et al.*, 1989). Fractals have non-integral dimensions, and are smaller than the dimension they are embedded in. In other words, the highest value that a fractal can have is three. In our case, an increase in the degree of heterogeneity on the biosensor surface would lead to an increase in the value of the fractal dimension. Another way of looking at the fractal dimension is its 'space filling' capacity. The more the space a surface fills, the higher is its fractal dimension. The fractal dimension cannot have a negative value, and very low values of the fractal dimension on the surface indicate that the surface exists as a Cantor-like dust.

Kopelman (1988) indicates that surface diffusion-controlled reactions that occur on clusters or islands are expected to exhibit anomalous and fractal-like kinetics. These kinetics exhibit anomalous reaction orders and time-dependent (e.g., binding) rate coefficients. As long as surface irregularities show scale invariance they can be characterized by a single number, the fractal dimension. Later on in the book we will characterize the surfaces of the biosensors used in the different examples by a fractal dimension. More specifically, we will characterize the heterogeneity present on these biosensor surfaces by a fractal dimension.

The fractal dimension is a global property, and it is insensitive to structural or morphological details (Pajkossy and Nyikos, 1989). Markel *et al.* (1991) indicate that fractals are scale self-similar mathematical objects that possess non-trivial geometrical properties. Furthermore, these authors indicate that rough surfaces, disordered layers on surfaces, and porous objects all possess fractal structure. A consequence of the fractal nature is a power law dependence of a correlation function (in our case the analyte–receptor on the biosensor surface) on a coordinate (e.g., time).

Pfeifer (1987) indicates that fractals may be used to track topographical features of a surface at different levels of scale. Lee and Lee (1995) indicate that the fractal approach permits a predictive approach for transport (diffusion-related) and reaction processes occurring on catalytic surfaces. This approach may presumably be extended to diffusion-limited analyte–receptor reactions occurring on biosensor surfaces.

The binding of an analyte in solution to a receptor attached to a solid (albeit flow cell or biosensor surface) is a good example of a low-dimension reaction system in which the distribution tends to be 'less random' (Kopelman, 1988), and a fractal analysis would provide novel physical insights into the diffusion-controlled reactions occurring at the surface. Also, when too many parameters are involved in a reaction, which is the case for these analyte-receptor reactions on a solid (e.g., biosensor surface), a fractal analysis provides a useful lumped parameter. It is appropriate to take particular care to the design of such systems and to explore new avenues by which further insight or knowledge may be obtained in these biosensor systems. The fractal approach is not new and has been used previously in analyzing different phenomena on lipid membranes.

Fatin-Rouge *et al.* (2004) have recently presented a summary of cases where the analysis of diffusion properties in random media have led to significant theoretical and experimental interest. These cases include soils (Sahimi, 1993), gels (Starchev *et al.*, 1997; Pluen *et al.*, 1999), bacteria cytoplasm (Berland *et al.*, 1995; Schwille *et al.*, 1999), membranes (Saffman and Delbruck, 1975; Peters and Cherry, 1982; Ghosh and Webb, 1988), and channels (Wei *et al.*, 2000). Coppens and Froment (1995) have analyzed the geometrical aspects of diffusion and reaction occurring in a fractal catalyst pore. In this chapter, and in this book as a whole, we are extending the analysis to analyte–receptor binding (and dissociation) on biosensor surfaces.

Fatin-Rouge *et al.* (2004) indicate that in most real systems disorder may exist over a finite range of distances. Harder *et al.* (1987) and Havlin (1989) indicate that in this range the diffusion process cannot be characterized by the classical Fick's law. In this range, anomalous diffusion applies. Fatin-Rouge *et al.* (2004) emphasize that at larger distances than in the above window range, the effects of disorder on diffusion may be very small due to statistical effects, and may cancel each other.

Prior to presenting the Havlin (1989) analysis modified for the analyte–receptor binding occurring on biosensor surfaces, it is appropriate to present briefly the analysis presented by Fatin-Rouge *et al.* (2004) on size effects on diffusion processes within agarose gels, and apply it to analyte–receptor binding and dissociation for biosensor kinetics. This analysis provides some insights into general fractal-related processes. Fatin-Rogue *et al.* (2004) have considered diffusion within a fractal network of pores. They indicate that fractal networks such as percolating clusters may be characterized by a power law distribution (Havlin, 1989):

$$M \propto (L)^{D_f} \tag{2.1}$$

Here M is the average number of empty holes in the (gel) space characterized by a linear size, L. The exponent D_f is the mass fraction dimension. Fatin-Rogue *et al.* (2004) emphasize that in the general case of fractals, D_f is smaller than the dimension of space of interest. Furthermore, the independence of D_f on scale is also referred to as self-similarity, and is an important property of rigorous fractals.

Havlin and Ben-Avraham (1987) indicate that the diffusion behavior of a particle within a medium can be characterized by its mean-square displacement, $r^2(t)$ versus time t, which is written as:

$$r^2(t) = \Gamma t^{(2/D_w)} \tag{2.2a}$$

Here Γ is the transport coefficient, and D_w is the fractal dimension for diffusion. Normal or regular diffusion occurs when D_w is equal to 2. In this case, $r^2(t)$ is equal to Γt. In other words,

$$r^2(t) = 2dDt \tag{2.2b}$$

Here d is the dimensionality of space, and D the diffusion coefficient.

Harder *et al.* (1987) and Havlin (1989) describe *anomalous diffusion* wherein the particles sense obstructions to their movement. This is within the fractal matrix, or in our case due to heterogeneities on the biosensor surface, perhaps due to irregularities on the biosensor surface. Fatin-Rogue *et al.* (2004) are careful to point out that anomalous diffusion may also occur due to non-elastic interactions between the network and the diffusing particles in a gel matrix (Saxton, 2001). Furthermore, Fatin-Rouge *et al.* (2004) indicate that *anomalous diffusion* is different from *trapped diffusion* wherein the particles are permanently trapped in holes, and are unable to come out of these holes. When the particles (analyte in our case) are in these trapped holes, then as time $t \rightarrow \infty$, the mean-square displacement $r^2(t)$ tends to a constant value. Fatin-Rouge *et al.* (2004) emphasize that in real heterogeneous porous media anomalous diffusion of particles occurs over a limited length- or time-scales since the structure is only fractal over a limited size scale. In other words, there is a lower bound and an upper bound over which the fractal structure applies. Similarly, in our case, the anomalous diffusion of the analyte on the biosensor surface occurs over a limited range of length- or time-scales.

For anomalous diffusion, one may combine the right-hand sides of eqs. (2.2a and b). Then, the diffusion coefficient D is given by (Fatin-Rouge *et al.*, 2004):

$$D(t) = (1/4)\Gamma t^{[(2/D_{\mathrm{w}})^{-1}]} \tag{2.3}$$

Due to the temporal nature of $D(t)$, it is better to characterize the diffusion of the analyte in our case by D_{w}. If we were still talking about the medium and gels, then D_{w} would refer to the diffusing medium.

We will now develop the theory for the analyte–receptor binding and dissociation on biosensor surfaces. We will use the Havlin (1989) approach.

2.2 THEORY

We present now a method of estimating fractal dimension values for analyte–receptor binding and dissociation kinetics observed in biosensor applications. The following chapters will present the different examples of data that have been modeled using the fractal analysis. The selection of the binding and dissociation data to be analyzed in the later chapters is constrained by whatever is available in the literature.

2.2.1 Variable rate coefficient

Kopelman (1988) has indicated that classical reaction kinetics are sometimes unsatisfactory when the reactants are spatially constrained on the microscopic level by either walls, phase boundaries, or force fields. Such heterogeneous reactions, for example, bioenzymatic reactions, that occur at interfaces of different phases, exhibit fractal orders for elementary reactions and rate coefficients with temporal memories. In such reactions, the rate coefficient exhibits a form given by:

$$k_1 = k' t^{-b} \qquad 0 \leqslant b \leqslant 1 (t \geqslant 1) \tag{2.4}$$

In general, k_1 depends on time whereas $k' = k_1\ (t=1)$ does not. Kopelman (1988) indicates that in three dimensions (homogeneous space) $b=0$. This is in agreement with the results obtained in classical kinetics. Also, with vigorous stirring, the system is made homogeneous and b again equals zero. However, for diffusion-limited reactions occurring in fractal spaces, $b \geqslant 0$; this yields a time-dependent rate coefficient.

Antibodies may form fractal clusters on biosensor surfaces. These antibodies or receptors on the biosensor surface may consist of islands of highly organized or disorganized antibodies. This has similarity to the growth of crystalline structures. It is quite possible that a cooperative effect may arise due to this tightly organized fractal structures. This is one possibility that could lead to an increase in the binding rate coefficient with an increase in the fractal dimension or the degree of heterogeneity on the biosensor surface.

The diffusion-limited binding kinetics of antigen (or antibody or analyte or substrate) in solution to antibody (or antigen or receptor or enzyme) immobilized on a biosensor surface has been analyzed within a fractal framework (Sadana and Beelaram, 1994; Sadana *et al.*, 1995). One of the findings, for example, is that an increase in the surface roughness or fractal dimension leads to an increase in the binding rate coefficient. Furthermore, experimental data presented for the binding of HIV virus (antigen) to the antibody immobilized on a surface displays characteristic ordered 'disorder' (Anderson, 1993). This indicates the possibility of a fractal-like surface.

A biosensor system (wherein either the antigen, antibody, analyte, or substrate is attached to the surface), along with its different complexities, which include heterogeneities on the surface and in solution, diffusion-coupled reaction, time-varying adsorption or binding rate coefficients, etc., can be characterized as a fractal system.

The diffusion of reactants toward fractal surfaces has been analyzed (De Gennes, 1982; Pfeifer *et al.*, 1984a,b; Nyikos and Pajkossy, 1986). Havlin (1989) has briefly reviewed and discussed these results. The diffusion is in the Euclidean space surrounding the fractal surface (Giona, 1992). Havlin (1989) presents an equation that may be utilized to describe the build-up of the analyte–receptor on a biosensor surface during the binding reaction. The receptor is immobilized on the biosensor surface. This equation is given below. In all fairness, at the outset, it is appropriate to indicate that the biosensor surface is assumed to be fractal, or possibly so.

Ideally, it is advisable to provide independent proof or physical evidence for the existence of fractals in the analysis of analyte–receptor reactions occurring on biosensor surfaces. Also, and as indicated earlier, if the diffusion effects can be separated from the heterogeneity effects, then one may better understand the effects of each of these on analyte–receptor reactions occurring on biosensor surfaces. In general, diffusion effects may be minimized either by increasing flow rates or by immobilizing less amounts of receptors on the biosensor surface.

In general, to demonstrate fractal-like behavior log–log plots of distribution of molecules $M(r)$ as a function of the radial distance (r) from a given molecule are required. This plot should be close to a straight line. The slope of log $M(r)$ *versus* log(r) plot determines the fractal dimension. In our case, one could try to obtain a log–log plot of two variables k and time t, and perform a least squares fit in this parameter space to find the slope of the curve. A regression coefficient at this stage could be beneficial in understanding the efficacy of this metric. However, an easier method, without the use of the required log–log plots, is presented below.

This is the equation developed by Havlin (1989) for diffusion of analytes toward fractal surfaces.

2.2.2 Single-fractal analysis

In the literature some authors refer to binding as comprising two phases, an association phase and a dissociation phase. In this chapter and in the book, we will refer to binding as just binding. The dissociation phase is separate.

Binding rate coefficient

Havlin (1989) indicates that the diffusion of a particle (analyte) from a homogeneous solution to a solid surface (e.g., receptor-coated surface) on which it reacts to form a product (analyte–receptor complex) is given by

$$(\text{Analyte.Receptor}) \sim \begin{cases} t^{(3-D_{\text{f,bind}})/2} = t^{p} & (t < t_c) \\ t^{1/2} & (t > t_c) \end{cases} \tag{2.5a}$$

where the analyte–receptor represents the association (or binding) complex formed on the surface. Here $p=-b$, and D_f is the fractal dimension of the surface. Havlin (1989) states that the crossover value may be determined by $r_c^2 \sim t_c$. Above the characteristic length r_c, the self-similarity of the surface is lost and the surface may be considered homogeneous. Eq. (2.5a) indicates that the concentration of the product [analyte–receptor] on a solid fractal surface scales at short and intermediate times as analyte–receptor ~ t^p with the coefficient $p=(3-D_f)/2$ at short time scales and $p=1/2$ at intermediate time scales. Note that D_f, $D_{f,assoc}$, and $D_{f,bind}$ are used interchangeably. This equation is associated with the short-term diffusional properties of a random walk on a fractal surface. Note that, in perfectly stirred kinetics on a regular (non-fractal) structure (or surface), the binding rate coefficient k_1 is a constant, i.e., it is independent of time. In other words, the limit of regular structures (or surfaces) and the absence of diffusion-limited kinetics lead to k_1 being independent of time. In all other situations, one would expect a scaling behavior given by $k_1 \sim k' t^{-b}$ with $-b=p < 0$. Also, the appearance of the coefficient p different from $p=0$ is the consequence of two different phenomena, that is, the heterogeneity (fractality) of the surface and the imperfect mixing (diffusion-limited) condition. Finally, for a homogeneous surface where $D_f = 2$, and when only diffusional limitations are present, $p=½$ as it should be. Another way of looking at the $p=1/2$ case (where $D_{f,bind} = 2$) is that the analyte in solution views the fractal object, in our case the receptor-coated biosensor surface, from a 'large distance'. In essence, in the association process, the diffusion of the analyte from the solution to the receptor surface creates a depletion layer of width $(Đt)^{1/2}$ where $Đ$ is the diffusion constant. This gives rise to the fractal power law,

$$(\text{Analyte.Receptor}) \sim t^{(3-D_{\text{f,bind}})/2}$$

The values of the parameters k (binding rate coefficient), p, and D_f in eq. (2.5a) may be obtained for analyte–receptor association kinetics data. This may be done by a regression analysis using, for example, Corel Quattro Pro (1997) along with eq. (2.5a) where (analyte.receptor) $=kt^p$ (Sadana and Beelaram, 1994; Sadana *et al.*, 1995). The fractal dimension may be obtained from the parameter p. Since $p=(3-D_{f,bind})/2$, $D_{f,bind} = (3-2p)$. In general, low values of p would lead to higher values of the fractal dimension $D_{f,bind}$. Higher values of the fractal dimension would indicate higher degrees of 'disorder' or heterogeneity or inhomogeneity on the surface.

Another way of looking at the diffusive process is that it inherently involves fluctuations at the molecular level that may be described by a random walk (Weiss, 1994). This author indicates that the kinetics of transport on disordered (or heterogeneous) media itself needs to be described by a random-walk model. When both these are present, that is the diffusion phenomena as well as a fractal surface, one needs to analyze the interplay of both of these fluctuations. In essence, the disorder on the surface (or a higher fractal dimension, D_f) tends to slow down the motion of a particle (analyte in our case) moving in such a medium. Basically, according to Weiss (1994) the particle (random walker analyte) is trapped in regions in space, as it oscillates for a long time before resuming its motion.

Havlin (1989) indicates that the crossover value may be determined by $r_c^2 \sim t_c$. Above the characteristic length, r_c, the self-similarity of the surface is lost. Above t_c, the surface may be considered homogeneous, and 'regular' diffusion is now present. One may consider the analysis to be presented as an intermediate 'heuristic' approach in that in the future one may also be able to develop an autonomous (and not time-dependent) model of diffusion-limited kinetics in disordered media.

It is worthwhile commenting on the units of the association and the dissociation rate coefficient(s) obtained for the fractal analysis. In general, for SPR biosensor analysis, the unit for the analyte–receptor complex on the biosensor surface is RU (resonance unit). One thousand resonance units is generally 1 ng/mm^2 (of surface), or one resonance unit is 1 pg/mm^2. Here, ng and pg are nanogram and picogram, respectively. Then, to help determine the units for the binding coefficient k from eq. (2.5a):

$$\text{(analyte.receptor), pg/mm}^2 = k\,t^p = k\,t^{(3-D_{f,bind})/2}$$

This yields a unit for the binding rate coefficient k as (pg)(mm)$^{-2}$(sec)$^{(D_{f,bind}-3)/2}$. Note that the unit of dependence in time exhibited by the association (or binding) rate coefficient, k changes slightly depending on the corresponding fractal dimension obtained in the binding phase, $D_{f,bind}$. The fractal dimension value is less than or equal to three. Three is the highest value of the fractal dimension, since the system is embedded in a three-dimensional system. k and k_{bind} and D_f and $D_{f,bind}$ are used interchangeably in this chapter and in the book.

It should be indicated that different laboratories use different technologies or different experimental designs to analyze the binding affinity of ligands to target proteins (or analytes) of interest (or to determine the rate coefficients for association and dissociation kinetics). The comparison of data between different technologies and experimental designs and conclusions thereof should be made with great caution. The fractal analysis is of value in that it provides the pros and cons of different *in vitro* technologies (or more precisely, in this case, analysis procedures). It makes the user of the technology aware of the quality of data generated and what can be done to improve the analysis.

One might very reasonably question the utility of the approach considering the different dimensions, and subsequently the units one may obtain even for the same interactions. It would be difficult to compare this technique with other approaches for different interactions. Nevertheless, the inclusion of the surface effects is essential, albeit difficult. This is especially true, if the rate coefficients for association and dissociation for binding are very significantly dependent on the nature of the surface. Unless, a simpler and alternate approach is suggested that includes the surface effects, it is reasonable for the present, to follow the present approach. Hopefully, modifications, to this approach may be suggested that permit the comparison for different interactions as well as with other approaches.

It would be useful to specify what the carrier of fractal properties is. It could either be the analyte surface, the receptor surface, or the immobilizing (in our case, the biosensor) surface. There is a considerable body of work on fractal surface properties of proteins (Li *et al.*, 1990; Federov *et al.*, 1999; Dewey and Bann, 1992; Le Brecque, 1992). Le Brecque (1992) indicates that the active sites (in our case the receptors on the biosensor surface) may themselves form a fractal surface. Furthermore, the inclusion of non-specific association sites on the surface would increase the degree of heterogeneity on the surface, thereby leading to an increase in the fractal dimension of the surface. At present, we are unable to specify what the carrier of the fractal properties is. This is exacerbated by our re-analysis of kinetic data available in the literature. Presumably, it is due to a composite of some or all of the factors mentioned above. No evidence of fractality is presented.

Dissociation rate coefficient

The diffusion of the dissociated particle (receptor or analyte) from the solid surface (e.g., analyte–receptor complex coated surface) into solution may be given as a first approximation by

$$\begin{aligned}(\text{Analyte.Receptor}) &\sim -t^{(3-D_{\text{f,diss}})/2} \qquad (t > t_{\text{diss}})\\ &= -k_{\text{diss}}t^{(3-D_{\text{f,diss}})/2}\end{aligned} \tag{2.5b}$$

Here $D_{\text{f,diss}}$ is the fractal dimension of the surface for the dissociation step. t_{diss} represents the start of the dissociation step. This corresponds to the highest concentration of the analyte–receptor complex on the surface. Henceforth, its concentration only decreases. $D_{\text{f,bind}}$ may or may not be equal to $D_{\text{f,diss}}$. k_{d} and k_{diss}, and $D_{\text{f,d}}$ and $D_{\text{f,diss}}$ are used interchangeably in this chapter and in this book.

One may obtain a unit for the dissociation rate coefficient k_{d} in a similar manner as done for the binding rate coefficient. In this case, the units for the binding and the dissociation rate coefficient are the same. The unit for the dissociation rate coefficient, k_{d} is (pg)(mm)$^{-2}$(sec)$^{(D_{f,diss}-3)/2}$. Once again, note that the unit dependence on time exhibited by k_d changes slightly due to the dependence on $D_{f,diss}$.

2.2.3 Dual-fractal analysis

Binding rate coefficient

The single-fractal analysis we have just presented is extended to include two fractal dimensions. At present, the time ($t=t_1$) at which the first fractal dimension "changes" to the second fractal dimension is arbitrary and empirical. For the most part it is dictated by the data analyzed and the experience gained by handling a single-fractal analysis. The r^2 (regression coefficient) value obtained is also used to determine if a single-fractal analysis is sufficient, or one needs to use a dual-fractal analysis to provide an adequate fit. Only if the r^2 value is less than 0.97 for a single-fractal analysis, do we use a dual-fractal model.

In this case, the analyte–receptor complex is given by

$$(\text{Analyte.Receptor}) \sim \begin{cases} t^{(3-D_{f1,bind})/2} = t^{p2} & (t < t_1) \\ t^{(3-D_{f2,bind})/2} = t^{p2} & (t_1 < t < t_2 = t_c) \\ t^{1/2} & (t > t_c) \end{cases} \tag{2.5c}$$

In analyte–receptor binding the analyte–receptor binds with the active site on the surface and the product is released. In this sense the catalytic surface exhibits an unchanging fractal surface to the reactant in the absence of fouling and other complications. In the case of analyte–receptor association the biosensor surface exhibits a changing fractal surface to the analyte in solution. This occurs since as each association reaction takes place, smaller and smaller amounts of 'association' sites or receptors are available on the biosensor surface to which the analyte may bind. Furthermore, as the reaction proceeds, there is an increasing degree of heterogeneity on the biosensor surface for some reaction systems. This is manifested by two degrees of heterogeneity, or two fractal dimensions on the biosensor surface. In the theoretical limit one might envisage a temporal fractal dimension wherein there is a continuous change in the degree of heterogeneity on the surface; though of course, such situations would be very rare, if at all.

Surfaces exhibit roughness, or a degree of heterogeneity at some scale. This degree of heterogeneity on the surface may be due to fracture or erosion. In our case of biosensors, this may arise due to (a) the inherent roughness of the biosensor surface, or (b) due to the immobilization or deposition of the receptors on the biosensor surface. The method of deposition of the receptors on the surface would also lead to different degrees of heterogeneity on the surface. The binding reaction takes place between the analyte in solution and the receptors on the surface through chemical bond formation and subsequent molecular association. The geometric nature (or parameter) of the surface will significantly influence these reactions. The influence of surface morphology and structure has been analyzed (Lee and Lee, 1994; Chaudhari *et al.*, 2002a,b). It would be of interest to determine the scale of these roughness heterogeneities. Are these at the Angstrom level or lower? With the current emphasis on nano-technology and nano-biotechnology these types of questions are becoming more and more relevant and of significance. The nature of surfaces in general, and of biosensors in particular (our case) should exhibit a fractal nature at the molecular level. Furthermore, one of the reasons for the emphasis on nano-technology is that as one goes down in scale, the properties of some substances change, sometimes for the better. It is these beneficial changes that one wishes to exploit in nano-technology and nano-biotechnology. Hopefully, similar parallels can be drawn on analyzing the fractal nature of biosensor surfaces. Do they exhibit self-similarity; and if they do what are their limits? In other words, what are their lower and upper bounds.

Furthermore, each binding event need not result in the formation of an analyte–receptor complex on the biosensor surface. All of the receptors on the biosensor surface are presumably not, and do not exhibit the same activity. In other words, their active sites should comprise presumably a probability distribution in 'activity.' *In lieu* of any prior information, it is reasonable to assume a bell-shaped Gaussian (or normal) distribution of active sites on the surface. A probabilistic approach is more realistic here. Such sort of analyses have presumably not been performed (at least this author is unaware of this) for analyte–receptor reactions occurring on biosensor surfaces. Thus, the fractal analysis is a convenient method of providing a lumped parameter analysis of analyte–receptor reactions occurring on biosensor surfaces.

Note that, at present, the dual-fractal analysis does not have a basis at the molecular level. This represents two different levels of heterogeneity on the biosensor surface. But, in some of the examples presented, a single-fractal analysis is clearly inadequate to model the data. Only in these cases does one resort to a dual-fractal analysis. The binding rate coefficients k_1 and k_2 in the dual-fractal analysis have the same units $(pg)(mm)^{-2}(sec)^{(D_{f1,bind}-3)/2}$ and $(pg)(mm)^{-2}(sec)^{(D_{f2,bind}-3)/2}$, respectively, as the association rate coefficient, k in the single-fractal analysis.

Dissociation rate coefficient

In this case the dissociation rate coefficient is given by

$$(\mathrm{Ab.Ag}) \approx \begin{cases} -t^{(3-D_{f1,bind})/2} & (t_{diss} < t < t_{d1}) \\ -t^{(3-D_{f2,diss})/2} & (t_{d1} < t < t_{d2}) \end{cases} \tag{2.5d}$$

Here $D_{f,diss}$ is the fractal dimension of the surface for the dissociation step. t_{diss} represents the start of the dissociation step. This corresponds to the highest concentration of the analyte–receptor on the surface. Henceforth, its concentration only decreases. $D_{f,bind}$ or $D_{f,assoc}$ may or may not be equal to $D_{f,diss}$. The dissociation rate coefficients, k_{d1} and k_{d2} in the dual-fractal analysis have the same units $(pg)(mm)^{-2}(sec)^{(D_{fd1}-3)/2}$ and

$(\text{pg})(\text{mm})^{-2}(\text{sec})^{(D_{\text{fd2}}-3)/2}$, respectively, as the dissociation rate coefficient, k_d, in the single-fractal analysis.

2.2.4 Triple-fractal analysis

One resorts to a triple-fractal analysis if the dual-fractal analysis does not provide a reasonable fit. As will be shown later on in the book, one resorts to a triple-fractal analysis when the dual-fractal analysis does not provide an adequate fit. The equation for the fractal analysis equation is generic in nature, and one may easily extend the single- and the dual-fractal analysis equations (eqs. 2.5a and 2.5c) to describe the binding (and/or the dissociation) kinetics for a triple-fractal analysis. In fact, in the extreme case, *n* fractal dimensions may be present. In this case, the degree of heterogeneity D_f or the fractal dimension is continuously changing on the biosensor surface, and the surface needs to be represented by D_{fi} where *i* goes from 1 to *n*. Similarly, we have *n* binding rate coefficients on the biosensor surface. A similar representation may also be made for the dissociation phase.

It is perhaps appropriate to at least mention one more approach that has been used to model the binding kinetics on surfaces.

2.2.5 Pfeifer's fractal binding rate theory

Pfeifer (1989) has suggested an alternate form of the binding rate theory. In the equation given in this reference *N* is the number of complexes, N_0 is the number of receptors on the solid surface, *D* is the diffusion coefficient of the analyte, *L* is the receptor diameter, and λ is the mean distance between two neighboring receptors. This equation may also be used to analyze the analyte–receptor binding kinetics. The problem, however, is that it may not be possible in all instances to estimate *a priori* all of the parameters described in the equation (not given here). In that case, one may have to approximate or assume certain values, and this will affect the accuracy and reliability of the analysis. The suggested equation does have an advantage compared to the fractal analysis described above in that it does include a pre-factor necessary to convert the time interval over which fractal scaling is observed into a length interval. It also provides an expression for $t_c(=L^2/D)$, which separates the short-term regime from the long-term regime. The short-term regime is the one in which the anomalous diffusion applies. At the end of the short-term interval $(t=t_c)$, the self-similarity of the system is lost, the surface is homogeneous, and regular diffusion applies.

Pfeifer (1989) states that the application of the above equation is contingent on

(a) The analyte is uniformly distributed in the solution at time $t=0$.
(b) Binding is irreversible and first-order (*N* equals the number of analyte particles that have reached the receptors).
(c) Binding occurs whenever an incoming analyte particle hits a receptor surface for the first time. In other words, the 'sticking' probability is one.

It is very difficult to imagine perhaps any one or all of these conditions being satisfied for analyte–receptor binding interactions occurring in continuous-flow reactors.

Given the extremely small volume of the flow channels there is a high probability that the mixing of the analyte is not proper. This in turn may lead to analyte depletion in the flow channel. Also, the binding cannot be assumed to be irreversible in all instances. There may be cases of extremely fast binding and dissociation, especially for analytes with low affinity which can dissociate in the continuously flowing buffer without any regeneration reagent.

Condition (c) may be satisfied, however, it does not include the 'sticking' probability, in that each collision leads to a binding event. Also, the presence of non-specific binding, avidity effects, and binding with reactions or binding of dissociated analytes may interfere with condition (c) being satisfied. Furthermore, the equation makes assumptions about the number of active sites, and the immobilized receptors. For example it states that the analyte binds to one specific active site. The receptor cannot bind to more than one analyte molecule at one time (1:1 binding).

The equilibrium dissociation rate coefficient, $K_D = k_{diss}/k_{assoc}$ can be calculated using the above models. The K_D value is frequently used in analyte–receptor reactions occurring on biosensor surfaces. The ratio besides providing physical insights into the analyte–receptor system is of practical importance since it may be used to help determine (and possibly enhance) the regenerability, reusability, stability, and other biosensor performance parameters. K_D has the unit $(\text{sec})^{[D_{f,diss}-D_{f,assoc}]/2}$. This applies to both the single- as well as the dual-fractal analysis. For example, for a single-fractal analysis, K_D has the units $(\text{sec})^{[D_{fd}-D_f]/2}$. Similarly, for a dual-fractal analysis, the affinity K_{D1} has the units (sec) $[D_{fd1}-D_{fassoc1}]/2$ and K_{D2} has the units $(\text{sec})^{[D_{fd2}-D_{fassoc\ 2}]/2}$. Note the difference in the units of the equilibrium dissociation rate coefficient obtained for the classical as well as the fractal-type kinetics. Though the definition of the equilibrium dissociation rate coefficient is the same in both types of kinetics (ratio of the dissociation rate coefficient to the association rate coefficient), the difference(s) in the units of the different rate coefficients eventually leads to a different unit for the equilibrium dissociation rate coefficient in the two types of kinetics. This is not entirely unexpected since the classical kinetic analysis does not include the characteristics of the surface in the definition of the equilibrium dissociation rate coefficient whereas the present fractal analysis does. Thus, one may not be able to actually compare the equilibrium dissociation rate coefficient affinities in these two types of systems. This is a significant difference in the kinetic analysis of binding and dissociation reactions on biosensor surfaces from what is available in the literature.

REFERENCES

Anderson, J. (1993). NIH Panel Review Meeting, Case Western Reserve University, Cleveland, Ohio, July.

Berland, K.M., So, P.T.C., and Gratton, E. (1995). Two-photon fluorescence correlation spectroscopy: method and application to the intracellular environment. *Biophysical Journal*, **68**, 694–701.

Chaudhari, A., Yan, C.C., and Lee, S.L. (2002a). Multifractal analysis of diffusion-limited reactions over surfaces of diffusion-limited aggregates. *Chemical Physics Letters*, **207**, 220–226.

Chaudhari, A., Yan, C.C., and Lee, S.L. (2002b). *Journal of Physics A*, **36**, 3757.

Coppens, M.O., and Froment, G.F. (1995). Diffusion and reaction in a fractal catalyst pore. 1. Geometrical aspects. *Chemical Engineering Science*, **50**(6), 1013–1026.

Corel Quattro Pro (1997). Corel Corporation Limited, Ottawa, Canada.

De Gennes, P.G. (1982). Diffusion-controlled reactions. In: *Polymer Melts. Radiat. Phys. Chem.*, **22**, 193.

Dewey, T.G., and Bann, J.G. (1992). Diffusion-controlled reaction in polymer melts. *Biophysical Journal*, **63**, 594.

Fatin-Rouge, N., Starchev, K., and Buffle, J. (2004). Size effects on diffusion process with agarose gels. *Biophysical Journal*, **86**, 2710–2719.

Federov, B.A., Federov, B.B., and Schmidt, P.W. (1999). An analysis of the fractal properties of globular proteins. *Journal of Chemical Physics,* **99**, 4076–4083.

Ghosh, R.N., and Webb, W.W. (1988). Results of automated tracking of low-density lipoprotein receptors on cell surfaces. *Biophysical Journal*, **53**, A352.

Giona, M. (1992). First-order reaction-diffusion in complex fractal media. *Chemical Engineering Science*, **47**, 1503–1515.

Harder, F.H., Havlin, S., and Bunde, A. (1987). Diffusion in fractals with singular waiting-time distribution. *Physics Reviews B*, **36**, 3874–3879.

Havlin, S. (1989). Molecular diffusion and reaction. In: *The Fractal Approach To Heterogeneous Chemistry: Surfaces, Colloids, Polymers* (ed. D. Avnir), Wiley, New York, pp. 251–269.

Havlin, S., and Ben-Avraham, D. (1987). Diffusion in disordered media. *Advance in Physics*, **36**, 695–798.

Henke, L., Nagy, N., and Krull, U.J. (2002). An AFM determination of the effects of surface roughness caused by cleaning of fused silica and glass substrates in the process of optical biosensor preparation. *Biosensors & Bioelectronics*, **17**, 547–555.

Kopelman, R. (1988). Fractal reaction kinetics. *Science*, **241**, 1620–1624.

Le Brecque, M. (1992). *Mosaic*, **23**, 12–15.

Lee, C.K., and Lee, S.L. (1994). Multifractal scaling analysis of autocatalytic and autopoisoning reactions over DLA surfaces. *Chemical Physics Letters*, **228**, 539-546.

Lee, C.K., and Lee, S.L. (1995). Multifractal scaling analysis of reactions over fractal surfaces. *Surface Science*, **325**, 294–310.

Li, H.L., Chen, S., and Zhao, H. (1990). Fractal mechanisms for the allosteric effects of proteins and enzymes. *Biophysical Journal*, **58**, 1313–1320.

Markel, V.A., Muratov, L.S., Stockman, M.I., and George, T.F. (1991). Theory and numerical simulation of optical properties of fractal clusters. *Physics Reviews B*, **43**(10), 8183.

Nyikos, L., and Pajkossy, T. (1986). Diffusion to fractal surfaces. *Electrochimica Acta*, **31**, 1347.

Pajkossy, T., and Nyikos, L. (1989). Diffusion to fractal surfaces. II.Verification of theory. *Electrochimica Acta*, **34**, 171.

Peters, R., and Cherry, R.J. (1982). Lateral and rotational diffusion of bacteriorhodopsin in lipid bilayers: experimental test of the Saffman-Delbruck equations. *Proceedings of the National Academy of Sciences, USA,* **79**, 4317–4321.

Pfeifer, P. (1987). Characterization of surface irregularity. Chapter 2. In: *Preparative Chemistry Using Supported Reagents*, Academic Press, San Diego.

Pfeifer, P., Avnir, D., and Farin, D.J. (1984a). Molecular fractal surfaces. *Nature*, **308**, 261.

Pfeifer, P., Avnir, D., and Farin, D.J. (1984b). Surface geometric irregularity of particulate materials. The fractal approach. *Journal of Colloid and Interface Science*, **103**(1), 112.

Pfeifer, P., and Obert, M. (1989). Fractals: Basic concepts and terminology. In: *The Fractal Approach to Heterogeneous Chemistry: Surfaces, Colloids, Polymers* (ed. D. Avnir), Wiley, New York, pp. 251–269.

Pluen, A., Netti, P.A., Rakesh, K.J., and Berk, D.A. (1999). Diffusion of macromolecules in agarose gels: comparison of linear and globular configurations. *Biophysical Journal*, **77**, 542–552.

Sadana, A., and Beelaram, A. (1994). Fractal analysis of antigen-antibody binding kinetics. Biosensor applications. *Biotechnology Progress*, **9**, 45.

Sadana, A., Alarie. J.P., and Vo-Dinh, T. (1995). A β-cyclodextrin based fiber-optic chemical sensor: a fractal analysis, *Talanta*, **42**, 1567.

Saffman, P.G., and Delbruck, M. (1975). Brownian motion in biological membranes. *Proceedings of the National Academy of Sciences, USA*, **72**, 3111–3113.

Sahimi, M. (1993). Flow phenomena in rocks: from continuum models to fractals, percolation, cellular automata and simulated annealing. *Reviews in Modern Physics*, **65**, 1393–1534.

Saxton, M.J. (2001). Anomalous diffusion due to binding: a Monte Carlo study. *Biophysical Journal*, **77**, 2251–2265.

Schwille, P., Haupts, U., Maiti, S., and Webb, W.W. (1999). Molecular dynamics in living cells observed by fluorescence correlation spectroscopy with one- and two-photon excitation. *Biophysical Journal*, **77**, 2251–2265.

Starchev, K., Sturm, J., Weill, G., and Brogen, C.H. (1997). Brownian motion and electrophoretic transport in agarose gels studied by epifluorescence microscopy and simple particle tracking analysis. *Journal of Physical Chemistry*, **101**, 5659–5663.

Wei, Q.H., Bechinger, C., and Leiderer, P. (2000). Single-file diffusion of colloids in one-dimensional channels. *Science*, **287**, 625–627.

Weiss, G.H, *Fractals in Science*, Springer-Verlag, Berlin, 1994, p. 119.

– 3 –

Fractal Analysis of Toxin and Pathogen Detection on Biosensor Surfaces

3.1 INTRODUCTION

The detection of biological toxins is important in the areas of food safety, industry, and in environmental monitoring. Events in the recent past have indicated the involvement of biological toxins as terrorist threats. They have been used in the military arena for quite a while. One of the earliest documented uses is perhaps the donation of blankets infected with small pox by a British general to American Indians to purposely infect them with this dreaded disease (History Channel on TV, 2005). Over the years, different countries in the West and elsewhere have developed 'weaponized toxins' that can annihilate large numbers of a population at a relatively low cost compared to other weapons of mass destruction (WMD).

Rubina *et al.* (2005) very recently emphasized the importance of developing fast, reliable, and sensitive methods to detect a wide variety of biological toxins that may be used as 'weapons' by individual groups or even a country. Some very lethal biological toxins include botulinum, cholera, ricin, and tetanus toxins. These authors emphasize that the lethal doses for these deadly toxins are in the nanogram range. Some other deadly toxins indicated by these authors with their sources given in brackets include viscumin [*viscum album* mistletoe] (Temyakov *et al.*, 1997; Jaggy *et al.*, 1995), staphylococcal enterotoxin B [*Staphylococcus aureus*] (Miyamoto *et al.*, 2003; Poli *et al.*, 2002), tetanus toxin [*Clostridia tetani*] (USAMRIID, 2001; Burkin *et al.*, 2004), and Anthrax toxin protective agent [*Bacillus anthracis*] (Leppla, 1995). In the recent past a couple of Anthrax-related incidents in the United States in the civilian arena (governmental facilities) have made news headlines, which has led to corrective and protective measures.

There is a need for the simultaneous detection of different biotoxins. Microarrays and microchips with sets of different probes (specific for a biotoxin) would permit the parallel and simultaneous detection of different pathogens. Rucker *et al.* (2005) indicate that they have recently developed an antibody-based microarray for the multiplexed detection of a variety of biological toxins that include cholera toxin β-subunit, anthrax lethal factor and protective antigen, tetanus toxin C fragment and *Staphylococcus aureus* enterotoxin B. Their antibody microarray was used in spiked samples. O'Brien *et al.* (2000) and Rowe *et al.* (1999) indicate the different advantages of the microarray platform that include portability, low rate of false positives (which helps minimize inconvenience(s), low analysis time, inexpensive components, and flexibility. Furthermore, Rucker *et al.* (2005) emphasize that the microarray platform would also be effectively used in the clinical area and for field-ready diagnostic devices.

Ligler *et al.* (2003) have developed an array biosensor that can rapidly detect multiple targets simultaneously on a single-waveguide surface. These authors have used sandwich and competitive fluoroimmunoassays to simultaneously detect both low- and high-molecular-weight toxins in complex samples. They were able to detect toxins such as ricin, cholera toxin, and trinitrotoluene at levels as low as 0.5 ng/M. Finally, this same group (Sapsford *et al.*, 2004) has reviewed fluorescence-based array biosensors for the detection of biohazards.

An analysis of the kinetics of binding (and dissociation) of biological toxins would significantly enhance the performance of the detection devices used. In this chapter fractal analysis is used to analyze the binding kinetics of (a) yessotoxin (YTX) in solution to phosphodiesterase 3′,5′ - cyclic-nucleotide specific to bovine brain (PDE) immobilized on an aminosilane surface (Pazos *et al.*, 2004), (b) of Cy3-labeled ricin in solution to gel-immobilized antibodies against ricin Rch 1 (Rubina *et al.*, 2005), and (c) lipopolysaccharide (LPS) in solution to cyclic decapeptide immobilized on a CM5 sensor chip surface (Thomas and Surolia, 1999). Binding rate coefficient and fractal dimension values are provided. The fractal analysis may be considered as an alternate analysis, which provides much needed physical insights into these types of interactions that in turn may lead to enhancement of the performance parameters involved in biosensing methods.

3.2 THEORY

Havlin (1989) has reviewed and analyzed the diffusion of reactants toward fractal surfaces. The details of the theory and the equations involved for the binding and the dissociation phases for analyte–receptor binding are available (Sadana, 2001). The details are not repeated here, except that just the equations are given to permit an easier reading. These equations have been applied to other biosensor systems (Ramakrishnan and Sadana, 2001; Sadana, 2001). Here we will attempt to apply these equations to the binding kinetics of (a) yessotoxin (YTX) in solution to phosphodiesterase 3′,5′-cyclic-nucleotide specific from bovine brain (PDE) immobilized on an aminosilane surface (Pazos *et al.*, 2004), (b) of Cy3-labeled ricin in solution to gel-immobilized antibodies against ricin Rch 1 (Rubina *et al.*, 2005), and (c) lipopolysaccharide in solution to cyclic decapeptide immobilized on a CM5 sensor chip surface (Thomas and Surolia, 1999). We recognize that these systems are very dilute. A triple-fractal analysis may be used only, if necessary, to accommodate this very dilute nature of these systems. For most applications, a single- or a dual-fractal analysis is often adequate to describe the binding and the dissociation kinetics. Peculiarities in the values of the binding and the dissociation rate coefficients as well as in the values of the fractal dimensions with regard to the dilute analyte systems being analyzed will be carefully noted, if applicable.

3.2.1 Single-fractal analysis

Binding rate coefficient

Havlin (1989) indicates that the diffusion of a particle (analyte [Ag]) from a homogeneous solution to a solid surface (e.g. receptor [Ab]-coated surface) on which it reacts to form a product (analyte–receptor complex; (Ab.Ag)) is given by

$$(\text{Ab.Ag}) \approx \begin{cases} t^{(3-D_{\text{f,bind}})/2} = t^{p}, & t < t_c \\ t^{1/2}, & t > t_c \end{cases} \tag{3.1a}$$

Here $D_{\text{f,bind}}$ or D_{f} (used later on in the chapter) is the fractal dimension of the surface during the binding step; t_{c} is the cross-over value. Havlin (1989) indicates that the cross-over value may be determined by using $r_{\text{c}}^{2} \sim t_{\text{c}}$. Above the characteristic length, r_{c}, the self-similarity of the surface is lost and the surface may be considered homogeneous. Above time t_{c}, the surface may be considered homogeneous, since the self-similarity property disappears and 'regular' diffusion is now present. For a homogeneous surface where $D_{\text{f}} = 2$ and when only diffusional limitations are present, $p = ½$ as it should be. Another way of looking at the $p = ½$ case (where $D_{\text{f,bind}} = 2$) is that the analyte in solution views the fractal object; in our case, the receptor-coated biosensor surface from a 'large distance.' In essence, in the association process, the diffusion of the analyte from the solution to the receptor surface creates a depletion layer of width $(Ɖt)^{½}$, where $Ɖ$ is the diffusion constant. This gives rise to the fractal power law, (Analyte·Receptor) ~ $t^{(3 - D_{\text{fbind}})/2}$. For the present analysis, t_{c} is arbitrarily chosen and we assume that the value of t_{c} is not reached. One may consider the approach as an intermediate 'heuristic' approach that may be used in the future to develop an autonomous (and not time-dependent) model for diffusion-controlled kinetics.

3.2.2 Dual-fractal analysis

Binding rate coefficient

Sometimes, the binding curve exhibits complexities and two parameters (k, D_{f}) are not sufficient to adequately describe the binding kinetics. This is further corroborated by low values of r^2 factor (goodness-of-fit). In that case, one resorts to a dual-fractal analysis (four parameters: k_1, k_2, D_{f1}, and D_{f2}) to adequately describe the binding kinetics. The single-fractal analysis presented above is thus extended to include two fractal dimensions. At present, the time ($t = t_1$) at which the 'first' fractal dimension 'changes' to the 'second' fractal dimension is arbitrary and empirical. For the most part, it is dictated by the data analyzed and experience gained by handling a single-fractal analysis. A smoother curve is obtained in the 'transition' region if care is taken to select the correct number of points for the two regions. In this case, the product (antibody–antigen, or analyte–receptor complex Ab.Ag, or analyte.receptor) is given by

$$(\text{Ab.Ag}) \approx \begin{cases} t^{(3-D_{\text{f1,bind}})/2} = t^{p1}, & t < t_1 \\ t^{(3-D_{\text{f2,bind}})/2} = t^{p2}, & t_1 < t < t_2 = t_c \\ t^{1/2}, & t > t_c \end{cases} \tag{3.1b}$$

In some cases, as mentioned above, a triple-fractal analysis with six parameters (k_1, k_2, k_3, D_{f1}, D_{f2}, and D_{f3}) may be required to adequately model the binding kinetics. This is when the binding curve exhibits convolutions and complexities in its shape due to the very dilute nature of the analyte (in some of the cases to be presented) or for some other reasons. Also, in some cases, a dual-fractal analysis may be required to describe the dissociation kinetics.

3.3 RESULTS

In this chapter we use fractals to analyze the binding kinetics of yessotoxin (YTX) in solution to PDE immobilized on an immunosilane surface (Pazos *et al.*, 2004), Cy3-labeled ricin in solution to gel-immobilized antibodies against ricin Rch 1 (Rubina *et al.*, 2005), and lipopolysaccharide (LPS) in solution to cyclic decapeptide immobilized on a CM5 sensor chip surface (Thomas and Surolia, 1999). Note that this is just one possible way of analyzing the binding kinetics.

Alternate expressions for fitting the data are available that include saturation, first-order reaction, and no diffusion limitations, but these expressions are apparently deficient in describing the heterogeneity that inherently exists on the surface. One might justifiably argue that appropriate modeling may be achieved by using a Langmuirian or other approach. The Langmuirian approach may be used to model the data presented if one assumes the presence of discrete classes of sites (e.g. double exponential analysis as compared with a single-fractal analysis). Lee and Lee (1995) indicate that the fractal approach has been applied to surface science, for example, adsorption and reaction processes. These authors indicate that the fractal approach provides a convenient means to represent the different structures and morphology on the reaction surface. These authors also emphasize using the fractal approach to develop optimal structures and as a predictive approach. Another advantage of the fractal technique is that the analyte–receptor association (as well as the dissociation reaction) is a complex reaction, and the fractal analysis via the fractal dimension and the rate coefficient provides a useful lumped parameter(s) analysis of the diffusion-limited reaction occurring on a heterogeneous surface.

To demonstrate fractality in a classical situation, one should make a log–log plot, and definitely have a large amount of data. It may be useful to compare the fit to some other forms, such as exponential or one involving saturation, etc. At present, we do not present any independent proof or physical evidence of fractals in the examples presented. It is a convenient means (since it is a lumped parameter) to make the degree of heterogeneity that exists on the surface more quantitative. Thus, there is some arbitrariness in the fractal model to be presented. The fractal approach provides additional information about interactions that may not be obtained by conventional analysis of biosensor data.

There is no nonselective adsorption of the analyte. The present system (pathogens) being analyzed is typically very dilute. Nonselective adsorption would skew the results obtained very significantly. In these types of systems, it is imperative to minimize this nonselective adsorption. We also recognize that, in some cases, this nonselective adsorption may not be a significant component of the adsorbed material and that this rate of association, which is of a temporal nature, would depend on surface availability. If we were to accommodate the nonselective adsorption in the model, there would be an increase in the heterogeneity on the surface, since, by its very nature, nonspecific adsorption is more homogeneous than specific adsorption. This would lead to higher fractal dimension values since the fractal dimension is a direct measure of the degree of heterogeneity that exists on the surface.

Pazos *et al.* (2004) have recently used the resonant mirror biosensor to kinetically analyze the binding interactions of 1 μM YTX in solution to phosphodiesterase 3′-5′-cyclic-nucleotide specific from bovine serum brain (PDE) immobilized on an aminosilane surface of a biosensor. These authors indicate that YTX is a polyether lipophilic marine toxin, and exhibits a potent mouse lethality of 0.1 mg/kg when administered intraperitoneally. However, they further state that its toxicological effects on human health are unknown. Shellfish get contaminated by YTX. Thus, the efforts by these authors are to detect and make the YTX concentrations quantitative.

Figures 3.1a–e show the binding of 1, 3.75, 7, 10, and 15 μM YTX in solution to PDE immobilized on an aminosilane surface of a biosensor (Pazos *et al.*, 2004). A dual-fractal analysis is required to adequately describe the binding kinetics. The values of (a) the binding rate coefficient, k, and the fractal dimension, D_f, for a single-fractal analysis, and (b) the binding rate coefficients, k_1 and k_2, and the fractal dimensions, D_{f1} and D_{f2}, for a dual-fractal analysis are given in Table 3.1.

Note the change in the binding mechanism as one goes from the 1 to the 3.75 μM YTX in solution. At the lower concentration (1 μM) a dual-fractal analysis is required to describe the binding kinetics, whereas at the higher concentration (3.75 μM) a single-fractal analysis is adequate to describe the binding kinetics.

It is of interest to note that at the intermediate range of YTX concentrations (3.75–10 μM) used in solution, a single-fractal analysis is adequate to describe the binding kinetics. This would indicate that the binding mechanism of YTX in solution in this concentration range to immobilized PDE is similar. At the lowest (1 μM YTX) and at the highest (15 μM YTX) concentration in solution used, the binding mechanisms are similar, since in both cases a dual-fractal analysis is required to describe the binding kinetics.

Table 3.1

Binding rate coefficients and fractal dimensions for the yessotoxin (YTX) in solution to phosphodiesterase 3′-5′-cyclic-nucleotide-specific from bovine serum brain (PDE) immobilized on an aminosilane surface (Pazos *et al.*, 2004)

YTX concentration (μM)	k	k_1	k_2	D_f	D_{f1}	D_{f2}
1	0.2380 ± 0.0700	0.1433±0.0361	0.9612 ± 0.0301	0.9088±0.1260	0.0766+0.2204	1.9320±0.0488
3.75	2.059 ± 0.115	NA	NA	1.7482±0.0264	NA	NA
7	4.057 ± 0.1414	NA	NA	2.0146±0.0168	NA	NA
10	7.9170 ± 0.5038	NA	NA	2.1702±0.0302	NA	NA
15	9.0557 ± 0.8609	7.3564±0.7545	12.0630 ± 0.1225	2.1122±0.044	1.7190±0.1164	2.3292±0.1022

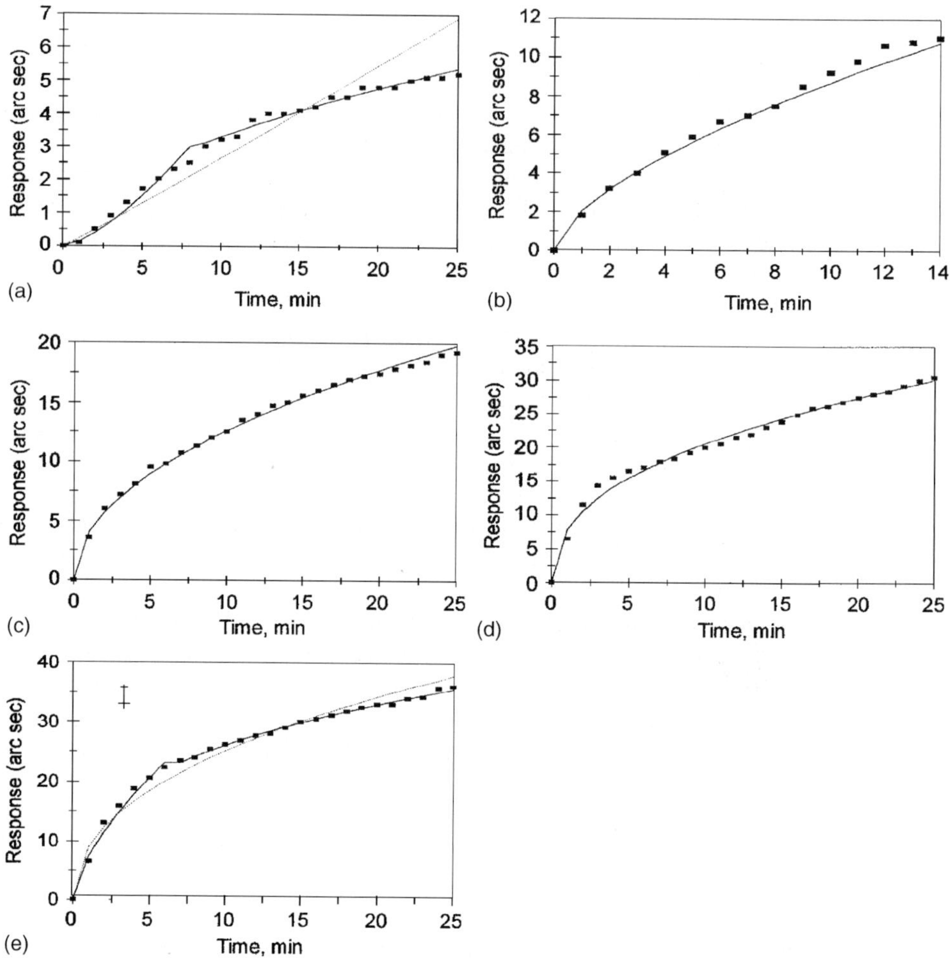

Figure 3.1 Binding of different concentrations (in μM) of yessotoxin (YTX) in solution to phosphodiesterase 3′-5′-cyclic-nucleotide-specific from bovine serum brain (PDE) immobilized on an aminosilane surface (Pazos *et al.*, 2004): (a) 1 (b) 3.75 (c) 7 (d) 10 (e) 15. When only a solid line is used then a single-fractal analysis applies. When both dashed (-------) and solid (_____) lines are used, then the dashed line is for a single-fractal analysis, and the solid line is for a dual-fractal analysis. In this case the solid line provides a better fit.

In the 3.75–10 μM YTX concentration range in solution where a single-fractal analysis is adequate to describe the binding kinetics, we observe the following: (1) Table 3.1 and Figure 3.2a show the increase in the binding rate coefficient, *k*, with an increase in the YTX concentration. The binding rate coefficient, *k*, is given by

$$k = (0.4152 \pm 0.0772)[\text{YTX}]^{1.2410 \pm 0.2260} \tag{3.2a}$$

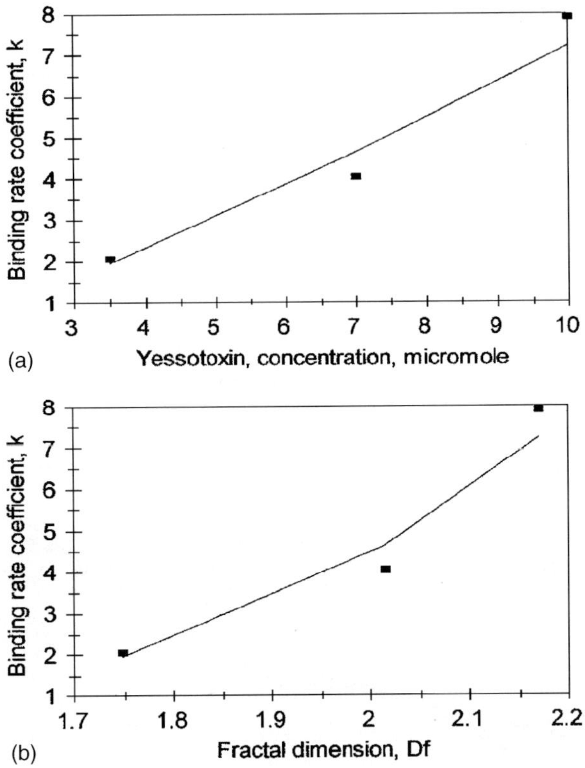

Figure 3.2 (a) Increase in the binding rate coefficient, k, with an increase in the YTX concentration in solution, (b) Increase in the binding rate coefficient, k, with an increase in the fractal dimension, D_f.

(2) Table 3.1 and Figure 3.2b show the increase in the binding rate coefficient, k, with an increase in the fractal dimension, D_f. The binding rate coefficient, k, is given by

$$k = (0.06752 \pm 0.0122) D_f^{6.0359 \pm 1.0693} \tag{3.2b}$$

The fit is good. Only three data points are available. The availability of more data points would provide for a more reliable fit. In case (1), the binding rate coefficient, k, exhibits an order of dependence between first and one and one-half (equal to 1.2410) on the YTX concentration in solution. The fractional order of dependence exhibited by the binding rate coefficient, k, in the 3.75–10 μM range reinforces the fractal nature of the system. In case (2) the binding rate coefficient, k, is very sensitive to the degree of heterogeneity on the aminosilane surface as noted by the greater than sixth order of dependence exhibited by k on the fractal dimension, D_f.

Figures 3.3a and b and Table 3.2 show the effect of temperature on the binding kinetics of yessotoxin in solution to PDE immobilized on an aminosilane surface. In this case, at higher temperature, a dual-fractal analysis is required to describe the binding kinetics, and

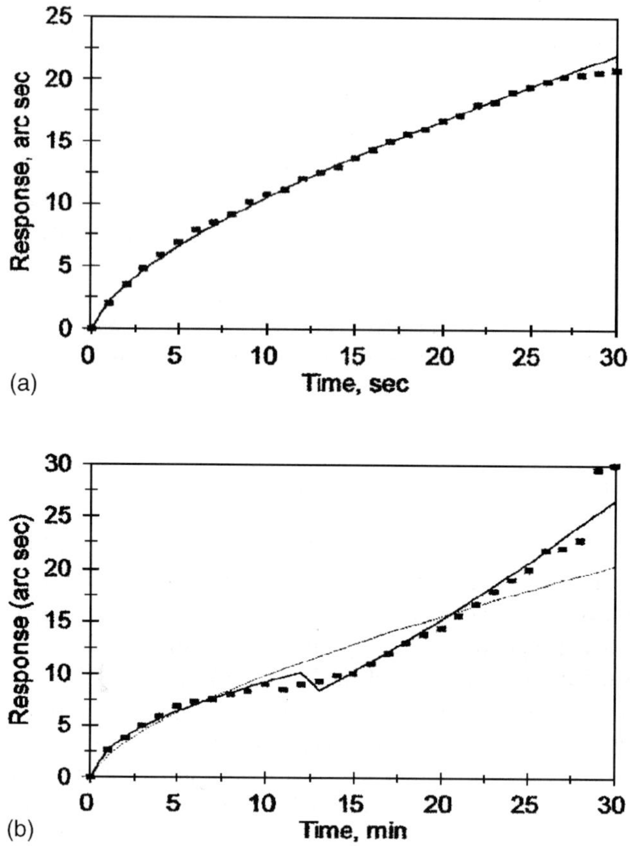

Figure 3.3 Binding of 7 μM YTX in solution at (a) 20 °C and (b) 37 °C to PDE immobilized on an aminosilane surface (Pazos *et al.*, 2004).

Table 3.2

Binding rate coefficients and fractal dimensions for the yessotoxin (YTX) in solution to phosphodiesterase 3′-5′-cyclic-nucleotide-specific from bovine serum brain (PDE) immobilized on an aminosilane surface. Effect of (a) temperature and (b) spiking YTX to a contaminated sample (Pazos *et al.*, 2004)

YTX concentration (μM)	T (°C)	k	k_1	k_2	D_f	D_{f1}	D_{f2}
(a) 7	22	2.2275 ± 0.0709	NA	NA	1.6540 ± 0.0139	NA	NA
7	37	2.1704 ± 0.4109	2.6967 ± 0.1254	0.2493 ± 0.0167	1.6838 ± 0.0756	1.9300 ± 0.0412	0.2548 ± 0.1212
(b) 3.75	NA	1.6839 ± 0.1070	1.1019 ± 0.1070	4.7532 ± 0.1455	1.7368 ± 0.0965	1.0110 ± 0.0994	2.5124 ± 0.0388

the results are given in Table 3.2. It is of interest to note that as one goes from the lower (22 °C) to the higher temperature (37 °C), there is a change in the binding mechanism since at the lower and higher temperatures a single- and a dual-fractal analysis is required, respectively, to describe the binding kinetics.

Also note that for the 37 °C case as the fractal dimension decreases by a factor of 7.57 from D_{f1} = 1.9300 to D_{f2} = 0.2548, the binding rate coefficient decreases by a factor of 10.82 from k_1 = 2.6967 to k_2 = 0.2493. Changes in the degree of heterogeneity on the aminosilane surface and in the binding rate coefficient are in the same direction.

Pazos *et al.* (2004) deliberately spiked a shellfish sample with 3.75 μM YTX. Their calculations yielded a YTX concentration of 3.85 μM. This indicated the sensitivity of their method. Figure 3.4 shows the binding curve for this spiked sample. A dual-fractal analysis is required to adequately describe the binding kinetics, and the results are given in Table 3.2b.

Note that in this case as the fractal dimension increases by a factor of 2.49 from D_{f1} = 1.0110 to D_{f2} = 2.5124, the binding rate coefficient increases by a factor of 4.31 of k_1 = 1.1019 to k_2 = 4.7532. Once again, changes in the degree of heterogeneity or the fractal dimension on the aminosilane surface and in the binding rate coefficient are in the same direction.

Rubina *et al.* (2005) have developed a hydrogel-based protein microchip to quantitatively detect a series of plant and bacterial toxins. These include ricin, viscumin, staphylococcal enterotoxin B, tetanus and diphtheria toxins, and the lethal factor of anthrax. Using their sandwich-type immunoassay they were able to obtain a highly sensitive biosensor that could detect ricin at levels as low as 0.1 ng/ml.

Figure 3.5a shows the binding of 0.1 mg/ml of ricin at 20 °C in PBS (phosphate-buffered saline) containing 0.15% PVA (polyvinyl alcohol), 0.15% PVP (polyvinyl pyrrolidone), at pH 7.2 (curve 1) to the gel-immobilized antibody. Figure 3.5b shows the binding of 0.1 mg/ml of ricin at 20 °C in PBS containing 4% bovine serum albumin at pH 7.2 (curve 2) to the gel-immobilized antibody. Figure 3.5c shows the binding of 0.1 mg/ml of ricin at 20 °C in PBS at pH 7.2 (curve 3) to the gel-immobilized antibody. A dual-fractal

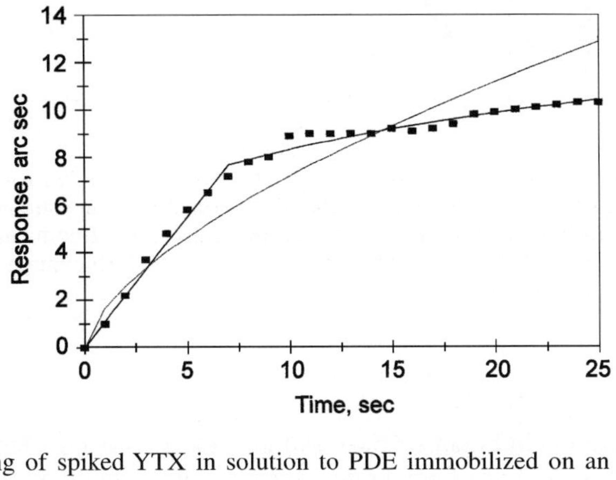

Figure 3.4 Binding of spiked YTX in solution to PDE immobilized on an aminosilane surface (Pazos *et al.*, 2004).

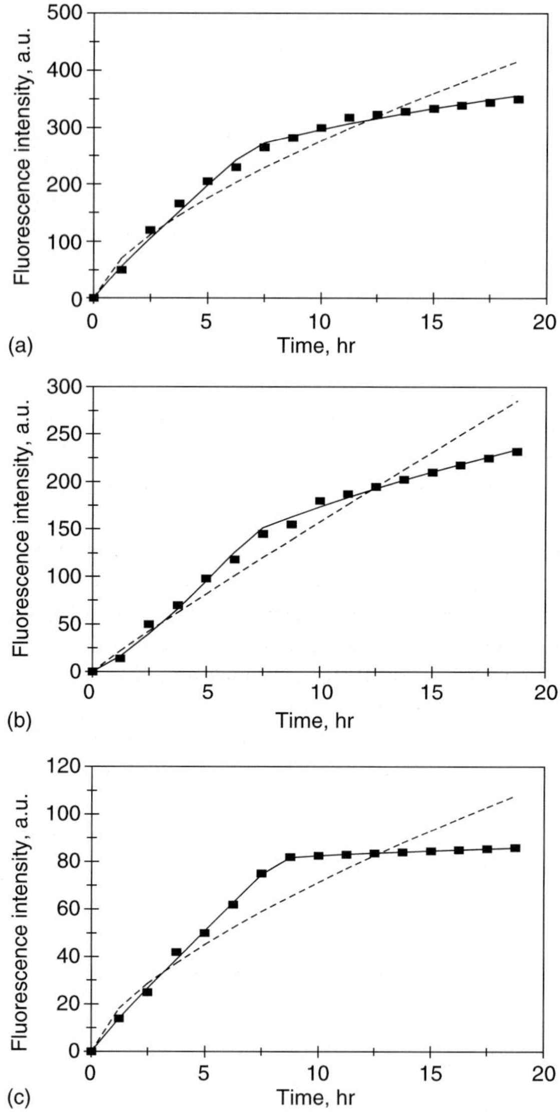

Figure 3.5 Binding of different concentrations of Cy3 labeled Ricin in solution to gel-immobilized antibodies against ricin Rch1 (0.1 mg/ml) under different conditions (Rubina *et al*., 2005): (a) curve 1: PBS containing 0.15% PVA, 0.115% PVP, pH 7.2, (b) curve 2: PBS containing 4% bovine serum albumin, pH 7.2, (c) curve 3: PBS, pH 7.2.

analysis is required in the above three cases to adequately describe the binding kinetics, and the results are given in Table 3.3.

For the data presented in Figures 3.5a–c and in Table 3.3 no apparent trend of the binding rate coefficient, k_1 with the fractal dimension D_{f1}, or k_2 with the fractal dimension D_{f2} is readily observable, and no figure(s) are presented.

Table 3.3

Binding rate coefficients and fractal dimensions for the different concentrations of Cy3-labeled Ricin to gel-immobilized against ricin Rch1 (0.1 mg/ml gel) at 20 °C under different conditions: (a) PBS containing 0.15% PVA, 0.15% PVP, pH 7.2 (curve 1), (b) PBS containing 4% bovine serum albumin, pH 7.2 (curve 2), (c) PBS, pH 7.2 (curve 3) (Rubina *et al.*, 2005)

Curve	k	k_1	k_2	D_f	D_{f1}	D_{f2}
1	61.336 ± 10.052	45.743 ± 5.261	151.61 ± 3.171	1.6928 ± 0.1037	1.1778 ± 0.1469	2.4172 ± 0.0452
2	17.8698 ± 3.8691	12.4577 ± 2.1103	58.005 ± 1.4902	1.1098 ± 0.1308	0.4746 ± 0.2112	2.0482 ± 0.0692
3	15.6463 ± 2.9702	11.2206 ± 0.6763	71.412 ± 0.126	1.6830 ± 0.1188	1.1210 ± 0.0921	2.8748 ± 0.0479

Thomas and Surolia (1999) have used the surface plasmon resonance biosensor to analyze the binding and dissociation kinetics of endotoxin with polymyxin B and its analogs. These authors indicate that lipopolysaccharide (LPS) is an invariant structural component of outer membranes in Gram-negative bacteria (Galanos *et al.* 1977; Raetz, 1990; Westphal, 1975). Thomas and Surolia (1999) indicate that one of the effects of LPS in circulation caused by infection with Gram-negative bacteria is 'endotoxic shock.' Fink (1990) indicates that hemodynamic and coagulation abnormalities are characteristics of endotoxic shock. These abnormalities lead to multiple system organ failure and account for about 60% fatalities in humans (Ziegler, 1991). Thomas and Surolia (1999) indicate that the removal of LPS by molecules that interact strongly with it is part of the strategy to combat endotoxic shock (Quezado *et al.*, 1995; Nelson *et al.*, 1995). Polymyxin B (PMB) and related family members of cyclic decapeptide antibiotics may be used to treat sepsis cases (Morrison and Jacobs, 1976).

Figure 3.6a shows the binding of 50 nM LPS in solution to cyclic decapeptide (PMB) immobilized on a CM5 sensor chip (Thomas and Surolia, 1999). A single-fractal analysis is adequate to describe the binding and the dissociation kinetics, and the results are given in Tables 3.4a and b.

Figures 3.6b–d show the binding of 75, 100, and 125 nM LPS in solution to cyclic decapeptide (PMB) immobilized on a CM5 sensor chip. A single-fractal analysis is still adequate to describe the binding kinetics; however, a dual-fractal analysis is required to adequately describe the dissociation kinetics. The results are given in Table 3.4a. Apparently, the dissociation mechanism changes as one goes from the 50 to the 75 nM LPS concentration in solution since a singe-fractal analysis is adequate for the lower (50 nM) LPS concentration whereas a dual-fractal analysis is required at the higher (75 nM) LPS concentration.

In the 50–125 nM LPS concentration range in solution where a single-fractal analysis is adequate to describe the binding kinetics, we observe the following: (1) Tables 3.4a and b and Figure 3.7a show the increase in the binding rate coefficient, k, with an increase in the LPS concentration in solution. The binding rate coefficient, k, is given by

$$k = (22.671 \pm 0.778)[\text{LPS}]^{0.331 \pm 0.0472} \tag{3.3a}$$

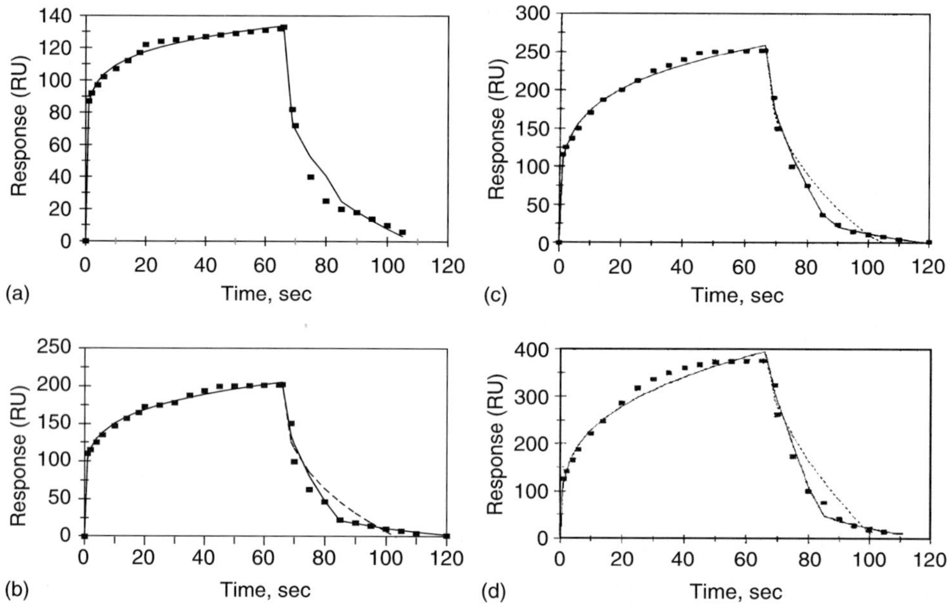

Figure 3.6 Binding of different concentrations (in nM) of lipopolysaccharide (LPS) in solution to cyclic decapeptide immobilized on a CM5 sensor chip surface (Thomas and Surolia, 1999): (a) 50 (b) 75 (c) 100 (d) 125. When only a solid line is used then a single-fractal analysis applies. When both dashed (-------) and solid (_____) lines are used, then the dashed line is for a single-fractal analysis, and the solid line is for a dual-fractal analysis. In this case the solid line provides a better fit.

(2) Tables 3.4a and b and Figure 3.7b show the decrease in the fractal dimension D_f with an increase in the LPS concentration in solution. The fractal dimension, D_f, is given by

$$D_f = (4.9697 \pm 0.092)[\text{LPS}]^{-0.1456 \pm 0.02676} \tag{3.3b}$$

(3) Tables 3.4a and b and Figure 3.7c show the decrease in the binding rate coefficient, k, with an increase in the fractal dimension, D_f. The binding rate coefficient, k, is given by

$$k = (781.857 \pm 47.261) D_f^{-2.1099 \pm 0.5694} \tag{3.3c}$$

The fit is good. Only three data points are available. The availability of more data points would lead to a more reliable fit. In case (1) above, the binding rate coefficient, k, exhibits a low (= 0.3331) order of dependence on the LPS concentration in solution. The fractional order of dependence exhibited by the binding rate coefficient, k, in the 50–125 nM concentration range reinforces the fractal nature of the system.

In case (2), the fractal dimension, D_f, exhibits a low negative (= −0.1456) order of dependence on the LPS concentration in solution.

Table 3.4a

Binding and dissociation rate coefficients for the interactions of different concentrations (in nM) of lipopolysaccharide (LPS) in solution to cyclic decapeptide immobilized on the CM5 sensor chip (Thomas and Surolia, 1999)

LPS (analyte) concentration (in nM) in solution/cyclic decapeptide immobilized on CM5 sensor chip	k	k_1	k_2	k_d	k_{d1}	k_{d2}
50	85.5679 ± 1.4356	NA	NA	42.045 ± 2.0160	NA	NA
75	103.687 ± 2.719	NA	NA	49.868 ± 9.927	38.323 ± 8.872	134.532 ± 0.449
100	108.181 ± 3.454	NA	NA	50.221 ± 7.693	39.324 ± 5.493	162.386 ± 1.386
125	117.506 ± 6.098	NA	NA	41.207 ± 11.300	24.704 ± 7.070	223.264 ± 2.015

In case (3), the binding rate coefficient, k, exhibits close to a negative second ($= -2.1099$) order of dependence on the degree of heterogeneity on the sensor chip surface. The binding rate coefficient, k, is quite sensitive to the degree of heterogeneity or the fractal dimension that exists on the sensor chip surface.

In the 50–125 nM LPS concentration range in solution where a dual-fractal analysis is adequate to describe the binding kinetics: (1) Tables 3.4a and b and Figure 3.8a show the increase in the dissociation rate coefficient, k_{d2}, with an increase in the LPS concentration in solution. The dissociation rate coefficient, k_{d2}, is given by

$$k_{d2} = (2.0096 \pm 0.1608)[\mathrm{LPS}]^{0.9673 \pm 0.2125} \tag{3.4a}$$

(2) Tables 3.4a and b and Figure 3.8b show the increase in the dissociation rate coefficient, k_{d2}, with an increase in the fractal dimension, D_{fd2}. The binding rate coefficient, k_{d2}, is given by

$$k_{d2} = (95.418 \pm 0.311) D_{fd2}^{-1.3146 \pm 0.1196} \tag{3.4b}$$

Table 3.4b

Fractal dimensions in the binding and in the dissociation phase for the interactions of different concentrations (in nM) of lipopolysaccharide (LPS) in solution to cyclic decapeptide immobilized on the CM5 sensor chip (Thomas and Surolia, 1999)

LPS (analyte) concentration (in nM) in solution/cyclic decapeptide immobilized on CM5 sensor chip	D_f	D_{f1}	D_{f2}	D_{fd}	D_{fd1}	D_{fd2}
50	2.7872 ± 0.00646	NA	NA	2.4050 ± 0.0562	NA	NA
75	2.6732 ± 0.0100	NA	NA	2.2274 ± 0.1188	1.9470 ± 0.2204	2.8021 ± 0.0139
100	2.5834 ± 0.0122	NA	NA	2.1092 ± 0.0933	1.8514 ± 0.1384	2.7796 ± 0.0261
125	2.4212 ± 0.0191	NA	NA	1.7648 ± 0.1586	1.2028 ± 0.3174	2.7378 ± 0.02754

The fit is good. Only three data points are available. The availability of more data points would lead to a more reliable fit. In case (1) above, the dissociation rate coefficient, k_{d2}, exhibits close to a first (equal to 0.9673) order of dependence on the LPS concentration in solution. The fractional order of dependence exhibited by the dissociation rate coefficient,

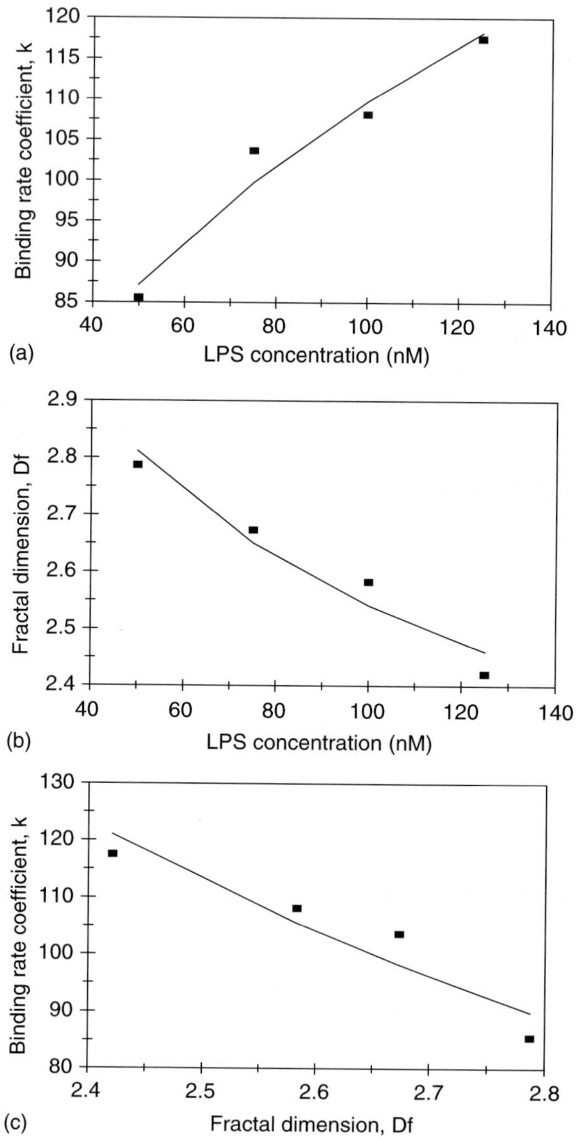

Figure 3.7 Increase in the (a) binding rate coefficient, k and (b) fractal dimension, D_f, with an increase in the LPS concentration in solution; Increase in the (c) binding rate coefficient, k with an increase in the the fractal dimension, D_f.

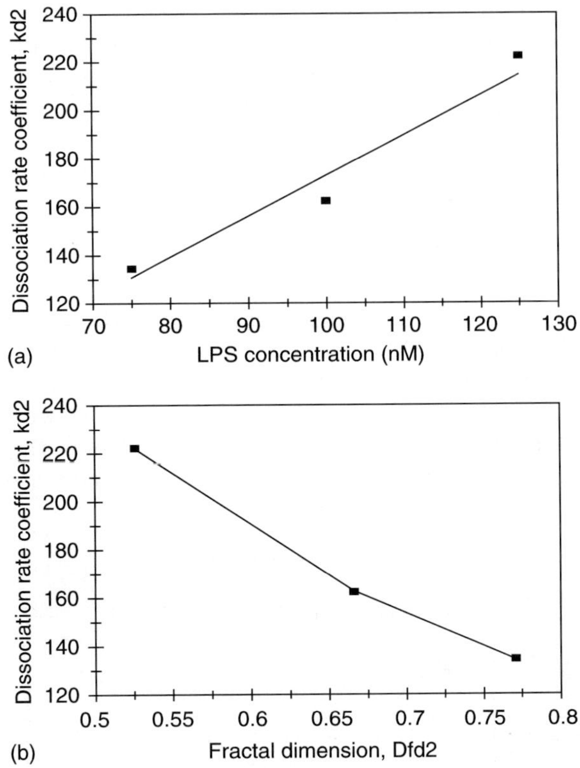

Figure 3.8 Increase in the dissociation rate coefficient, k_{d2}, with an increase in the (a) LPS concentration in solution and (b) fractal dimension, D_{fd2}.

k_{d2}, on the LPS concentration in solution in the 50–125 nM concentration range reinforces the fractal nature of the system.

In case (2), the dissociation rate coefficient, k_{d2}, exhibits a negative order of dependence between first and one and half ($= -1.3146$) on the degree of heterogeneity that exists on the sensor chip surface.

Affinity values are of interest to practicing biosensorists. Define affinity, K_1 as $= k/k_{d1}$. In the 50–125 nM LPS concentration range in solution where a single-fractal analysis is adequate to describe the binding kinetics: (1) Tables 3.4a and b and Figure 3.9a show the increase in the affinity, K_1, with an increase in the LPS concentration in solution. The affinity, K_1, is given by

$$K_1 = (0.00206 \pm 0.000337)[\text{LPS}]^{1.5897 \pm 0.4185} \tag{3.5a}$$

(2) Tables 3.4a and b and Figure 3.9b show the decrease in the affinity, K_2, with an increase in LPS concentration in solution. The affinity, K_2, is given by

$$K_2 = (18.9729 \pm 1.0957)[\text{LPS}]^{-0.7373 \pm 0.1550} \tag{3.5b}$$

(3) Tables 3.4a and b and Figure 3.10a show the increase in the affinity, K_1, with an increase in the fractal dimension ratio, D_f/D_{fd1}. The affinity, K_1, is given by

$$K_1 = (1.3038 \pm 0.2522)(D_f/D_{fd1})^{0.8578 \pm 0.5779} \tag{3.6a}$$

(4) Tables 3.4a and b and Figure 3.10b show the increase in the affinity, K_2, with an increase in the fractal dimension ratio, D_f/D_{fd2}. The affinity, K_2, is given by

$$K_2 = (0.9708 \pm 0.0117)(D_f/D_{fd2})^{5.0150 \pm 0.2199} \tag{3.6b}$$

The fit is very good. Only three data points are available. The availability of more data points would lead to a more reliable fit. In case (1) above, the affinity, K_1, exhibits close to a one and half (= 1.5897) order of dependence on the LPS concentration in solution.

In case (2) the affinity, K_2, exhibits an order of dependence between negative one half and first (= −0.7373) order on the LPS concentration in solution. The fractional order of dependence by the affinity, K_1 and K_2, on the LPS concentration in solution once again lends support to the fractal nature of the system.

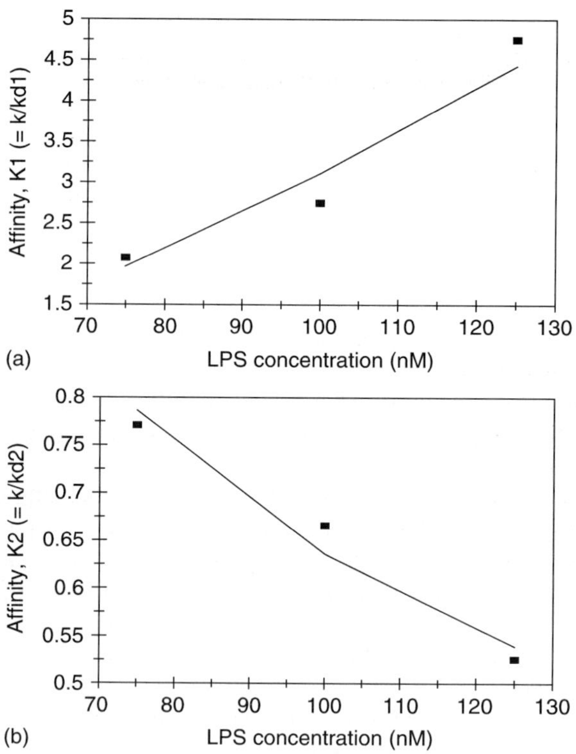

Figure 3.9 Increase in the affinity (a) $K_1(=k_1/k_d)$ and (b) $K_2(=k_2/k_d)$ with an increase in the LPS concentration in solution.

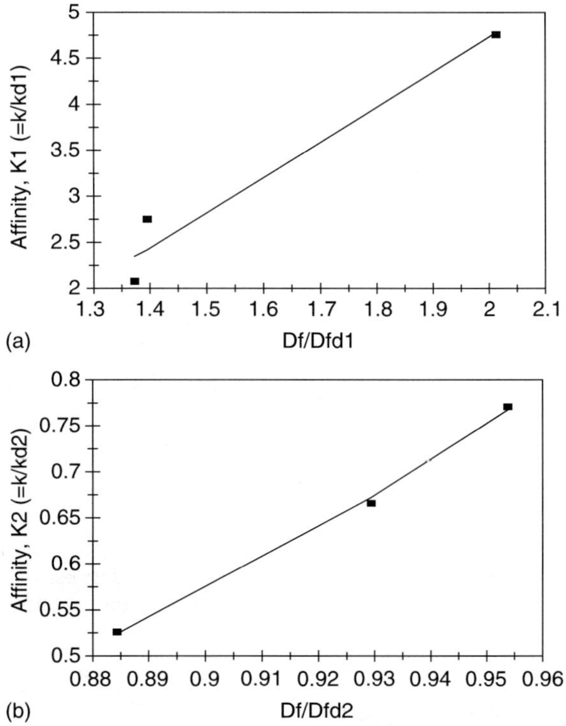

Figure 3.10 Increase in the affinity (a) $K_1(=k/k_{d1})$ with an increase in the fractal dimension ratio, D_f/D_{fd1} and (b) $K_2(=k/k_{d2})$ with an increase in the fractal dimension ratio, D_f/D_{fd2}.

In case (3), the affinity, K_1, exhibits close to a second (= 1.8578) order of dependence on the ratio of fractal dimensions, D_f/D_{fd1}. This is one way of possibly controlling the affinity, by changing the degree of heterogeneity on the sensor chip surface. Of course, one has to be careful in that one should increase the degree of heterogeneity in the binding phase selectively compared to that in the dissociation phase if one is interested in obtaining higher affinity values.

In case (4), the affinity, K_2, is very sensitive to the degree of heterogeneity that exists on the surface as noted by the very high (= 5.0150) order of dependence exhibited.

Thomas and Surolia (1999) also compared the binding and the dissociation curves for LPS in solution to different cyclic peptides immobilized on a CM5 sensor chip. Figures 3.11a–c show the binding and dissociation of LPS in solution to (a) polymyxin B (PMB) immobilized on a sensor chip surface, (b) cyclic decapeptide immobilized on a sensor chip surface, and (c) polymyxin B nonapeptide immobilized on a sensor chip surface, respectively. A dual-fractal analysis is required to adequately describe the binding and dissociation kinetics. The results are given in Tables 3.5a and b.

It is of interest to note that as one goes from the Polymyxin B to the cyclic decapeptide immobilized on the sensor chip surface, the fractal dimension, D_{f1}, increases by 6.92% from

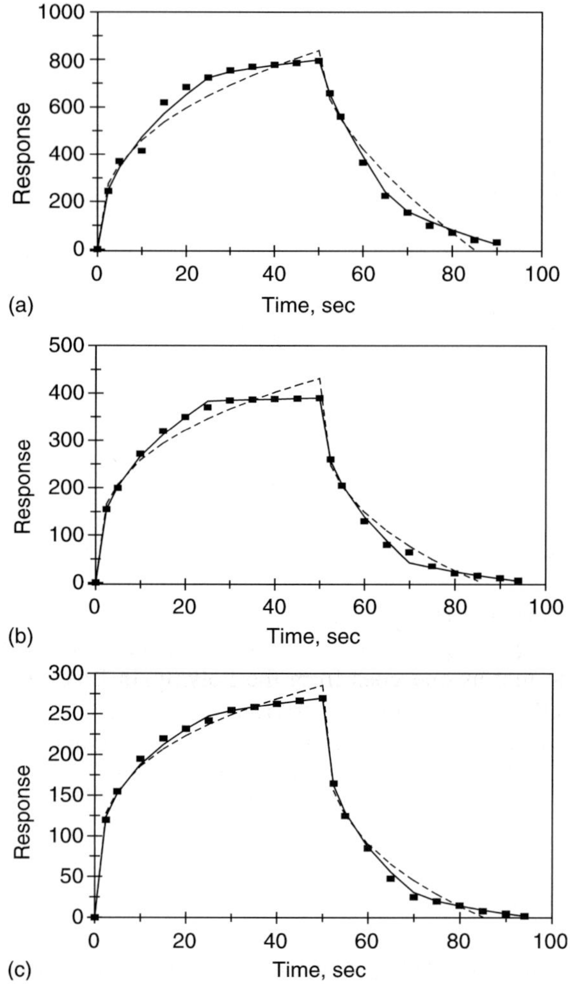

Figure 3.11 Binding of LPS (lipopolysaccharide) in solution to different compounds immobilized on a CM5 sensor chip (Thomas and Surolia, 1999): (a) polymyxin B (PMB), (b) cyclic decapeptide, (c) polymyxin B nonapeptide.

a value of 2.0772 to 2.2210, and the binding rate coefficient, k_1 decreases by 33.63% from a value of 164.06 to 108.885. Changes in the binding rate coefficient, k_1, and in the degree of heterogeneity on the sensor chip surface during the first phase of binding are in opposite directions. Similarly, a 7.34% increase in the fractal dimension D_{f2} from 2.7494 to 2.9511 leads to a decrease in the binding rate coefficient k_2 by 27.5% from a value of 489.146 to 354.48. Once again, changes in the fractal dimension or the degree of heterogeneity on the

Table 3.5a

Binding and dissociation rate coefficients for the interaction of lipopolysaccharide (LPS) in solution to polymyxin B (PMB), cyclic decapeptide, and polymyxin B nonapeptide (PMBN) immobilized on a CM5 sensor chip (Thomas and Surolia, 1999)

Analyte in solution/receptor immobilized on CM5 sensor chip	k	k_1	k_2	k_d	k_{d1}	k_{d2}
LPS/polymyxin B	196.44 ± 6.34	164.06 ± 14.09	489.15 ± 3.84	91.14 ± 13.49	68.22 ± 4.60	271.26 ± 6.38
LPS/cyclic decapeptide	124.43 ± 9.21	108.69 ± 2.56	354.48 ± 0.21	101.07 ± 7.54	87.86 ± 4.02	231.36 ± 1.70
LPS/polymyxin B nonapeptide	100.605 ± 4.60	94.23 ±3.03	170.35±0.12	84.85 ± 5.76	75.89±2.76	166.29 ± 0.68

surface (fractal dimension, D_{f2}) and in the binding rate coefficient, k_2, are in opposite directions.

The trends are similar to the ones noted above and are not repeated here to avoid repetition. Only one example is given to demonstrate the similarity in the trend observed. It is of interest to note that as one goes from the Polymyxin B to the Polymyxin B nonapeptide immobilized on the sensor chip surface, the fractal dimension D_{f1} increases by 15.5% from 2.0772 to 2.3990, and the binding rate coefficient k_1 decreases by 42.56% from 164.06 to 94.23. Changes in the binding rate coefficient, k_1, and in the degree of heterogeneity on the sensor chip surface during the first phase of binding are in opposite directions.

Table 3.5b

Fractal dimensions for the binding and the dissociation phase for the interaction of lipopolysaccharide (LPS) in solution to polymyxin B (PMB), cyclic decapeptide, and polymyxin B nonapeptide (PMBN) immobilized on a CM5 sensor chip (Thomas and Surolia, 1999)

Analyte in solution/receptor immobilized on CM5 sensor chip	D_f	D_{f1}	D_{f2}	D_{fd}	D_{fd1}	D_{fd2}
LPS/polymyxin B	2.2576±0.0786	2.0772±0.0740	2.7494±0.02772	1.7812±0.0968	1.4554±0.0772	2.4318±0.0489
LPS/cyclic decapeptide	2.3630±0.0470	2.2210±0.02364	2.9511±0.4892	2.2436±0.0506	2.0904±0.0530	2.7296±0.0324
LPS/polymyxin B nonapeptide	2.4666±0.0294	2.3990±0.02852	2.7644±0.0053	2.3508±0.0460	2.2289±0.0362	2.7470±0.0284

For the data presented in Tables 3.5a and b, Figure 3.12a shows the decrease in the binding rate coefficient, k_1, with an increase in the fractal dimension, D_{f1}. For the data presented in Figure 3.12a and in Tables 3.5a and b, the binding rate coefficient, k_1, is given by

$$k_1 = (2946.92 \pm 334.48)D_{f1}^{-3.799\pm1.233} \tag{3.7a}$$

For the data presented in Tables 3.5a and b, Figure 3.12b shows the increase in the ratio of the binding rate coefficients, k_2/k_1, with an increase in the fractal dimension, ratio, D_{f2}/D_{f1}. For the data presented in Figure 3.12b and in Tables 3.5a and b, the ratio of the binding rate coefficients, k_2/k_1, is given by

$$k_2/k_1 = (1.0403 \pm 0.0565)D_{f2}/D_{f1}^{3.8908\pm0.4610} \tag{3.7b}$$

The fit is good. Only three data points are available. The binding rate coefficient, k_1, is quite sensitive to the degree of heterogeneity that exists on the biosensor chip surface as noted by the order of dependence between negative three and half and fourth ($= -3.799$) exhibited. The availability of more data points would lead to a more reliable fit. The ratio of the binding rate coefficients, k_2/k_1, is quite sensitive to the degree of heterogeneity that exists on the biosensor chip surface as noted by the order of dependence close to fourth ($= 3.8908$) order exhibited.

For the data presented in Tables 3.5a and b, Figure 3.12c shows the decrease in the dissociation rate coefficient, k_{d2}, with an increase in the fractal dimension, D_{fd2}. For the data presented in Figure 3.12c and in Tables 3.5a and b, the dissociation rate coefficient, k_{d2} is given by

$$k_{d2} = (3402.95 \pm 841.66)D_{fd2}^{-2.837\pm2.2779} \tag{3.7c}$$

The fit is reasonable. Only three data points are available. The availability of more data points would lead to a more reliable fit. The dissociation rate coefficient, k_{d2}, is quite sensitive to the degree of heterogeneity that exists on the biosensor chip surface as noted by the order of dependence between negative two and half and three ($= -2.837$) exhibited.

The data presented in Tables 3.5a and b and in Figure 3.13a show the increase in the affinity, $K_1(= k_1/k_{d1})$ with an increase in the ratio of the fractal dimensions, D_{f1}/D_{fd1}. The affinity, K_1, is given by

$$K_1(= k_1/k_{d1}) = (1.0629 \pm 0.0187)(D_{f1}/D_{fd1})^{2.293\pm0.0734} \tag{3.8a}$$

The fit is good. Only three data points are available. The availability of more data points would lead to a more reliable fit. The affinity, K_1, exhibits slightly higher than second ($= 2.293$) order of dependence on the ratio of fractal dimensions, D_{f1}/D_{fd1}. The affinity, K_1, is quite sensitive to the ratio of fractal dimensions, D_{f1}/D_{fd1}. Once again, this is one way of controlling the affinity (K_1 in this case) by selectively changing the degree of heterogeneity on the sensor chip surface of the binding phase compared to the dissociation phase.

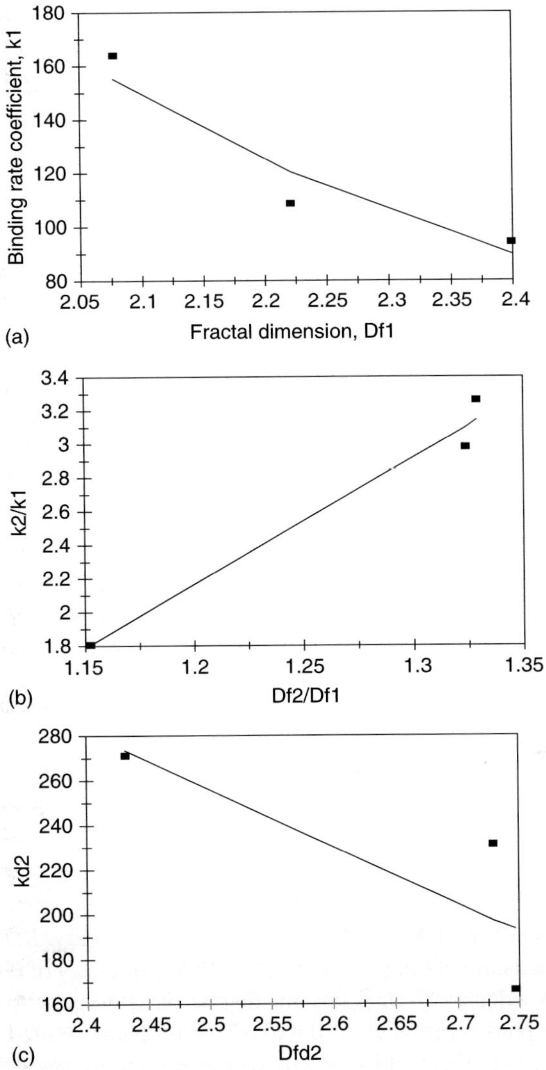

Figure 3.12 (a) Decrease in the binding rate coefficient, k_1, with an increase in the fractal dimension, D_{f1}, (b) Increase in the ratio of the binding rate coefficients, k_2/k_1, with an increase in the ratio of fractal dimensions, D_{f2}/D_{f1}, (c) Decrease in the dissociation rate coefficient, k_{d2}, with an increase in the fractal dimension, D_{fd2}.

The data presented in Tables 3.5a and b and in Figure 3.13b show the increase in the affinity, $K_2(= k_2/k_{d2})$ with an increase in the ratio of the fractal dimensions, D_{f2}/D_{fd2}. The affinity, K_2, is given by

$$K_2(= k_2/k_{d2}) = (1.00688 \pm 0.04529)(D_{f2}/D_{fd2})^{4.9257 \pm 0.5295} \tag{3.8b}$$

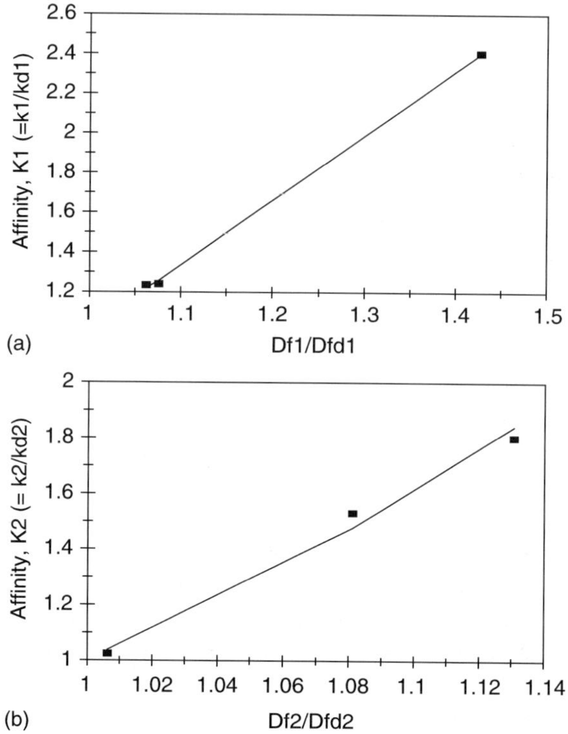

Figure 3.13 Increase in the affinity (a) $K_1(=k_1/k_{d1})$ with an increase in the ratio of fractal dimensions, D_{f1}/D_{fd1}, and (b) $K_2(=k_2/k_{d2})$ with an increase in the ratio of fractal dimensions, D_{f2}/D_{fd2}.

The fit is good. Only three data points are available. The availability of more data points would lead to a more reliable fit. The affinity, K_2, exhibits close to fifth (= 4.9257) order of dependence on the ratio of fractal dimensions, D_{f2}/D_{fd2}. The affinity, K_2, is very sensitive to the ratio of fractal dimensions, D_{f2}/D_{fd2}. Once again, this is one way of controlling the affinity (K_2 in this case) by selectively changing the degree of heterogeneity on the sensor chip surface of the binding phase compared to the dissociation phase. Note that the affinity, K_2, is much more sensitive than the affinity, K_1, to the corresponding fractal dimension ratio.

3.4 CONCLUSIONS

A fractal analysis is presented for (a) the binding of yessotoxin (YTX) in solution to phosphodiesterase 3′-5′-cyclic-nucleotide (PDE) immobilized on an aminosilane surface (Pazos *et al.*, 2005), (b) Cy3-labeled Ricin in solution to gel-immobilized antibodies against ricin Rch1 (Rubina *et al.*, 2005), and (c) lipopolysaccharide (LPS) in solution to cyclic decapeptides immobilized on a sensor chip surface (Thomas and Surolia, 1999). The fractal analysis provides a quantitative indication of the state of disorder (fractal dimension) and the binding (and dissociation) rate coefficients on the sensor chip surface.

Both types of examples are given wherein either a single- or a dual-fractal analysis is used. The dual-fractal analysis is used only when the single-fractal analysis does not provide an adequate fit. This was done by the regression analysis provided by Corel Quattro Pro 8.0 (1997). In accord with the prefactor analysis for fractal aggregates (Sorenson and Roberts, 1997), quantitative (predictive) relations are developed for (a) the binding rate coefficient, k, as a function of the YTX concentration in solution, and as a function of the fractal dimension, D_f (Pazos *et al.*, 2004), (b) the binding rate coefficient, k as a function of lipopolysaccharide (LPS) in solution and as a function of the fractal dimension, D_f (Thomas and Surolia, 1999), (c) the dissociation rate coefficient, k_{d2}, as a function of the LPS concentration in solution and as a function of the fractal dimension in the dissociation phase, D_{fd2}, and (d) the affinities, K_1 and K_2, as a function of the ratio of the fractal dimensions present in the binding and in the dissociation phases, respectively (Thomas and Surolia, 1999).

The binding rate coefficient, k, for YTX (Pazos *et al.*, 2004) is very sensitive to the fractal dimension, D_f, or the degree of heterogeneity present on the chip surface as noted by the slightly higher than sixth (= 6.0359) order of dependence on D_f exhibited. Similarly, the binding rate coefficient, k, for LPS (Thomas and Surolia, 1999) exhibits close to a negative second (= −2.1099) order of dependence on the degree of heterogeneity on the sensor chip surface. The above two examples emphasize the importance of the degree of heterogeneity on the sensor chip surface, and its influence on the binding (and dissociation) rate coefficients. More such examples are provided in later chapters throughout the book. In fact, one of the main aims of this book is to emphasize the influence of heterogeneity on the sensor chip surface on the binding and (the dissociation) rate coefficients and affinity values.

It is suggested that the fractal surface (roughness) leads to turbulence, which enhances mixing, decreases diffusional limitations, and leads to an increase in the binding (and dissociation) rate coefficient (Martin *et al.*, 1991). For this to occur, the characteristic length of this turbulent boundary layer may have to extend a few monolayers above the sensor surface to affect the bulk diffusion to and from the surface. Considering the extremely laminar flow regimes in most biosensors this may not be possible. However, due to the fractal nature of the surface that involves, for example, grooves and ridges, the surface morphology may contribute substantially toward the presence of eddy diffusion. This eddy diffusion enhances mixing and helps extend the characteristic length of the boundary layer to affect the bulk diffusion to and from the surface.

The characterization of the surface by a fractal dimension provides extra flexibility and an avenue whereby the nature of the surface may be modulated in desired directions, and thereby simultaneously affecting or changing the binding and the dissociation rate coefficients in required directions. This predictive approach is of considerable value in the design of biosensor experiments. More such studies are required and are presented throughout the book to determine whether the binding and dissociation rate coefficients are sensitive to their respective fractal dimensions or the degree of heterogeneity that exists on the biosensor surface. If this is correct, then experimentalists may find it worth their effort to pay a little more attention to the nature of the surface, and how it may be manipulated to control the relevant parameters and biosensor performance in desired directions. Also, in a more general sense the treatment should also be applicable to non-biosensor applications wherein further physical insights could also be obtained.

REFERENCES

Burkin, M.A., Sviridov, V.V., and Perelygina, O.V. (2004). Determination of tetanus toxin by ELISA using monoclonal antibodies. *Prikl Biokhim. Microbiology*, **40**, 478–484.

Corel Corporation (1997). Corel Quattro Pro 8.0, Ottawa, Canada.

Fink, P.F. (1990). Sepsis syndrome. In: *Handbook of Critical Care*, 3rd edn. (eds J.L. Berk and J.E. Sampliner), Little, Brown and Company, Boston, Massachussetts.

Galanos, C., Luderitz, O., Rietschel, E.T., and Westphal, O.C. (1977). *International Reviews in Biochemistry*, **14**, 239–334.

Havlin, S. (1989). Molecular diffusion and reaction. In: *The Fractal Approach to Heterogeneous Chemistry: Surfaces, Colloids, Polymers* (ed. D. Avnir), Wiley, New York, pp. 251–269.

History Channel, Channel 46, American Television (2005). Biological warfare, June 28.

Jaggy, C., Musielski, H., Urech, K., and Schaller, G. (1995). Quantitative determination of lectins in mistletoe preparations. *Arzneimittelforschung*, **45**, 905–909.

Leppla, S. (1995). Anthrax toxins. In: *Bacterial toxins and Virulence Factors in Disease* (ed. J. Moss), Marcel Dekker, New York, pp. 543–572.

Lee, C.K., and Lee, S.L. (1995). Multi-fractal scaling analysis of reactions over fractal surfaces. *Surface Science*, **325**, 294–310.

Ligler, F.S., Taitt, C.R., Shriver-Lake, L.C., Sapsford, K.E., Shubin, Y., and Golden, J.P. (2003). Array biosensor for detection of toxins. *Analytical and Bioanalytical Chemistry*, **377**, 469–477.

Martin, S.J., Granstaff, V.E., and Frye, G.C. (1991). Effect of surface roughness on the response of thickness-shear mode resonators in liquids. *Analytical Chemistry*, **65**, 2910–2922.

Miyamoto, T., Kamikado, H., Kobayashi, H., Honjoh, K., and Iio, M. (2003). Immunomagnetic flow cytometric detection of staphylococcal enterotoxin B in raw and dry milk. *Food Protection*, **66**, 1222–1226.

Morrison, D.C., and Jacobs, D.M. (1976). Binding of polymyxin B to the lipid portion of bacterial lipopolysaccharides. *Immunochemistry*, **13**, 813–818.

Nelson, D., Kuppermann, N., and Fleischer, G.R., et al. (1995). *Critical Care in Medicine*, **23**, 56–63.

O'Brien, T., Johnson III, L.H., Aldrich, J.L., Allen, S.G., Liang, L.-T., Plummer, A.L., Krak, S.J., and Boiarski, A.A. (2000). The development of immunoassays to four biological threat agents in a bidiffractive grating biosensor. *Biosensors & Bioelectronics*, **14**, 815–828.

Pazos, M.J., Alfonso, A., Vieytes, M.R., Yasumoto, T., Vietes, J.M., and Botana, L.M. (2004). Resonant mirror biosensor detection method based on yessotoxin-phosphodiesterase interactions. *Analytical Biochemistry*, **335**, 112–118.

Poli, M.A., Rivera, V.R., and Neal, D. (2002). Sensitive and specific colorimetric ELISAs for *Staphylococcus aureus* enterotoxins A and B in urine and buffer. *Toxicon*, **40**, 1723–1726.

Quezado, S.M., Zenaide, M.N., and Nutanson, C. (1995). New Strategies for combatting sepsis: The magic bullet missed the mark but the search continues. *Trends in Biotechnology*, **13**, 56–63.

Raetz, C.R.H. (1990). Biochemistry of endotoxins. *Annual Reviews in Biochemistry*, **59**, 129–170.

Ramakrishnan, A., and Sadana, A. (2001). A single-fractal analysis of cellular analyte-receptor binding kinetics utilizing biosensors. *Biosystems*, **59**(1), 35–45.

Rowe, C.A., Tender, L.M., Feldstein, M.J., Golden, J.P., Scruggs, S.B., MacCraith, B.D., Cras, J.J., and Ligler, F.S. (1999). Array biosensor for simultaneous identification of bacterial, viral, and protein analytes. *Analytical Chemistry*, **71**, 3846–3852.

Rubina, A. Yu., Dyokora, V.I., Dementieva, E.I., Stomakhin, A.A., Nesmeyanov, V.A., Grishin, E.V., and Zasedatelev, A.S. (2005). Quantitative immunoassay of biotoxins on hydrogel-based protein microchips. *Analytical Biochemistry*, **340**, 317–329.

Rucker, V.C., Havenstrite, K.L., and Herr, A.E. (2005). Antibody microarrays for native toxin detection. *Analytical Biochemistry*, **339**, 262–270.

Sadana, A. (2001). A fractal analysis approach for the evaluation of hybridization kinetics in biosensors. *Journal of Colloid and Interface Science,* **234**, 9–18.

Sapsford, K.E., Shubin, Y.S., Delehanty, J.B., Golden, J.P., Taitt, C.R., Shriver-Lake, L.C., and Ligler, F.S. (2004). Fluorescence-based array biosensors for the detection of biohazards. *Journal of Applied Microbiology*, **96**(1), 47–58.

Sorenson, C.M., and Roberts, G.C. (1997). The prefactor of fractal aggregates. *Journal of Colloid and Interface Science*, **186**, 447–452.

Temyakov, D.E., Agapov, I.I., Moisenovch, M.M., Egorova, S.G., Pfueller, U., Zinke, H., and Tonevitskii, A.G. (1997). Heterogeneity of mistletoe lectin catalytic subunits assessed with monoclonal antibodies. *Molecular Biology (Moscow)*, **31**, 448–454.

Thomas, C.J., and Surolia, A. (1999). Kinetics of interaction of endotoxin with polymyxin B and its analogs: A surface plasmon resonance analysis. *FEBS Letters*, **45**, 420–424.

USAMRIID's medical management of biological casualties handbook, Fort Detrick, Frederic, Maryland, 2001. Available from: http://www.usamriid.army.mil./education/bluebook/html.

Westphal, O. (1975). *Transactions Coll. Int. Aller.*, **49**, 1–43.

Ziegler, E.T., Fisher, C.J., Sprung, C.L., Straube, R.C., Sadoff, J.C., Foulke, G.E., Wortel, C.H., Fink, M.P., Dellinger, R.P., and Teng, N.N. (1991). Treatment of gram-negative bacteremia and septoic shock with HA-1A human monoclonal antibody against endotoxin. A randomized, double-blind, placebo-controlled trail. The HA-1A sepsis study group. *New England Journal of Medicine*, **324**, 429–436.

– 4 –

Fractal Binding and Dissociation Kinetics of Heart-Related Compounds on Biosensor Surfaces

4.1 INTRODUCTION

The early detection of atherosclerosis is an important area of medical investigation. Gaus and Hall (1999) emphasize the early detection of risk and risk factors for this ailment. For example, Ross (1993) has indicated that the accumulation of low-density lipoproteins (LDL) in microphages may be considered as the first step of atherosclerosis. Gaus and Hall (1999), however, did indicate that the precise mechanism of LDL deposition is still not clear. Brown and Goldstein (1983) had initially suggested that the oxidative modification of LDL could lead to atherosclerotic LDL. Thus, a method to detect the extent of the presence of oxidative LDL is required.

Jin *et al.* (1999) indicate that cholesterol is converted to cholesterol ester by lecithin cholesterol acyltransferase (LCAT). LCAT catalyzes the transfer of *sn*-2 fatty acid of phosphatidylcholine to the 3-OH group of cholesterol, thereby converting cholesterol to cholesterol ester. This is a key step in reverse cholesterol transport, which is the pathway by which free cholesterol is returned to the liver for catabolism (Fielding and Fielding, 1995). Glomset (1972) and Jonas (1991) indicate that the conversion of cholesterol to cholesterol ester occurs preferentially on high-density lipoproteins (HDL). Jonas (1998) indicates that just as other lipases LCAT is an interfacially activated enzyme. It has a lid region (between cys 50 and 74), which mediates the binding to lipoprotein substrates (Adimoolam and Jonas, 1997; Peelman *et al.*, 1999). Fielding and Fielding (1995) indicate that the apolipoprotein A-I (apo-I) on the HDL is the main physiological activator of LCAT. Singh and Poirier (1993) have developed a fiber-optic biosensor to measure the amount of cholesterol and lipoproteins in blood samples in real time.

Serious heart problems are caused by acute myocardiac infarction (AMI). The early detection of AMI has been emphasized by medical clinics (Apple *et al.*, 1999; Delrey *et al.*, 1998; Mair, 1997). Cardiac troponin (cTnI) is a good biochemical marker in the bloodstream that indicates damage or death of the cardiac muscle. Liu *et al.* (2003) have attempted to detect cTnI by monoclonal antibodies that recognize the antigenic epitopes of cTnI. This would facilitate the early detection of AMI, since Liu *et al.* (2003) indicate that cTnI has a wide diagnostic window with time and has excellent specificity and sensitivity. These authors further indicate that the detection of cTnI for the diagnosis of AMI is becoming a 'gold standard.'

In this chapter, we use fractal analysis to analyze the binding of (a) different LCAT concentrations in solution to egg-white apoA-I rHDL immobilized on a biosensor chip (Jin *et al.*, 1999), (b) native, mildly oxidized, and strongly oxidized LDL in solution to heparin-modified Au-surface of a surface plasmon resonance (SPR) biosensor (Gaus and Hall, 1999), (c) TRITC (tetramethylrhodamine-5-(and-6)-isothiocyanate)-labeled HDL in solution to a bare optical fiber (Singh and Poirier, 1993), and (d) interactions of other heart-related analytes. Binding and dissociation rate coefficients and fractal dimension values are provided. The analysis may be considered as an alternate analysis to the conventional analysis provided by the software that comes along with the SPR biosensor (when that applies).

4.2 THEORY

Havlin (1989) has reviewed and analyzed the diffusion of reactants toward fractal surfaces. The details of the theory and the equations involved for the binding and dissociation phases of analyte–receptor binding are available (Sadana, 2001). The details are not repeated here, except that just the equations are given to permit an easier reading. These equations have been applied to other biosensor systems (Ramakrishnan and Sadana, 2001; Sadana, 2001). For most applications, a single- or a dual-fractal analysis is often adequate to describe the binding and dissociation kinetics. Peculiarities in the values of the binding and dissociation rate coefficients, as well as in the values of the fractal dimensions with regard to the dilute analyte systems being analyzed will be carefully noted, if applicable. In this chapter, we analyze the binding and dissociation kinetics of LCAT in solution to egg-white apo-I rHDL immobilized on a biosensor chip surface (Jin *et al.*, 1999), native, mildly oxidized, and strongly oxidized LDL in solution to a heparin-modified Au-surface of an SPR biosensor (Gaus and Hall, 1999), TRITC-labeled HDL in solution to a bare optical fiber (Singh and Poirier, 1993), and interactions of other heart-related analytes.

4.2.1 Single-fractal analysis

Binding rate coefficient

Havlin (1989) indicates that the diffusion of a particle (analyte [Ag]) from a homogeneous solution to a solid surface (e.g., receptor [Ab]-coated surface) on which it reacts to form a product (analyte–receptor complex (Ab·Ag)) is given by

$$(\text{Analyte}\cdot\text{Receptor}) \sim \begin{cases} t^{(3-D_{\text{f,bind}})/2} = t^{p} & (t < t_{\text{c}}) \\ t^{1/2} & (t < t_{\text{c}}) \end{cases} \tag{4.1a}$$

Here $D_{\text{f,bind}}$ or D_{f} (used later on in the chapter) is the fractal dimension of the surface during the binding step. t_{c} is the cross-over value. Havlin (1989) indicates that the cross-over value may be determined by $r_{\text{c}}^{2} \sim t_{\text{c}}$. Above the characteristic length, r_{c}, the self-similarity of the surface is lost and the surface may be considered homogeneous. Above time t_{c}, the surface may be considered homogeneous since the self-similarity property disappears, and 'regular' diffusion is now present. For a homogeneous surface where $D_f = 2$, and when only diffusional limitations are present, $p = ½$, as it should be. Another way of looking at the $p = ½$ case (where $D_{f,bind}=2$) is that the analyte in solution views the fractal object, in our case, the receptor-coated biosensor surface, from a 'large distance.' In essence, in the association process, the diffusion of the analyte from the solution to the receptor surface creates a depletion layer of width $(Ɖt)^{½}$, where $Ɖ$ is the diffusion constant. This gives rise to the fractal power law, (Analyte · Receptor) ~ $t^{(3-D_{f,bind})/2}$. For the present analysis, t_c is arbitrarily chosen and we assume that the value of t_c is not reached. One may consider the approach as an intermediate 'heuristic' approach that may be used in the future to develop an autonomous (and not time-dependent) model for diffusion-controlled kinetics.

Dissociation rate coefficient

The diffusion of the dissociated particle (receptor [Ab] or analyte [Ag]) from the solid surface (e.g., analyte [Ag]–receptor [Ab]) complex coated surface) into solution may be given, as a first approximation by

$$\text{(Analyte·Receptor)} \sim -k' t^{(3-D_{f,diss})/2} \quad (t > t_{diss}) \tag{4.1b}$$

Henceforth, its concentration only decreases. The dissociation kinetics may be analyzed in a manner 'similar' to the binding kinetics.

4.2.2 Dual-fractal analysis

Binding rate coefficient

Sometimes, the binding curve exhibits complexities and two parameters (k, D_f) are not sufficient to adequately describe the binding kinetics. This is further corroborated by low values of r^2 factor (goodness-of-fit). In that case, one resorts to a dual-fractal analysis (four parameters: k_1, k_2, D_{f1}, and D_{f2}) to adequately describe the binding kinetics. The single-fractal analysis presented above is thus extended to include two fractal dimensions. At present, the time ($t = t_1$) at which the 'first' fractal dimension 'changes' to the 'second' fractal dimension is arbitrary and empirical. For the most part, it is dictated by the data analyzed and experience gained by handling a single-fractal analysis. A smoother curve is obtained in the 'transition' region, if care is taken to select the correct number of points for the two regions. In this case, the product (antibody–antigen; or analyte–receptor complex, Ab·Ag or analyte·receptor) is given by

$$(\text{Ab}\cdot\text{Ag}) \approx \begin{cases} t^{(3-D_{f1,bind})/2} = t^{p1} & (t < t_1) \\ t^{(3-D_{f2,bind})/2} = t^{p2} & (t_1 < t < t_2) = t_c \\ t^{1/2} & (t > t_c) \end{cases} \tag{4.1c}$$

In some cases, as mentioned above, a triple-fractal analysis with six parameters (k_1, k_2, k_3, D_{f1}, D_{f2}, and D_{f3}) may be required to adequately model the binding kinetics. This is when the binding curve exhibits convolutions and complexities in its shape due to perhaps the very dilute nature of the analyte (in some of the cases to be presented) or for some other reasons. Also, in some cases, a dual-fractal analysis may be required to describe the dissociation kinetics.

4.3 RESULTS

In this chapter, we use fractals to analyze the binding and dissociation kinetics of LCAT in solution to egg-white apo-I rHDL immobilized on a biosensor chip surface (Jin *et al.*, 1999), native, mildly oxidized, and strongly oxidized LDL in solution to heparin-modified Au-surface of an SPR biosensor (Gaus and Hall, 1999), TRITC-labeled HDL in solution to a bare optical fiber (Singh and Poirier, 1993), and interactions of other heart-related analytes. Note that this is just one possible way of analyzing the binding and dissociation kinetics.

In order to obtain a better understanding of the reverse cholesterol process, Jin *et al.* (1999) have analyzed the binding and dissociation kinetics of LCAT to lipoprotein surfaces using an SPR biosensor. They obtained binding and dissociation rate coefficients as well as affinity values using the software provided by Biacore, the SPR manufacturer. This type of analysis as indicated above and elsewhere in this book does not take into account either the presence of external diffusional limitations or the degree of heterogeneity present on the sensor chip surface.

Figures 4.1a, b, and c–e show the binding and dissociation of 0.04, 0.08, and 0.16–0.64 μM wild-type LCAT in solution to egg-white apoA-I rHDL immobilized on a sensor chip surface containing streptavidin (Jin *et al.*, 1999). A dual-fractal analysis is required to adequately describe the binding and dissociation kinetics. Table 4.1 shows the values of (a) the binding rate coefficient k and the fractal dimension D_f for a single-fractal analysis, (b) the binding rate coefficients, k_1 and k_2, and the fractal dimensions, D_{f1} and D_{f2}, for a dual-fractal analysis, (c) the dissociation rate coefficient k_d and the fractal dimension D_{fd} for a single-fractal analysis, and (d) the dissociation rate coefficients, k_{d1} and k_{d2}, and the fractal dimensions, D_{fd1} and D_{fd2}, for a dual-fractal analysis. All values were obtained by using *Corel Quattro Pro* (1997).

The values of the parameters presented in Table 4.1 are within 95% confidence limits. For example, the value of k reported for the binding of 0.04 μM LCAT concentration in solution is 3.6837±0.389. The 95% confidence limits indicates that 95% of the k values lie between 3.2946 and 4.0727.

Note that a doubling of the LCAT concentration in solution from 0.04 to 0.08 μM has led to (a) an increase in the binding rate coefficient k_1 by 5.1% from a value of 2.4561 to 2.5814, and (b) to an increase in the binding rate coefficient k_2 by 4.82% from a value of 12.225 to 12.814. Note the almost similar amount of changes (5.1% and 4.82%) for the binding rate coefficients, k_1 and k_2.

It is of interest to note the increase in the binding rate coefficients, k_1 and k_2, as the LCAT concentration in solution increases from 0.04 to 0.64 μM.

Figures 4.2a and b show the increase in the binding rate coefficients, k_1 and k_2, with an increase in the LCAT concentration in solution from 0.04 to 0.64 μM. For the data

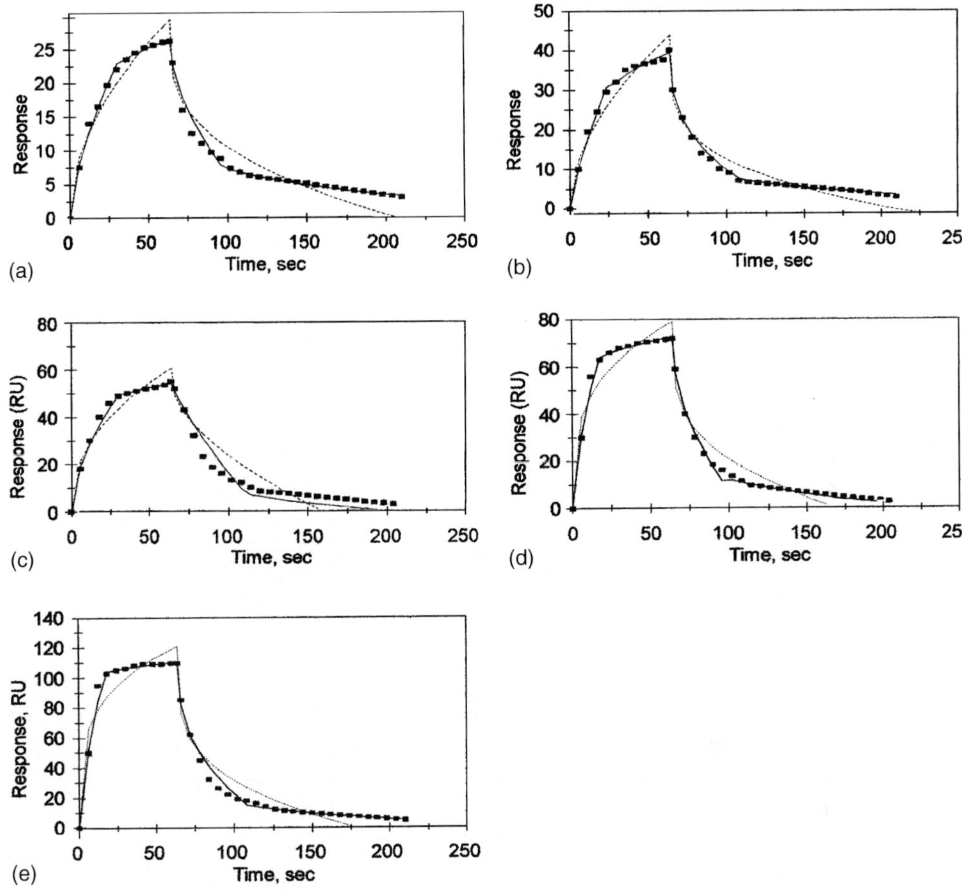

Figure 4.1 Binding and dissociation of different wild-type LCAT concentrations (in μM) in solution to egg-white apoA-I rHDL immobilized on a biosensor chip containing streptavidin (Jin *et al.*, 1999): (a) 0.04, (b) 0.08, (c) 0.16, (d) 0.32, (e) 0.64. When only a solid line is used then a single-fractal analysis applies. When both a dashed (------) and a solid (_____) line are used, then the dashed line is for a single-fractal analysis, and the solid line is for a dual-fractal analysis. In this case, the solid line provides the better fit.

presented in Table 4.1 and in Figures 4.2a and b, the binding rate coefficients, k_1 and k_2, are given by

$$k_1 = (20.610 \pm 4.942)[\mathrm{LCAT}]^{0.7058 \pm 0.0980} \tag{4.2a}$$

$$k_2 = (118.086 \pm 27.088)[\mathrm{LCAT}]^{0.7655 \pm 0.0942} \tag{4.2b}$$

Table 4.1

(a) Binding and dissociation rate coefficients and (b) fractal dimensions for the binding and the dissociation phases for different wild-type LCAT in solution to egg-white apoA-I rHDL immobilized on a biosensor chip containing streptavidin (Jin *et al.*, 1999)

(a)

LCAT concentration (μM)	k	k_1	k_2	k_d	k_{d1}	k_{d2}
0.04	3.6837 ± 0.389	2.4561 ± 0.1931	12.225 ± 0.088	4.1041 ± 0.6945	2.5348 ± 0.3686	11.494 ± 0.0677
0.08	4.7452 ± 0.5910	2.5814 ± 0.1964	12.814 ± 0.338	10.289 ± 0.961	8.5019 ± 0.3988	22.860 ± 0.063
0.16	10.006 ± 1.2108	6.7078 ± 0.5654	30.201 ± 0.244	3.8827 ± 1.0229	2.1976 ± 0.4588	28.981 ± 0.087
0.32	22.700 ± 3.181	8.8928 ± 1.194	48.117 ± 0.435	16.776 ± 2.816	10.484 ± 1.507	39.951 ± 0.476
0.64	41.958 ± 6.322	15.276 ± 2.473	89.869 ± 3.548	27.277 ± 3.250	20.596 ± 1.838	69.159 ± 0.137

(b)

	D_f	D_{f1}	D_{f2}	D_{fd}	D_{fd1}	D_{fd2}
0.04	2.0030 ± 0.0856	1.6854 ± 0.1191	2.6308 ± 0.0292	2.2526 ± 0.0622	1.8568 ± 0.1002	2.7222 ± 0.0077
0.08	1.9322 ± 0.100	1.4414 ± 0.0141	2.4600 ± 0.0570	2.4524 ± 0.0356	2.2686 ± 0.0296	2.8130 ± 0.0055
0.16	2.1336 ± 0.0974	1.8238 ± 0.1092	2.7184 ± 0.0238	1.8360 ± 0.0932	1.3990 ± 0.1168	2.7412 ± 0.0068
0.32	2.400 ± 0.116	1.6010 ± 0.3206	2.9034 ± 0.0147	2.3764 ± 0.0620	1.9874 ± 0.0990	2.7754 ± 0.0136
0.64	2.4912 ± 0.1196	1.6266 ± 0.3820	2.9018 ± 0.0098	2.4134 ± 0.0448	2.1972 ± 0.0536	2.8324 ± 0.0039

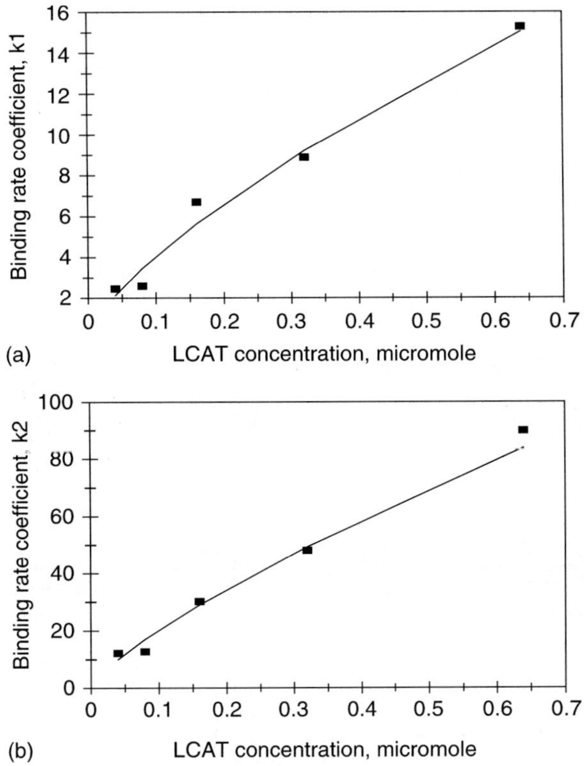

Figure 4.2 (a) Increase in the binding rate coefficient, k_1, with an increase in the LCAT concentration (in μM) in solution, and (b) increase in the binding rate coefficient, k_2, with an increase in the LCAT concentration (in μM) in solution.

The fit is very good. The binding rate coefficients exhibit an order of dependence between one-half and first (k_1=0.7058 and k_2=0.7655) order on the LCAT concentration in solution. The non-integer dependence indicates the fractal nature of the system.

It is of interest to note that the binding rate coefficients, k_1 and k_2, exhibit orders of dependence on the LCAT concentration that are close to each other (within 86%).

Figure 4.3a shows the increase in the dissociation rate coefficient k_{d2} with an increase in the LCAT concentration in solution from 0.04 to 0.64 μM. For the data presented in Table 4.1 and in Figure 4.3a, the dissociation rate coefficient, k_{d2}, is given by

$$k_{d2} = (87.263 \pm 12.134)[\mathrm{LCAT}]^{0.5983 \pm 0.0594} \tag{4.3a}$$

The fit is very good. The dissociation rate coefficient, k_{d2}, exhibits an order of dependence slightly more than one-half (=0.5983) order on the LCAT concentration in solution. The non-integer dependence indicates the fractal nature of the system.

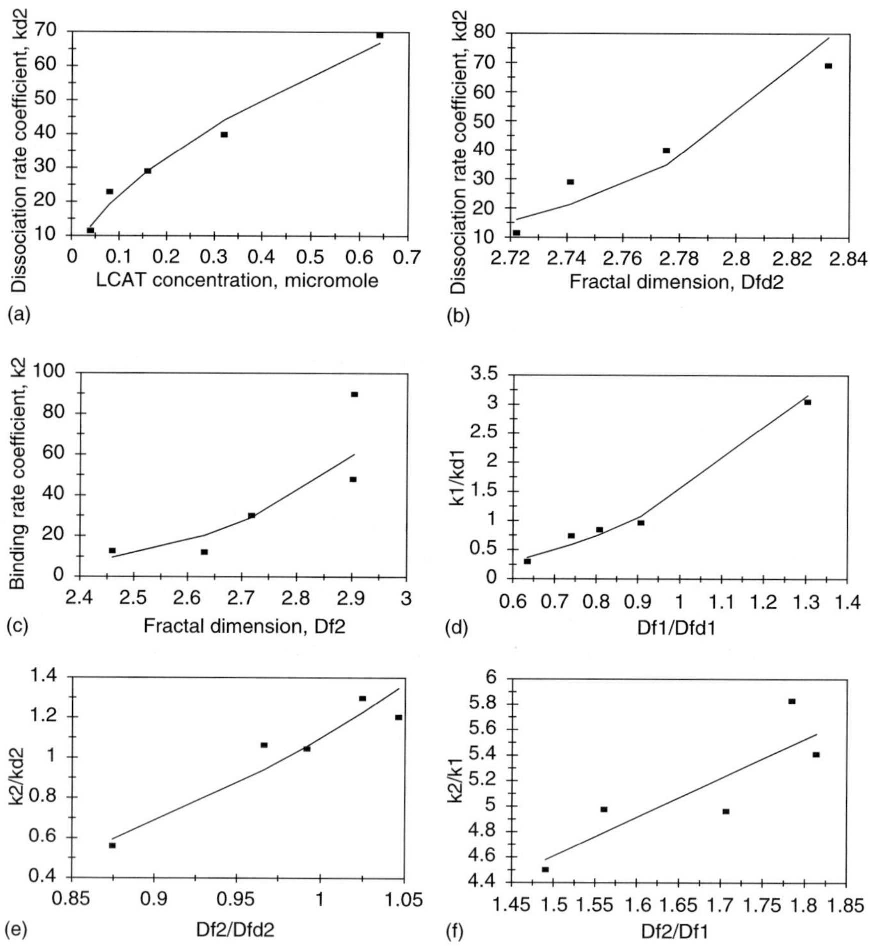

Figure 4.3 (a) Increase in the dissociation rate coefficient, k_{d2}, with an increase in the LCAT concentration (in μM) in solution, (b) increase in the dissociation rate coefficient, k_{d2}, with an increase in the fractal dimension, D_{fd2}, (c) increase in the binding rate coefficient, k_2, with an increase in the fractal dimension, D_{f2}, (d) increase in the affinity, K_1 (= k_1/k_{d1}), with an increase in the ratio of fractal dimensions in the binding and in the dissociation phase, D_{f1}/D_{fd1}, (e) increase in the affinity, K_2 (= k_2/k_{d2}), with an increase in the ratio of fractal dimensions in the binding and in the dissociation phase, D_{f2}/D_{fd2}, and (f) increase in the ratio of binding rate coefficients, k_2/k_1, with an increase in the ratio of fractal dimensions, D_{f2}/D_{f1}.

Figure 4.3b shows the increase in the dissociation rate coefficient, k_{d2}, with an increase in the fractal dimension in the dissociation phase, D_{fd2}. For the data presented in Table 4.1 and in Figure 4.3b, the dissociation rate coefficient, k_{d2}, is given by

$$k_{d2} = (6.6\text{E}-17 \pm 2.7\text{E}-17) D_{fd2}^{\,39.979 \pm 11.592} \tag{4.3b}$$

The fit is reasonable. The availability of more data points would lead to a more reliable fit. The dissociation rate coefficient, k_{d2}, is extremely sensitive to the degree of heterogeneity present on the sensor chip surface during the dissociation phase, as noted by the inordinately high order (=39.979) of dependence exhibited.

Figure 4.3c shows the increase in the binding rate coefficient, k_2, with an increase in the fractal dimension in the second binding phase, D_{f2}. For the data presented in Table 4.1 and in Figure 4.3c, the binding rate coefficient, k_2, is given by

$$k_2 = (1.5415 + 2.177) D_{f2}^{\,2.657+5.1542} \tag{4.3c}$$

There is scatter in the fit of the data. This is reflected in the error in the coefficient and in the order also. Only the positive errors in each case are presented. This is especially true for the rate coefficient which cannot be negative. The power (order) is also positive as noted by the shape of the curve in Figure 4.3c. The availability of more data points would lead to a more reliable fit. The binding rate coefficient, k_2, is sensitive to the degree of heterogeneity present on the sensor chip surface during the second binding phase, as noted by the order exhibited that lies between second and third (=2.657).

Affinity, K, values are of interest to practicing biosensorists. The affinity, K_1, may be defined as K_1 (= k_1/k_{d1}). Figure 4.3d shows the increase in the affinity, K_1, with an increase in the fractal dimension ratio, D_{f1}/D_{fd1}. For the data presented in Table 4.1 and in Figure 4.3d, the affinity, K_1 (= k_1/k_{d1}), is given by

$$K_1 = (1.4354 \pm 0.3179)(D_{f1}/D_{fd1})^{2.9775 \pm 0.3692} \tag{4.3d}$$

The fit is very good. The affinity, K_1, exhibits very close to third-order dependence on the ratio of fractal dimensions, D_{f1}/D_{fd1}. This indicates that the affinity, K_1, is very sensitive to the ratio of fractal dimensions.

The affinity, K_2, may be defined as K_2 (= k_2/k_{d2}). Figure 4.3e shows the increase in the affinity, K_2, with an increase in the fractal dimension ratio, D_{f2}/D_{fd2}. For the data presented in Table 4.1 and in Figure 4.3e, the affinity, K_2 (= k_2/k_{d2}), is given by

$$K_2 = (1.0975 \pm 0.1259)(D_{f2}/D_{fd2})^{4.572 \pm 0.771} \tag{4.3e}$$

The fit is very good. The affinity, K_2, exhibits an order of dependence between 4 and 5 (=4.572) on the ratio of fractal dimensions, D_{f2}/D_{fd2}. This indicates that the affinity, K_2, is very sensitive to the ratio of fractal dimensions, D_{f2}/D_{fd2}.

For a dual-fractal analysis, it is of interest to note the change in the degree of heterogeneity from the first phase of binding (D_{f1}) to the second phase of binding (D_{f2}), and the changes resulting therein in the binding rate coefficients, k_1 and k_2. For the data presented in Table 4.1 and in Figure 4.3f, the ratio of the binding rate coefficients, k_2/k_1, is given by

$$k_2/k_1 = (3.0755 \pm 0.1771)(\mathrm{D}_{f2}/\mathrm{D}_{f1})^{0.9973 \pm 0.3284} \tag{4.3f}$$

There is scatter in the fit of the data. The availability of more data points would lead to a more reliable fit. The ratio of binding rate coefficients, k_2/k_1, exhibits very close to a first (=0.9973) order of dependence on the ratio of fractal dimensions, D_{f2}/D_{f1}. One might easily consider this as a linear dependence, though this is fortuitious, and applies presumably for this case only.

Gaus and Hall (1999) have used the SPR biosensor to detect native LDL, mildly oxidized LDL, and strongly oxidized LDL. Brown and Goldstein (1983) had initially suggested that it is the oxidative modification of LDL that leads to atherosclerotic LDL. They further indicate that both the state of the LDL and the receptors could be causal in atherosclerosis. Fogelman *et al.* (1980) and Steinbrecher (1991, 1996) suggest that presumably the oxidation of LDL occurs through the action of aldehyde and oxidation byproducts of lipid peroxidation. Penn and Chisolm (1994) and Aviram (1993) have shown the link between oxidized LDL and atherogenesis. Gaus and Hall (1999) have used a heparin-modified biosensor chip surface as a modulator of LDL binding, according to its binding state. Heparin binding sites may be regarded as identical to LDL–receptor binding sites (Hirose *et al.*, 1987; Weisgraber and Rall, 1987). Thus, Gaus and Hall (1999) indicate that the oxidative modification of the lipoprotein would apparently lead to a decrease in the binding of LDL to the heparinized surface.

Figure 4.4a shows the binding of native LDL in solution to a heparinized Au-surface of an SPR sensor chip (Gaus and Hall, 1999). The binding kinetics is adequately described by a single-fractal analysis. The values of the binding rate coefficient, k, and the fractal dimension, D_f, are given in Table 4.2a.

Figure 4.4b shows the binding of mildly oxidized LDL in solution to a heparinized Au-surface of an SPR sensor chip. The values of the binding rate coefficient, k, and the fractal dimension, D_f, are given in Table 4.2a. It is of interest to note that as one goes from the native LDL to the mildly oxidized LDL in solution there is (a) an increase in the fractal dimension, D_f, or the degree of heterogeneity on the sensor chip surface by 11.3%, and (b) a subsequent increase in the binding rate coefficient, k, by 8.72%. Note that changes in the fractal dimension and in the binding rate coefficient are in the same direction. This result is contrary to what is indicated by Gaus and Hall (1999), in that a mildly oxidized LDL should exhibit a decrease in the binding rate coefficient when compared with native LDL. In our case, the reverse holds true wherein the mildly oxidized state leads to an increase in the degree of heterogeneity on the surface, and subsequently to a higher binding rate coefficient for the mildly oxidized state when compared with the native LDL state.

Figure 4.4c shows the binding of strongly oxidized LDL in solution to a heparinized Au-surface of an SPR sensor chip (Gaus and Hall, 1999). In this case, a dual-fractal analysis is required to describe the binding kinetics. The values of (a) the binding rate coefficient, k, and the fractal dimension, D_f, for a single-fractal analysis, and (b) the binding rate coefficients, k_1 and k_2, and the fractal dimensions, D_{f1} and D_{f2}, are given in Tables 4.2a and b. Note that as one goes from the native and mildly oxidized LDL to the strongly oxidized LDL, there is a change in the binding mechanism, since for the first two cases the binding kinetics is adequately described by a single-fractal analysis, whereas for the third case a dual-fractal analysis is required to describe the binding kinetics. The binding is more complex for the strongly oxidized LDL since a dual-fractal analysis applies here.

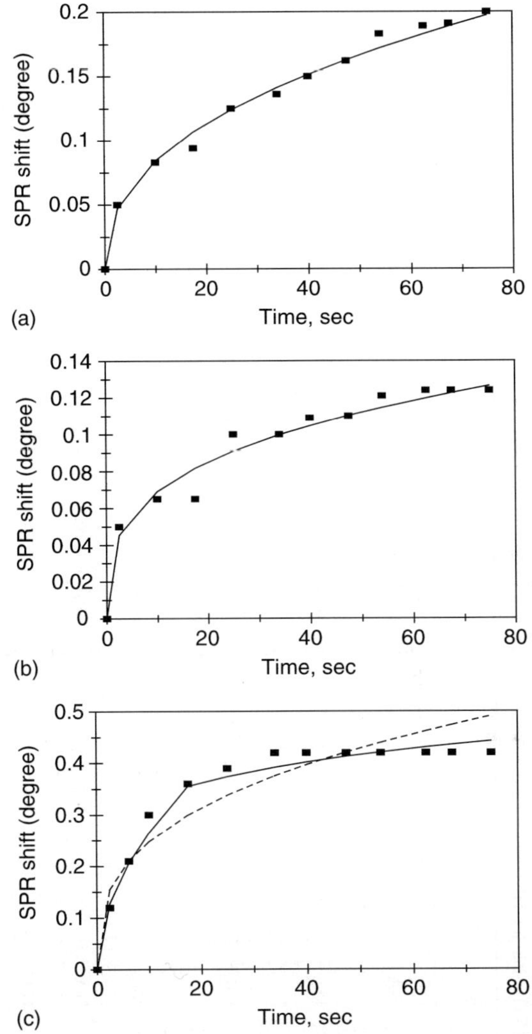

Figure 4.4 Binding of different forms of LDL in solution to heparin-modified Au-surface of an SPR biosensor (Gaus and Hall, 1999): (a) native, (b) mildly oxidized, (c) strongly oxidized. When only a solid line is used then a single-fractal analysis applies. When both a dashed (-------) and a solid (_____) line are used, then the dashed line is for a single-fractal analysis, and the solid line is for a dual-fractal analysis. In this case, the solid line provides the better fit.

One possible explanation could be the existence of a heterogeneous population in the oxidized state, as suggested by Gaus and Hall (1999), and may in fact apply here. Presumably, one can go one step further in this line of thought. There may be some (very minor) degree of heterogeneity in the native LDL. There may be a higher (but still small)

Table 4.2

Binding rate coefficients and fractal dimensions for (a) native, mildly oxidized, and strongly oxidized LDL in solution to heparin-modified Au-surface of an SPR biosensor (Gaus and Hall, 1999), and (b) different concentrations of TRITC-labeled HDL in solution to a bare optical fiber (Singh and Poirier, 1993)

Analyte in solution/ receptor on surface	k	k_1	k_2	D_f	D_{f1}	D_{f2}	Reference
(a) Native LDL/ heparinized Au-surface	0.03186 ± 0.00181	NA	NA	2.1554 ± 0.0342	NA	NA	Gaus and Hall (1999)
Mildly oxidized LDL/heparinized Au-surface	0.03464 ± 0.00342	NA	NA	2.3992 ± 0.0584	NA	NA	Gaus and Hall (1999)
Strongly oxidized LDL/heparinized Au-surface	0.1135 ± 0.0182	0.07902 ± 0.0082	0.2270 ± 0.0117	2.3214 ± 0.0839	1.9498 ± 0.1086	2.6906 ± 0.0516	Gaus and Hall (1999)
(b) 0.1 μg/ml TRITC-HDL	0.3263 ± 0.0277	NA	NA	1.1910 ± 0.1098	NA	NA	Singh and Poirier (1993)
0.5 μg/ml TRITC-HDL	8.0171 ± 0.8286	NA	NA	1.6801 ± 0.1917	NA	NA	Singh and Poirier (1993)
1.0 μg/ml TRITC-HDL	132.88 ± 3.747	NA	NA	2.5108 ± 0.0344	NA	NA	Singh and Poirier (1993)
5.0 μg/ml TRITC-HDL	523.74 ± 54.157	412.851 ± 54.157	1264.355 ± 1.787	2.5610 ± 0.1315	2.4512 ± 0.2094	2.8798 ± 0.0053	Singh and Poirier (1993)

degree of heterogeneity in the LDL in solution for the mildly oxidized LDL. However, in both these cases the degree of heterogeneity of the LDL in solution does not affect the binding mechanism, which is adequately described by a singe-fractal analysis. However, for the strongly oxidized LDL in solution, there is presumably a high enough degree of heterogeneity present in the LDL that subsequently leads to a change in the binding mechanism. Thus, in this case a dual-fractal analysis is required to describe the binding kinetics. This is offered as one possible mechanism to describe the change in the binding kinetics, based on the initial assumption outlined by Gaus and Hall (1999). Other explanations that are consistent with the data presented and analyzed are also possible.

Singh and Poirier (1993) have developed a fiber-optic biosensor to quantitatively measure cholesterol and lipoproteins in real time. They indicate that the intracellular transport of lipids (cholesterol, cholesterol esters, and triacylglycerides) in the aqueous environment of the body is accomplished by water-soluble lipoproteins. These authors indicate that the apolipoproteins provide the lipoproteins with enough hydrophobic character that they are soluble in water. The LDL transport the cholesterol to peripheral tissues (Rininger and Pittman, 1988), and are responsible for the formation of atherosclerotic plaques on arterial walls (Lawn, 1992). The HDL transport the cholesterol, as indicated earlier, to the liver for reutilization.

Figures 4.5a–c show the binding of (1) 0.1, (2) 0.5, and (3) 1.0 μg/ml TRITC-labeled HDL in solution to a bare optic (glass) fiber (Singh and Poirier, 1993). The binding kinetics is adequately described by a single-fractal analysis. The values of the binding rate coefficient, k, and the fractal dimension, D_f, for a single-fractal analysis are given in Table 4.2. In case (1), the concentration of the TRITC-HDL is rather low and almost a linear plot is obtained.

In case (2), note that as the TRITC-HDL concentration in solution increases by a factor of 5 from 0.1 to 0.5 μg/ml (a) the fractal dimension increases by 29.1% from a value of 1.1910 to 1.6801, and (b) the binding rate coefficient, k, increases by a factor of 24.57 from a value of 0.3263 to 8.0171. The fractal dimension is very sensitive to the degree of heterogeneity on the bare optical fiber surface.

In case (3), note that as the TRITC-HDL concentration in solution increases by an order of magnitude from 0.1 to 1.0 μg/ml (a) the fractal dimension increases by a factor of 2.108 from a value of 1.1910 to 2.5108, and (b) the binding rate coefficient, k, increases by a factor of 407.23 from a value of 0.3263 to 132.88. Once again the fractal dimension is very sensitive to the degree of heterogeneity on the bare optical fiber surface.

Figure 4.5d shows the binding of 5.0 μg/ml TRITC-labeled HDL in solution to a bare optic fiber. In this case, a dual-fractal analysis is required to adequately describe the binding kinetics. The values of (a) the binding rate coefficient, k, and the fractal dimension, D_f, for a single-fractal analysis, and (b) the binding rate coefficients, k_1 and k_2, and the fractal dimensions, D_{f1} and D_{f2}, for a dual-fractal analysis are given in Table 4.2b. Note that there is a change in the binding mechanism when compared with the 0.1–1.0 μg/ml TRITC-labeled HDL in solution, since for the former a dual-fractal analysis is required to adequately describe the binding kinetics, whereas for the latter a single-fractal analysis is adequate to describe the binding kinetics.

In the 0.1–1.0 μg/ml TRITC-HDL concentration range in solution, Table 4.2 and Figure 4.6a show for a single-fractal analysis, the increase in the binding rate coefficient,

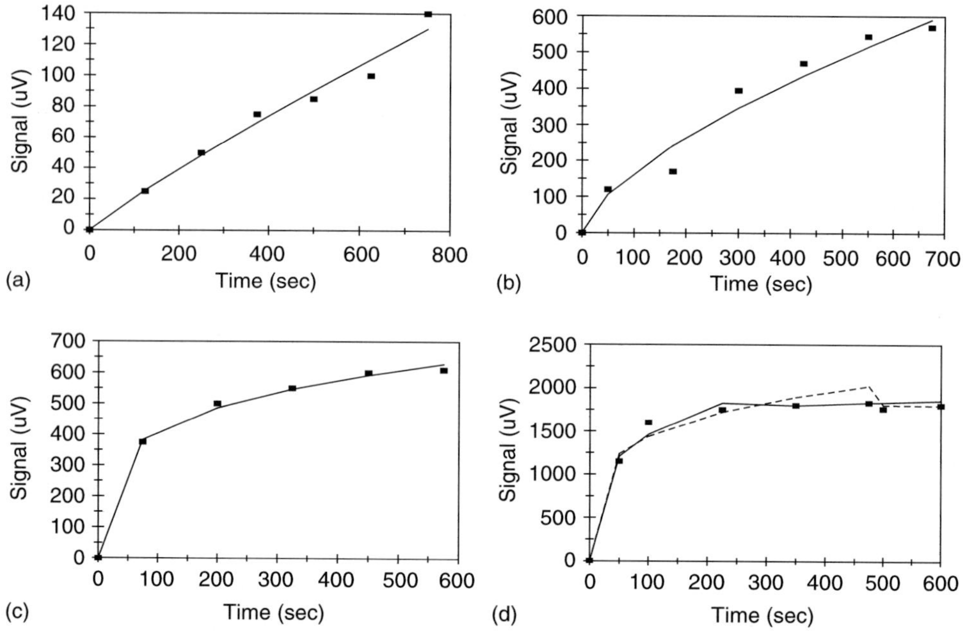

Figure 4.5 Binding of different concentrations (in μg/ml) of TRITC-labeled HDL in solution to a bare optical fiber (Singh and Poirier, 1993): (a) 0.1, (b) 0.5, (c) 1.0, (d) 5.0. When only a solid line is used then a single-fractal analysis applies. When both a dashed (-------) and a solid (_____) line are used, then the dashed line is for a single-fractal analysis, and the solid line is for a dual-fractal analysis. In this case, the solid line provides the better fit.

k, with an increase in the TRITC-HDL concentration in solution. In this concentration range, the binding rate coefficient, k, is given by

$$k = (85.39 + 103.70)[\text{TRITC-HDL}]^{2.50 \pm 0.476} \quad (4.4a)$$

There is scatter in the data. This is reflected in the error in the coefficient. Only the positive error is presented since the binding rate coefficient, k, cannot be negative. The availability of more data points would lead to a more reliable fit. The binding rate coefficient, k, exhibits a two and half order of dependence on the TRITC-labeled HDL concentration in solution. The non-integer dependence of the binding rate coefficient on the TRITC-HDL concentration in solution lends support to the fractal nature of the solution.

In the 0.1–1.0 μg/ml TRITC-HDL concentration range in solution, Table 4.2 and Figure 4.6b show for a single-fractal analysis the increase in the binding rate coefficient, k, with an increase in the fractal dimension, D_f. In this concentration range, the binding rate coefficient, k, is given by

$$k = (0.0935 + 0.0392)D_f^{8.0279 \pm 0.6632} \quad (4.4b)$$

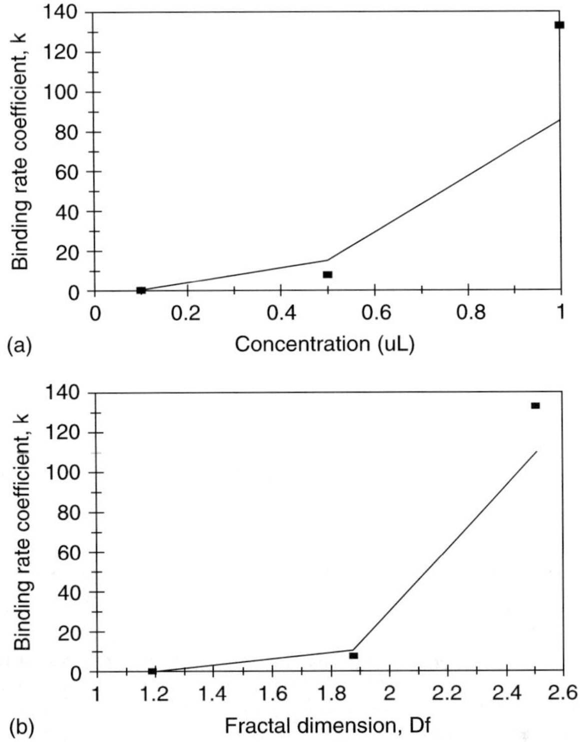

Figure 4.6 (a) Increase in the binding rate coefficient, k, with an increase in the TRITC-labeled HDL concentration (in μg/ml) in solution, and (b) increase in the binding rate coefficient, k, with an increase in the fractal dimension, D_f.

The fit is reasonable. The availability of more data points would lead to a more reliable fit. The binding rate coefficient, k, is extremely sensitive to the degree of heterogeneity on the surface as noted by the slightly more than eighth (equal to 8.0279) order of dependence exhibited.

Singh and Poirier (1993) analyzed the initial binding rates of 5 μg/ml HDL or LDL in solution to a bare optical fiber and to a cholesterol-coated fiber. Figures 4.7a and b show the binding of 5 μg/ml TRITC-labeled HDL in solution to (1) a bare optical fiber and (2) an optical fiber coated with 100mg/ml cholesterol. A single-fractal analysis is adequate to describe the binding kinetics. The values of the binding rate coefficient, k, and the fractal dimension, D_f, are given in Table 4.3.

It is of interest to note that as one goes from the case (1) to case (2), (a) the fractal dimension, D_f, decreases by 45.4% from a value of 1.7232 to 0.9414, and (b) the binding rate coefficient, k, decreases by a factor of 6.052 from a value of 18.491 to 3.0549. Note that changes in the binding rate coefficient and in the fractal dimension or the degree of heterogeneity on the surface are in the same direction.

Singh and Poirier (1993) also analyzed the influence of the concentration of cholesterol coating on the optical fiber on the initial binding kinetics of LDL in solution. Figures 4.7c

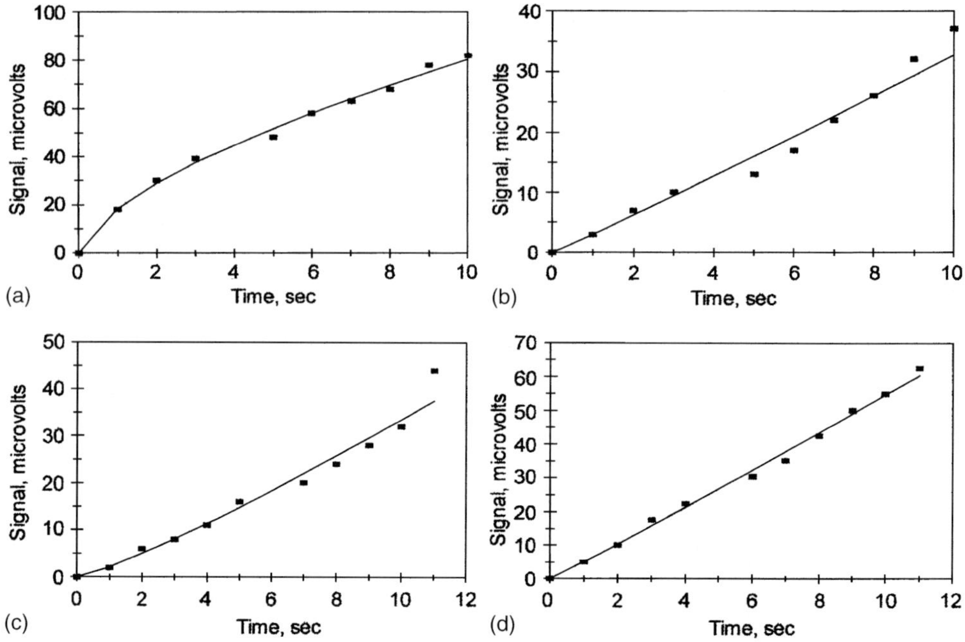

Figure 4.7 (a) Binding of 5 μg/ml TRITC-labeled HDL in solution to a bare optical fiber (Singh and Poirier, 1993), (b) binding of 5 μg/ml TRITC-labeled HDL in solution to a 100 μg/ml cholesterol-coated optical fiber (Singh and Poirier, 1993), (c) binding of 5 μg/ml TRITC-labeled LDL in solution to a 5 μg/ml cholesterol-coated optical fiber (Singh and Poirier, 1993), and (d) binding of 5 μg/ml TRITC-labeled LDL in solution to a 100 μg/ml cholesterol-coated optical fiber (Singh and Poirier, 1993).

and d show the binding of 5 μg/ml TRITC-labeled LDL in solution to an optical fiber coated with (1) 5 μg/ml cholesterol and (2) 100 μg/ml cholesterol. A single-fractal analysis is adequate to describe the binding kinetics. The values of the binding rate coefficient, k, and the fractal dimension, D_f, are given in Table 4.3.

It is of interest to note that as one increases the concentration of the cholesterol coating from 5 to 100 μg/ml, (a) the fractal dimension, D_f, decreases by 32.5% from a value of 0.9384 to 0.6336, and (b) the binding rate coefficient, k, deceases by a factor of 2.3 from a value of 5.079 to 2.2090. Once again, changes in the fractal dimension and in the binding rate coefficient are in the same direction. Similar behavior was also observed for the binding of HDL in solution to the uncoated and cholesterol-coated fiber. This result is consistent with the results presented by Singh and Poirier (1993) in that the cholesterol coating decreases the rate of binding of either LDL or HDL in solution to the optical fiber.

Finally, it is of interest to compare the binding of 5 μg/ml of either HDL or LDL in solution to a 100 μg/ml cholesterol-coated fiber. Note that as one goes from the HDL to the LDL in solution (a) the fractal dimension, D_f, decreases by 32.7% from a value of 0.9414 to 0.6336, and (b) the binding rate coefficient, k, decreases by 33.6% from a value of

Table 4.3

Binding rate coefficients and fractal dimensions for the binding of (a) TRITC-labeled HDL in solution to a bare and a cholesterol-coated (100 μg/ml) fiber, and (b) TRITC-labeled LDL in solution to an optical fiber coated with different concentrations of cholesterol (5 and 100 μg/ml) (Singh and Poirier, 1993)

Analyte in solution/receptor on surface	D_f	k
5 μg/ml TRITC-labeled HDL/bare optical fiber	1.7232 ± 0.03850	18.491 ± 0.7982
5 μg/ml TRITC-labeled HDL/100 μg/ml cholesterol-coated optical fiber	0.9414 ± 0.1098	3.0549 ± 0.03887
5 μg/ml TRITC-labeled LDL/5μg/ml cholesterol-coated optical fiber	0.9384 ± 0.04856	5.079 ± 0.3003
5 μg/ml TRITC-labeled LDL/100 μg/ml cholesterol-coated optical fiber	0.6336 ± 0.0926	2.2090 ± 0.2537

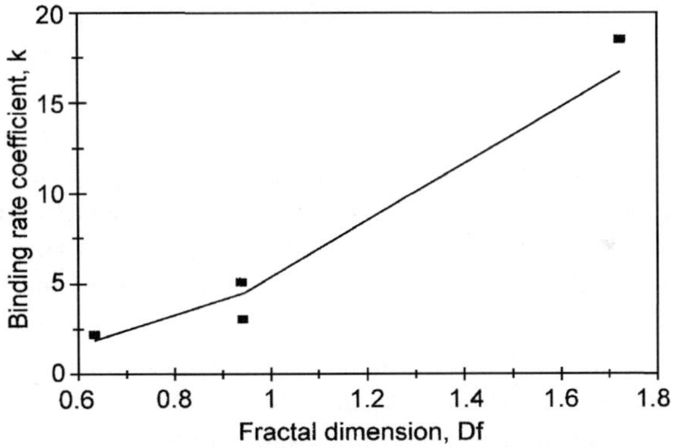

Figure 4.8 Increase in the binding rate coefficient, k, with an increase in the fractal dimension, D_f, for the binding of either HDL or LDL in solution to an uncoated or cholesterol-coated fiber.

3.0549 to 2.202. These results are consistent with those presented by Singh and Poirier (1993). Once again, changes in the fractal dimension or the degree of heterogeneity on the surface and in the binding rate coefficient are in the same direction. Also, note in the 100 μg/ml cholesterol-coated case, the changes (decreases) in the fractal dimension and in the binding rate coefficient are almost linear. This is, however, just a coincidence.

The data presented in Table 4.3 and in Figure 4.8 show for a single-fractal analysis the increase in the binding rate coefficient, k, with an increase in the fractal dimension, D_f. The binding rate coefficient, k, is given by

$$k = (5.116 \pm 1.8941) D_f^{2.1756 \pm 0.4403} \tag{4.5}$$

The fit is good. More data points are required to provide a better fit. Note that the data for the binding of LDL and HDL binding in solution to the uncoated or cholesterol-coated fiber are presented together, due to the lack of availability of points for each case separately. In that case the fit is remarkable. This indicates that there is similarity in the way by which LDL or HDL in solution affects the nature of the degree of heterogeneity or the fractal dimension of the optical fiber surface (uncoated or cholesterol coated). The binding rate coefficient, k, exhibits an order of dependence greater than 2 (=2.1756) on the degree of heterogeneity or the fractal dimension, D_f, on the surface.

Steinlund *et al.* (2003) have recently analyzed the binding (and dissociation) activity of G protein-coupled receptors (GPCRs) on a biosensor surface. These authors state that the GPCRs are involved in a variety of physiological functions, for example, control of cardiac function and neuronal transmission. The GPCRs are important targets for therapeutic intervention. These authors further state that the GPCR has seven transmembrane-spanning α-helices, and represents the largest group of membrane-spanning cell-surface receptors (Karlsson and Lofas, 2002). This receptor is extremely hydrophobic and requires a lipid environment to keep its native conformation. Steinlund *et al.* (2003) have developed a 'capture and reconstitute' method to assemble GPCRs on a biosensor chip surface. They were able to use the SPR biosensor method to analyze the binding (and dissociation) interactions of GPCRs by tethering one biomolecule to the surface and the partner in solution.

Figures 4.9a–c show the binding and dissociation of the anti-CCR5 linear epitope-recognizing antibody, 3A9, in solution to the human chemokine coupled receptor CCR5 (1) captured, reconstituted, and assembled on an L1 sensor chip, (2) captured on an L1 sensor chip, and (3) captured on a CM5 sensor chip (Steinlund *et al.*, 2003). A dual-fractal analysis is required to adequately describe the binding kinetics. A single-fractal analysis is adequate to describe the dissociation kinetics. The values of (a) the binding rate coefficient, k, and the fractal dimension, D_f, for a single-fractal analysis, (b) the binding rate coefficients, k_1 and k_2, and the fractal dimensions, D_{f1} and D_{f2}, for a dual-fractal analysis, and (c) the dissociation rate coefficient, k_d, and the fractal dimension for dissociation, D_{fd}, for a single-fractal analysis are given in Tables 4.4a and b.

It is of interest to note that as one goes from the CCR5 assembled on an L1 sensor chip to the CCR5 captured on an L1 chip, (a) the fractal dimension, D_{f1}, decreases by 16.1% from a value of 1.7936 to 1.5052, and the binding rate coefficient, k_1, decreases by 32.8% from a value of 0.02605 to 0.01751, and (b) the fractal dimension, D_{f2}, increases by 2.7% from a value of 2.587 to 2.6572 and the binding rate coefficient, k_2, increases by 18.55% from a value of 0.08031 to 0.09521. Note that changes in the binding rate coefficient and in fractal dimension are in the same direction for both cases. In other words, for example, an increase in the degree of heterogeneity or fractal dimension on the surface leads to an increase in the binding rate coefficient, k_1 or k_2.

It is of interest to note that as one goes from the CCR5 assembled on an L1 sensor chip to the CCR5 captured on a CM5 chip (a) the fractal dimension, D_{f1} decreases by 19% from a value of 1.7936 to 1.4526, and the binding rate coefficient k_1 decreases by 24.3% from a value of 0.02605 to 0.01972, and (b) the fractal dimension D_{f2} decreases by 6.05% from a value of 2.587 to 2.4308 and the binding rate coefficient k_2 increases by 2.4% from a value of 0.08031 to 0.08224. Note that changes in the binding rate coefficient and in the fractal dimension are in the same direction for one case, and in the opposite direction in the other

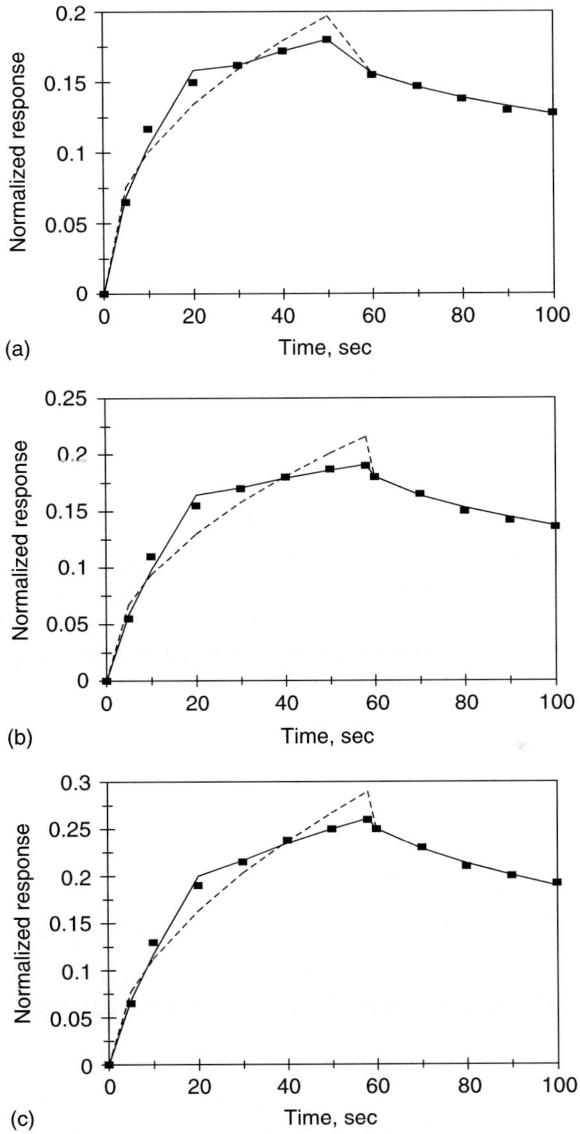

Figure 4.9 Binding and dissociation of anti-CCR5 linear epitope-recognizing antibody, 3A9 in solution to the human chemokine-coupled receptor CCR5 (Steinlund *et al.*, 2003): (a) captured and reconstituted and assembled on an L1 sensor chip, (b) captured on an L1 chip, and (c) captured on a CM5 chip. When only a solid line is used then a single-fractal analysis applies. When both a dashed (-----------) and a solid (_________) line are used, then the dashed line is for a single-fractal analysis, and the solid line is for a dual-fractal analysis. In this case, the solid line provides the better fit.

case. In other words, for example, an increase in the degree of heterogeneity or fractal dimension on the surface leads to an increase in the binding rate coefficient, k_1, but a decrease in the binding rate coefficient, k_2.

Table 4.4 and Figure 4.10a show for a dual-fractal analysis the increase in the binding rate coefficient, k_1, with an increase in the fractal dimension, D_{f1}, for the binding of 3A9 in solution to the receptor CCR5 assembled or captured on either an L1 or a CM5 chip. The binding rate coefficient, k_1, is given by

$$k_1 = (0.009929 \pm 0.001338) D_{f1}^{\,1.6232 \pm 0.77921} \tag{4.6a}$$

There is scatter in the data. Only three data points are available. The availability of more data points would lead to a better and more reliable fit. The binding rate coefficient, k_1, exhibits an order of dependence between 1 and 2 (=1.6232) on the degree of heterogeneity or the fractal dimension, D_{f1}, present on the chip surface.

Table 4.4

(a) Binding and dissociation rate coefficients and (b) fractal dimensions for the binding and dissociation phase for the anti-CCR5 linear epitope-recognizing antibody, 3A9, in solution to the human chemokine coupled receptor, CCR5, assembled on an L1 chip, captured on an L1 chip, and captured on a CM5 chip (Steinlund *et al.*, 2003)

(a)

Analyte in solution/ receptor on surface	k	k_1	k_2	k_d
3A9/CCR5 assembled on an L1 chip	0.03862±0.00532	0.02605±0.00387	0.08031±0.00004	0.007617±0.000628
3A9/CCR5 captured on an L1 chip	0.03150±0.00539	0.01751±0.00269	0.09521±0.00060	0.006617±0.000388
3A9/CCR5 captured on a CM5 chip	0.03280±0.00482	0.01972±0.00269	0.08224±0.00090	0.006374±0.000362

(b)

	D_f	D_{f1}	D_{f2}	D_{fd}
3A9/CCR5 assembled on an L1 chip	2.1676±0.1312	1.7936±0.2826	2.5874±0.0025	2.0100±0.1247
3A9/CCR5 captured on an L1 chip	2.0526±0.1430	1.5052±0.2916	2.6572±0.0249	1.8884±0.0469
3A9/CCR5 captured on a CM5 chip	1.9276±0.1242	1.4526±0.2612	2.4308±0.0436	1.7132±0.0454

Figure 4.10b shows the increase in the ratio of the binding rate coefficients, k_2/k_1, with an increase in the fractal dimension ratio, D_{f2}/D_{f1}. For the data presented in Tables 4.4a and b and in Figure 4.10b, the ratio of the binding rate coefficients, k_2/k_1, is given by

$$k_2/k_1 = (2.1034 \pm 0.0891)(D_{f2}/D_{f1})^{1.3910 \pm 0.3703} \quad (4.6b)$$

The fit is good. Only three data points are available. The availability of more data points would provide for a more reliable fit. The ratio of binding rate coefficients, k_2/k_1, exhibits an order of dependence between one and one-half order (=1.391) on the ratio of fractal dimensions, D_{f2}/D_{f1}, present on the sensor chip surface. If higher binding rate coefficients, such as k_2, are required then means or methods are required to selectively increase the degree of heterogeneity (or fractal dimension) during the second phase of binding, D_{f2}, while simultaneously not increasing or increasing only slightly the degree of heterogeneity present in the first

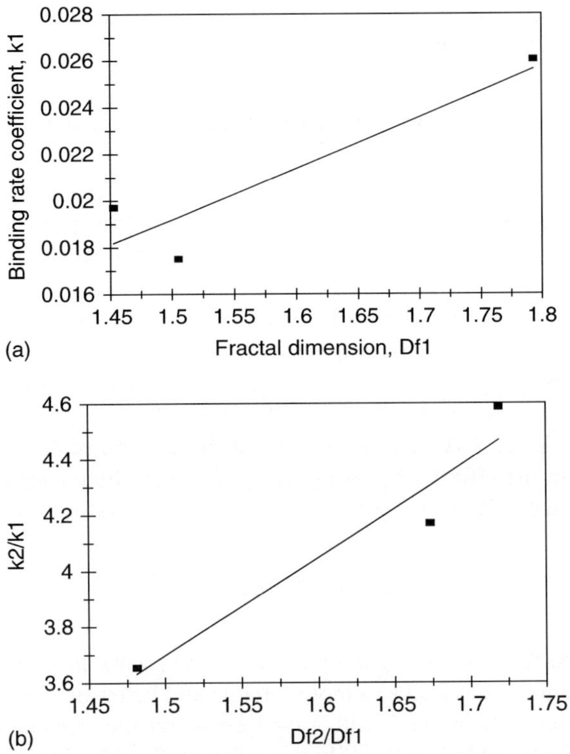

Figure 4.10 (a) Increase in the binding rate coefficient, k_1, with an increase in the fractal dimension, D_{f1}, for the binding of 3A9 in solution to the receptor CCR5 assembled, or captured on either an L1 or a CM5 chip, and (b) increase in the ratio of binding rate coefficients, k_2/k_1, with an increase in the fractal dimension ratio, D_{f2}/D_{f1}, for the binding of 3A9 in solution to the receptor CCR5 assembled or captured on either an L1 or a CM5 chip.

phase of binding. This is bound to be tricky and sure to test the abilities of experimentalists (biosensorists). Perhaps, a trial-and-error solution would be required here.

Table 4.4 and Figure 4.11a show for a single-fractal analysis the increase in the dissociation rate coefficient, k_d, with an increase in the fractal dimension in the dissociation phase, D_{fd}. For the data given in Table 4.4 and in Figure 4.11a, the dissociation rate coefficient, k_d, is given by

$$k_d = (0.003554 \pm 0.000211) D_{fd}^{\ 1.0511 \pm 0.5061} \tag{4.7a}$$

There is some scatter in the data. Once again, only three data points are available. The availability of more data points would provide for a more reliable fit. The dissociation rate coefficient, k_d, exhibits close to a first (=1.0511) order of dependence on the degree of heterogeneity of fractal dimension, D_{fd}, in the dissociation phase present on the chip surface.

Affinity values are of interest to practicing biosensorists. Define affinity, K_1, as k_1/k_d. Figure 4.11b and the data presented in Tables 4.4a and b show the increase in the affinity, K_1, as the ratio of fractal dimensions, D_{f1}/D_{fd}, increases. For the data presented in Table 4.4 and in Figure 4.11b, the affinity, K_1, is given by

$$K_1 = (4.4577 \pm 0.0579)(D_{f1}/D_{fd})^{2.2775 \pm 0.1614} \tag{4.7b}$$

The fit is very good. Only three data points are available. The availability of more data points would lead to a more reliable and better fit. The affinity, K_1, exhibits an order of dependence between 2 and 3 (=2.2775) on the ratio of fractal dimensions, D_{f1}/D_{fd}, that exists on the chip surface. If higher affinity, K_1, values are required, then this provides a method by which this may be obtained. As indicated earlier for eq. 4.6b, this could be tricky, and some ingenuity may be required to selectively change the degree of heterogeneity during the binding phase and not changing or changing to a much smaller extent the degree of heterogeneity in the dissociation phase.

Assume affinity, K_2, as k_2/k_d. Figure 4.11c and the data presented in Tables 4.4a and b show the increase in the affinity, K_2, as the ratio of fractal dimensions, D_{f2}/D_{fd}, increases. For the data presented in Table 4.4 and in Figure 4.11c, the affinity, K_2, is given by

$$K_2 = (6.1491 \pm 0.5855)(D_{f2}/D_{fd})^{2.2930 \pm 1.1921} \tag{4.7c}$$

There is scatter in the fit of the data. Only three data points are available. The availability of more data points would lead to a more reliable and better fit. The affinity, K_2, exhibits an order of dependence between 2 and 3 (=2.2930) on the ratio of fractal dimensions, D_{f2}/D_{fd}, that exists on the chip surface. If higher affinity, K_2, values are required, and as indicated above, then this provides a method by which this may be obtained. As indicated earlier for eq. 4.6b, this could be tricky, and some ingenuity may be required to selectively change the degree of heterogeneity during the binding phase and not changing or changing to a much smaller extent the degree of heterogeneity in the dissociation phase.

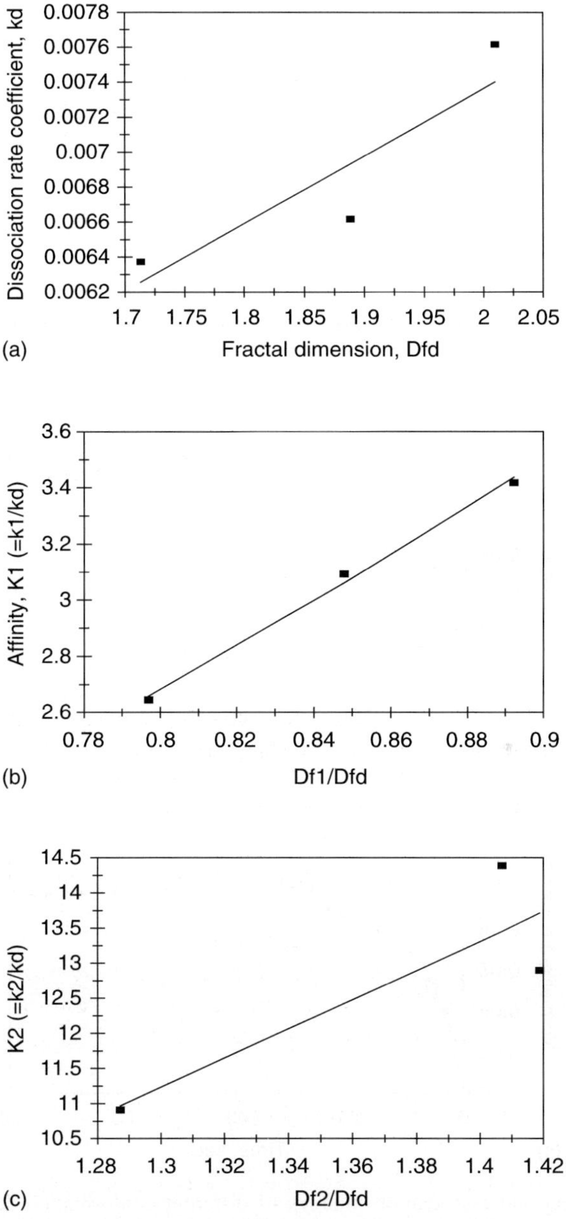

Figure 4.11 (a) Increase in the dissociation rate coefficient, k_d, with an increase in the fractal dimension, D_{fd}, for the binding of 3A9 in solution to the receptor CCR5 assembled, or captured on either an L1 or a CM5 chip, (b)increase in the affinity, K_1 $(= k_1/k_d)$, with an increase in the ratio of fractal dimensions, D_{f1}/D_{fd}, and (c) increase in the affinity, K_2 $(= k_2/k_d)$, with an increase in the ratio of fractal dimensions, D_{f2}/D_{fd}.

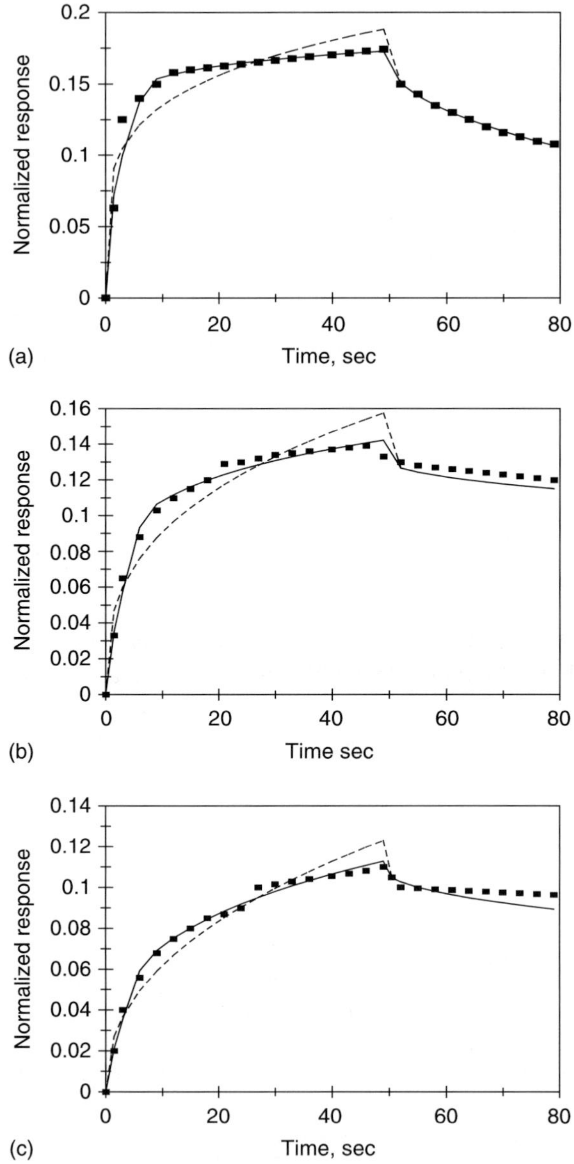

Figure 4.12 Binding and dissociation kinetics of different conformation-dependent monoclonal antibodies in solution to different human chemokine receptors immobilized on a sensor chip (Steinlund *et al.*, 2003): (a) 12G5 to captured and reconstituted CXCR4 on an L1 chip surface, (b) 44716.111 to captured and reconstituted CXCR4 on an L1 chip surface, and (c) 44717.111 to captured and reconstituted CXCR4 on an L1 chip surface. When only a solid line is used then a single-fractal analysis applies. When both a dashed (------------) and a solid (________) line are used, then the dashed line is for a single-fractal analysis, and the solid line is for a dual-fractal analysis. In this case, the solid line provides the better fit.

Figures 4.12a–c show the binding and dissociation of the conformation-dependent monoclonal antibody (1) 12G5 (2) 44716.111, and (3) 44717.111 in solution to the chemokine receptor CXCR4 captured and reconstituted on an L1 chip surface (Steinlund *et al.*, 2003). A dual-fractal analysis is required to adequately describe the binding kinetics. The dissociation kinetics are adequately described by a single-fractal analysis. The values of (a) the binding rate coefficient, k, and the fractal dimension, D_f, for a single-fractal analysis, (b) the binding rate coefficients, k_1 and k_2, and the fractal dimensions, D_{f1} and D_{f2}, for a dual-fractal analysis, and (c) the dissociation rate coefficient, k_d, and the fractal dimension, D_{fd}, for a single-fractal analysis are given in Tables 4.4a and 4.4b.

Note that as one goes from case (1) to case (2), (a) the fractal dimension, D_{f1}, decreases by 23.7% from a value of 2.0774 to 1.5850, and the binding rate coefficeint, k_1, decreases by a factor of 2.28 from a value of 0.0602 to 0.02637, and (b) the fractal dimension, D_{f2}, decreases by 7.05% from a value of 2.8620 to 2.6602, and the binding rate coefficient, k_2, decreases by 44.5% from a value of 0.1322 to 0.07339. Note that the changes in the fractal dimension or the degree of heterogeneity on the sensor chip surface and changes in their respective binding rate coefficients as noted above are in the same direction.

Table 4.5 and Figures 4.13a and b show the increase in the binding rate coefficients, k_1 and k_2, with an increase in the fractal dimensions, D_{f1} and D_{f2}, on the sensor chip for the binding of conformation-dependent monoclonal antibodies 12G5, 44716.111, and 44717.111 in solution to the captured and reconstituted human chemokine receptor, CCXR4, assembled on an L1 or a CM5 chip. For the data presented in Table 4.5 and Figures 4.13a and b, the binding rate coefficients, k_1 and k_2, are given by

$$k_1 = (0.003748 \pm 0.001041) D_{f1}^{3.8341 \pm 1.0151} \tag{4.8a}$$

$$k_2 = (4.1\mathrm{E}-05 \pm 0.1\mathrm{E}-05) D_{f2}^{7.6788 \pm 0.1806} \tag{4.8b}$$

The fit is good. Only three data points are available. The availability of more data points would lead to a more reliable fit. The binding rate coefficient, k_1, is very sensitive to the fractal dimension, D_{f1}, or the degree of heterogeneity present on the sensor chip surface as noted by the nearly fourth (=3.8341) order of dependence exhibited. The binding rate, k_2, is very sensitive to the fractal dimension, D_{f2}, or the degree of heterogeneity present on the sensor chip surface as noted by the order exhibited between 7 and 8 (=7.6788).

Table 4.5 and Figure 4.13c show the increase in the ratio of binding rate coefficients, k_2/k_1, with an increase in the fractal dimension ratio, D_{f2}/D_{f1}, on the sensor chip surface for the binding of conformation-dependent monoclonal antibodies 12G5, 44716.111, and 44717.111 in solution to the captured and reconstituted human chemokine receptor, CCXR4, assembled on an L1 or a CM5 chip. For the data presented in Table 4.5 and Figure 4.13c, the ratio of binding rate coefficients, k_2/k_1, is given by

$$k_2/k_1 = (1.5593 \pm 0.1471)(D_{f2}/D_{f1})^{1.0154 \pm 0.6188} \tag{4.8c}$$

Table 4.5

(a) Binding and dissociation rate coefficients and (b) fractal dimensions for the binding and dissociation phase for the conformation-dependent monoclonal antibodies 12G5, 44716.111, and 44717.111 in solution to the captured and reconstituted human chemokine receptor, CXCR4, assembled on an L1 chip, captured on an L1 chip, and captured on a CM5 chip (Steinlund *et al.*, 2003)

(a)

Analyte in solution/ receptor on on surface	k	k_1	k_2	k_d
12G5/CXCR4	0.08366±0.01148	0.00602±0.0133	0.1322±0.0006	0.014475±0.000275
44716.111/ CXCR4	0.0409±0.0060	0.02637±0.00436	0.07339±0.00226	0.003843±0.000127
44717.111/ CXCR4	0.02283±0.00282	0.01571±0.00241	0.03680±0.0011	0.004398±0.003811

(b)

	D_f	D_{f1}	D_{f2}	D_{fd}
12G5/CXCR4	2.5834±0.0644	2.0774±0.2938	2.8620±0.00588	2.0952±0.0170
44716.111/ CXCR4	2.3066±0.06822	1.5850±0.3122	2.6602±0.0322	2.0942±0.02960
44717.111/ CXCR4	2.1340±0.10583	1.5146±0.2970	2.4236±0.0322	2.0938±0.4104

There is scatter in the data. Only three data points are available. The availability of more data points would lead to a more reliable fit. The ratio of binding rate coefficients, k_2/k_1, exhibits close to a first-order (=1.1054) dependence on the ratio of fractal dimensions, D_{f2}/D_{f1}.

Loo *et al.* (2005) have recently used an enzyme-amplified diffraction-based immunoassay to detect picogram levels of antidigoxin (50–2000 pg/ml) in solution to digoxin immobilized on a glass substrate surface. These authors indicate that in this technique receptor molecules of interest are patterned on a substrate surface. This pattern will diffract light that hits it. When the patterned surface is exposed to binding analyte molecules, then the intensity of the diffracted light increases. Loo *et al.* (2005) emphasize that this technique has been used previously for the detection of whole cells (St. John *et al.*, 1998; Morhard *et al.*, 2000), small molecules (Mines *et al.*, 2002; Bailey and Hupp, 2002a,b), and proteins (Goh *et al.*, 2002; Tsay *et al.*, 1991).

Figures 4.14a–d show the binding of (1) 0.2, (2) 0.5, (3) 2.0, and (4) 5.0 μg/ml antidigoxin in solution to digoxin immobilized on a glass substrate surface (Loo *et al.*, 2005). A single-fractal analysis is adequate to describe the binding kinetics. The values of the binding rate coefficient, k, and the fractal dimension, D_f, are given in Table 4.6.

Note that as the digoxin concentration in solution increases from 0.2 to 0.5 μg/ml, (a) the binding rate coefficient, k, increases from a value of 0.6023 to 0.7538, and (b) the fractal dimension, D_f, or the degree of heterogeneity on the glass substrate surface decreases from 1.9828 to 1.8018. In this case, changes in the degree of heterogeneity on the glass

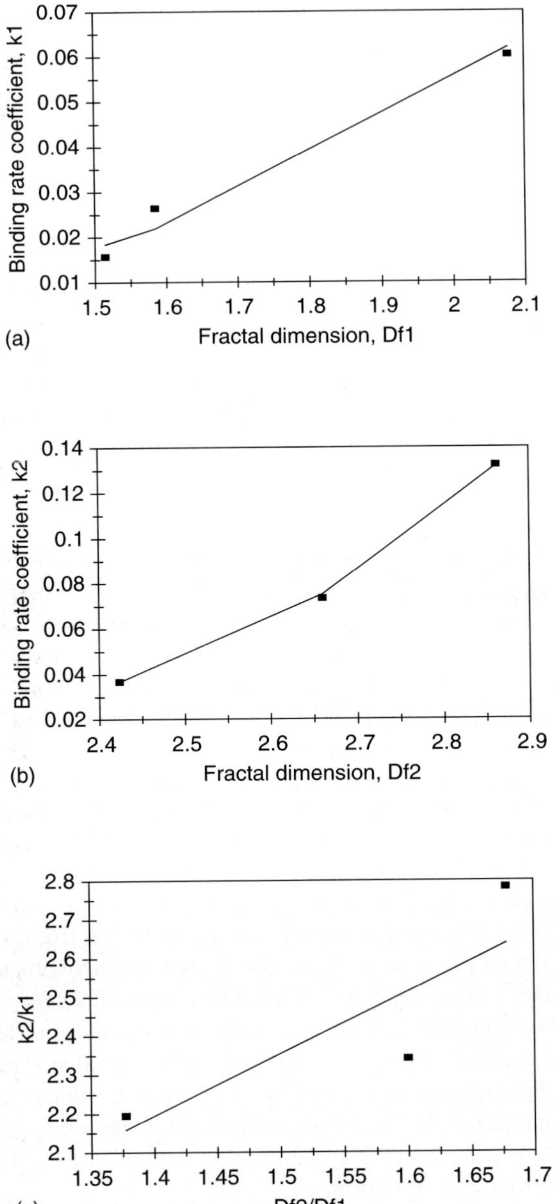

Figure 4.13 (a) Increase in the binding rate coefficient, k_1, with an increase in the fractal dimension, D_{f1}, or the degree of heterogeneity present on the sensor chip surface, (b) increase in the binding rate coefficient, k_2, with an increase in the fractal dimension, D_{f2}, or the degree of heterogeneity present on the sensor chip surface, and (c) increase in the ratio of binding rate coefficients, k_2/k_1, with an increase in the ratio of fractal dimensions, D_{f2}/D_{f1}, in the binding phase.

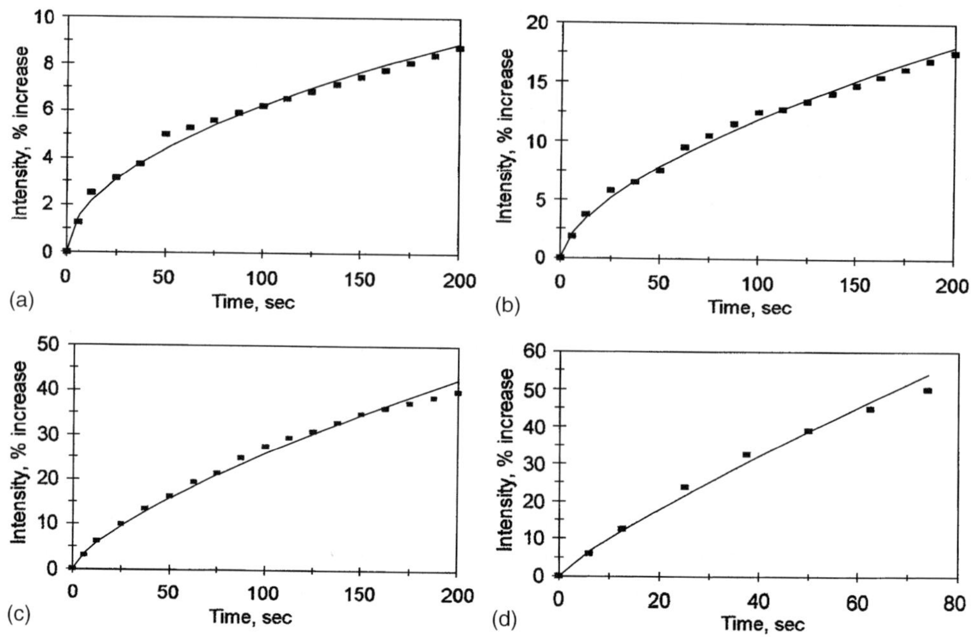

Figure 4.14 Binding of different concentrations of antidigoxin (in μg/ml) in solution to digoxin immobilized on a glass substrate surface (Loo *et al.*, 2005): (a) 0.2, (b) 0.5, (c) 2, (d) 5.

substrate surface and in the binding rate coefficient are in opposite directions. A 9.09% decrease in the fractal dimension from a value of 1.982 to 1.8018 leads to an increase in the binding rate coefficient by 25.05% from a value of 0.6023 to 0.7538.

Note that as the digoxin concentration in solution increases by an order of magnitude from 0.2 to 2.0 μg/ml, (a) the binding rate coefficient, k, increases from a value of 0.6023 to 0.9768, and (b) the fractal dimension, D_f, or the degree of heterogeneity on the glass substrate surface decreases from 1.9828 to 1.5712. Once again, changes in the degree of heterogeneity on the glass substrate surface and in the binding rate coefficient are in opposite directions. A 20.72% decrease in the fractal dimension from a value of 1.982 to 1.5712 leads to an increase in the binding rate coefficient by 62.2% from a value of 0.6023 to 0.9768.

The trend in the changes of the fractal dimension and in the binding rate coefficient with a change in the analyte concentration in solution is similar to the ones noted above, and are not repeated here to avoid repetition.

Table 4.6 and Figures 4.15a–c show for a single-fractal analysis (1) the increase in the binding rate coefficient k, and (2) the decrease in the fractal dimension, D_f, with an increase in the antidigoxin concentration (in μg/ml) in solution. In case (1), the binding rate coefficient, k, is given by

$$k = (0.9835 \pm 0.0724)[\text{antidigoxin}]^{0.2571 \pm 0.0315} \tag{4.9a}$$

Table 4.6

Binding rate coefficients and fractal dimension values for the binding of different concentrations of antidigoxin in solution to digoxin immobilized on a glass surface

Antidigoxin in solution/ digoxin immobilized on glass surface (μg/ml)	Fractal dimension (D_f)	Binding rate coefficient (k)
(a) 0.2	1.9828 ± 0.0366	0.6023 ± 0.0458
0.5	1.8018 ± 0.0312	0.7538 ± 0.0485
2	1.5712 ± 0.0250	0.9768 ± 0.0501
5	1.3150 ± 0.06678	1.4371 ± 0.1119
(b) 0.05[a]	2.5334 ± 0.0703	4.4044 ± 0.7648
0.25[a]	2.1336 ± 0.0692	3.3668 ± 0.5777
5[a]	2.2260 ± 0.05330	7.5627 ± 0.9765
2[a]	1.7034 ± 0.1283	6.435 ± 0.727

[a]An enzyme-amplified diffraction-based assay used (Loo *et al.*, 2005).

The fit is good. Only four data points are available. The availability of more data points would provide for a more reliable fit. The binding rate coefficient, k, is only very slightly dependent on the antidigoxin concentration in solution as noted by the very low (=0.2571) order exhibited. The fractional order of dependence exhibited lends support to the fractal nature of the system.

In case (2), the fractal dimension D_f is given by

$$D_f = (1.6483 \pm 0.0576)[\text{antidigoxin}]^{-0.1231 \pm 0.01388} \tag{4.9b}$$

The fit is good. Only four data points are available. The availability of more data points would provide for a more reliable fit. The fractal dimension, D_f, is only very slightly dependent on the antidigoxin concentration in solution as noted by the very low (=−0.1231) order exhibited. Note the negative order of dependence exhibited. The changes are very small, but this is not entirely unexpected since the fractal dimension as indicated elsewhere in this book is based on a log scale, and even very small changes in the fractal dimenson, D_f, indicate significant changes in the degree of heterogeneity on the glass substrate surface.

Figure 4.15 shows for a single-fractal analysis the decrease in the binding rate coefficient, k with an increase in the fractal dimension, D_f. The binding rate coefficient is given by:

$$k = (2.5426 \pm 0.0387) D_f^{-2.0927 \pm 0.04887} \tag{4.9c}$$

The fit is good. Only four data points are available. The availability of more data points would provide for a more reliable fit. The binding rate coefficient, k, exhibits close to a second (=−2.0927) order of dependence on the degree of heterogeneity or fractal dimension, D_f, on the glass substrate surface. Once again, note the negative order of dependence exhibited. In this case, contrary to the cases generally presented in this chapter or book, an increase in the degree of heterogeneity on the glass substrate surface leads to a decrease in the binding rate coefficient, k.

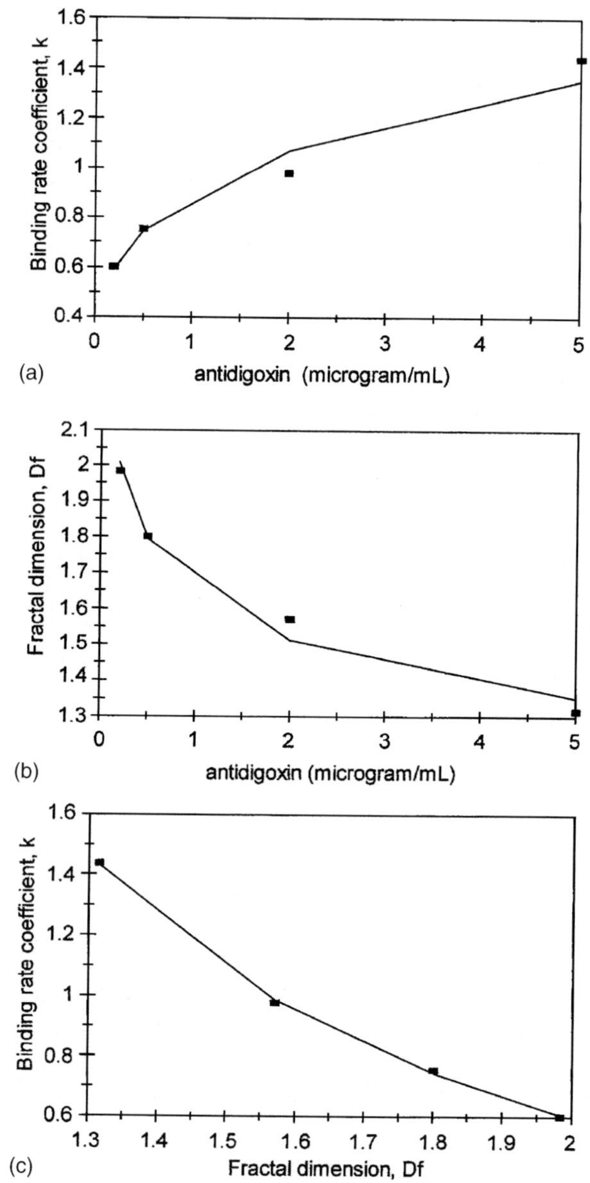

Figure 4.15 (a) Increase in the binding rate coefficient, k, with an increase in the antidigoxin concentration in solution, (b) decrease in the fractal dimension, D_f, with an increase in the antidigoxin concentration in solution, and (c) decrease in the binding rate coefficient, k, with an increase in the antidigoxin concentration in solution.

Goh *et al.* (2002) have shown that the detection limit in a diffraction-based immunoassay may be reduced by a factor of 40 if a gold-conjugated secondary antibody is used in a sandwich-assay format. Loo *et al.* (2005) have used a secondary enzymatic amplification with a precipitating substrate to improve the limit of detection.

A combination of peroxidase-conjugated secondary label and a precipitating substrate (TMB; 3,3′5,5′-tetramethyl benzidine) was used to reduce the limit of detection by more than 3 orders of magnitude.

Figure 4.16 shows the real-time turnover of TMB by HRP (horseradish peroxidase) conjugated on top of antidigoxin to digoxin immobilized on a glass substrate surface in a sandwich immunoasay format. Figures 4.16a–d show the binding of (1) 0.05, (2) 0.25, (3) 0.5, and (4) 2.0 ng/ml antidigoxin in solution in the enzyme-amplified diffraction-based immunoassay format. A single-fractal analysis is adequate to describe the binding kinetics. The values of the binding rate coefficient, k, and the fractal dimension, D_f, are given in Table 4.6b.

It is of interest to note that as the antidigoxin concentration in solution increases by a factor of 5 from 0.05 to 0.25 ng/ml, the fractal dimension, D_f, and the binding rate coefficient, k, both decrease.

Also, note that as the antidigoxin concentration in solution increases by an order of magnitude from 0.05 to 0.5 ng/ml, (a) the fractal dimenson, D_f, decreases from a value of 2.5334 to 2.2260, and (b) the binding rate coefficient, k, increases from a value of 4.4044 to 7.5627.

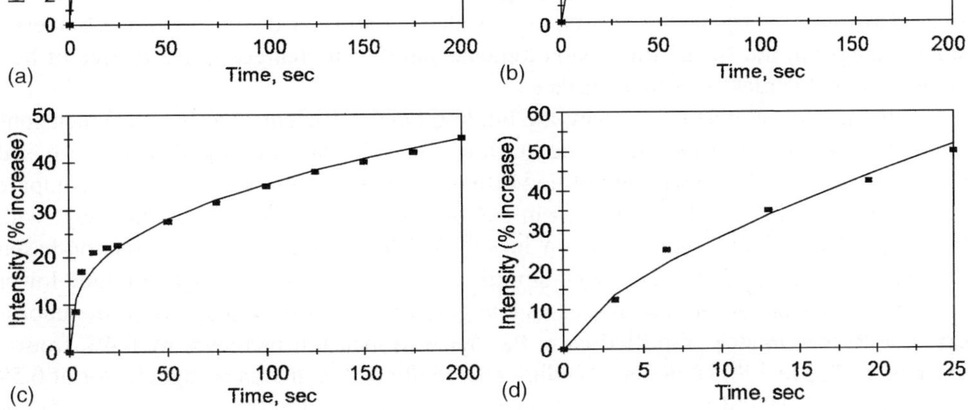

Figure 4.16 Binding of different concentrations of antidigoxin (in ng/ml) in solution to digoxin immobilized on a glass substrate surface in an enzyme-amplified diffraction-based immunoassay format (Loo *et al.*, 2005): (a) 0.05, (b) 0.25, (c) 0.5, (d) 2.

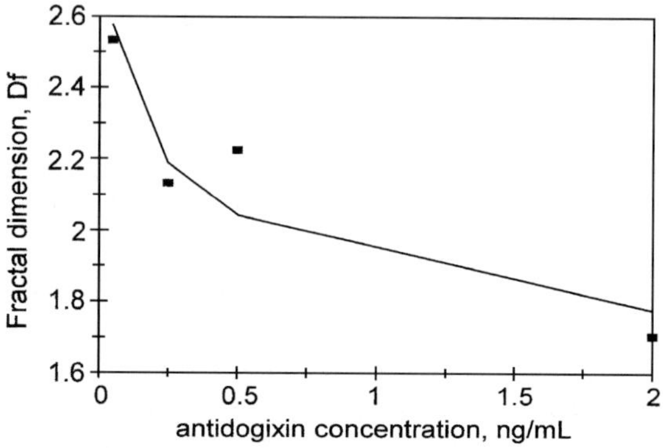

Figure 4.17 Decrease in the fractal dimension, D_f, with an increase in the antidigoxin concentration (in ng/ml) in solution for the enzyme-amplified diffraction-based immunoassay format.

Figure 4.17 and Table 4.6 show for a single-fractal analysis the decrease in the fractal dimension, D_f, with an increase in the antidigoxin in solution. In the 0.05–2 ng/ml antidigoxin concentration range in solution, the fractal dimension, D_f, is given by

$$D_f = (1.9050 \pm 0.1404)[\text{antidigoxin}]^{-0.100 \pm 0.0268} \tag{4.10}$$

The fit is good. Only four data points are available. The availability of more data points would lead to a more reliable fit. The fractal dimension, D_f, is only very slightly dependent on the antidigoxin concentration in solution as noted by the very low ($=-0.100$) order exhibited. Note the negative order exhibited. A similar negative order was exhibited by the fractal dimension for the diffraction-based immunoassay without enzymatic amplification (see eq. 4.9b). Once again, the changes in the fractal dimension, D_f, exhibited are very small. However, the fractal dimension, as indicated earlier is based on a log scale, and even small changes in the fractal dimension indicate significant changes in the degree of heterogeneity on the glass substrate surface.

Finally, it is of interest to compare the binding rate coefficients and fractal dimensions obtained for the antidigoxin concentrations in solution for the diffraction-based immunoassay with the enzyme-amplified diffraction-based immunoassay. Let us compare the 2 μg/ml antidigoxin concentration in solution (diffraction-based immunoassay) with the 2 ng/ml antidigoxin concentration in solution (enzyme-amplified diffraction-based immunoassay). Note that the enzyme-amplified case is for 3 orders of magnitude lower antidigoxin concentration in solution. As one goes from the diffraction-based immunoassay case to the enzyme-amplified case, the fractal dimension increases by 8.4% from a value of 1.5712 to 1.7034, and the binding rate coefficient, k, increases by a factor of 6.59 from a value of 0.9768 to 6.435. Once again, an increase in the fractal dimension, D_f, or the degree of heterogeneity on the surface leads to an increase in the binding rate coefficient, k. Changes in the binding rate coefficient are very sensitive to the changes in the degree of heterogeneity on the glass substrate surface.

4.4 CONCLUSIONS

A fractal analysis is presented for the binding and dissociation kinetics for (a) different wild-type LCAT concentrations in solution to egg-white apoA-I rHDL immobilized on a sensor chip surface (Jin *et al.*, 1999), (b) native, mildly oxidized, and strongly oxidized LDL in solution to a modified Au-surface of an SPR biosensor (Gaus and Hall, 1999), (c) different concentrations of TRITC-labeled HDL in solution to a bare optical fiber (Singh and Poirier, 1993), (d) the binding and dissociation of a G protein-coupled receptor (human chemokine-coupled receptor CCR5) on a biosensor surface (Steinlund *et al.*, 2003), (e) conformation-dependent monoclonal antibody 12G5 in solution to the chemokine receptor CXCR4 on a sensor chip surface (Steinlund *et al.*, 2003), (f) antidigoxin in solution to digoxin immobilized on a glass substrate surface in a diffraction-based immunoassay along with enzyme amplification (Loo *et al.*, 2005). The fractal analysis is used to provide a better physical understanding into these different types of reactions occurring on biosensor surfaces.

The fractal analysis provides a quantitative indication of the state of disorder (fractal dimension) and the binding (and dissociation) rate coefficients on the sensor chip surface. Both types of examples are given wherein either a single- or a dual-fractal analysis was used. The dual-fractal analysis was used only when the single-fractal analysis did not provide an adequate fit. This was done by the regression analysis provided by *Corel Quattro Pro 8.0* (1997).

In accordance with the prefactor analysis for fractal aggregates (Sorenson and Roberts, 1997), quantitative (predictive) equations are developed, for example, for (a) the binding rate coefficients, k_1 and k_2, as a function of the LCAT concentration in solution, (b) the dissociation rate coefficient, k_{d2}, as a function of the LCAT concentration in solution and as a function of the fractal dimension, D_{fd2}, or the degree of heterogeneity present in the dissociation phase on the sensor chip surface, (c) the binding rate coefficient, k_2, as a function of the degree of heterogeneity or fractal dimension, D_{f2}, present on the sensor chip surface, and (d) the affinity, $K_2(= k_2/k_{d2})$, as a function of the ratio of fractal dimensions, D_{f2}/D_{fd2} (Jin *et al.*, 1999). Predictive relations are also developed for the binding rate coefficient, k, as a function of the TRITC-labeled HDL concentration (in μg/ml) in solution, and as a function of the fractal dimension, D_f, or the degree of heterogeneity present on a bare optical fiber (Singh and Poirier, 1993). Predictive relationships are also presented for the affinities, K_1 and K_2, as a function of the ratio of fractal dimensions, D_{f1}/D_{fd} and D_{f2}/D_{fd}, respectively for the anti-CCR5 linear epitope-recognizing antibody, 3A9, in solution to the human chemokine-coupled receptor CCR5 (Steinlund *et al.*, 2003). Affinities are of particular interest to practicing biosensorists, and they should find these types of equations of interest. Finally, predictive relations are also presented for the binding rate coefficient, k, as a function of the antidigoxin concentration in solution, and as a function of the fractal dimension present on the sensor chip surface (Loo *et al.*, 2005).

The fractal dimension is not a classical independent variable such as analyte (antigen, antibody, or other biological molecule) concentration in solution. Nevertheless, the expressions obtained for the binding (and the dissociation) rate coefficients for a single- and a dual-fractal analysis as a function of the fractal dimension indicate a high sensitivity of these rate coefficients on their respective fractal dimensions on the SPR sensor chip surface.

For example, for the binding of LCAT in solution to egg-white apo-I rHDL immobilized on a sensor chip surface (Jin *et al.*, 1999), the binding rate coefficient, k_2, exhibits a 2.657 order of dependence on the fractal dimension, D_{f2}, or the degree of heterogeneity present on the sensor chip surface. Also, the affinity, K_1 ($= k_1/k_{d1}$), exhibits close to a third (=2.9775) order of dependence on the ratio of fractal dimensions, D_{f1}/D_{fd1}, present on the sensor chip surface. This emphasizes the importance of the extent of heterogeneity on the SPR biosensor chip surface and its impact on the binding rate coefficient.

Generally, it is seen that an increase in the fractal dimension of the surface or the degree of heterogeneity leads to an increase in the binding rate coefficient. One possible explanation of the observed increase could be due to the fact that the fractal surface (roughness) leads to turbulence, which enhances mixing, decreases diffusional limitations, and leads to an increase in the binding rate coefficient (Martin *et al.*, 1991). Granted, that for this to occur, the characteristic length of this turbulent boundary layer may have to extend a few monolayers above the sensor surface to affect the bulk diffusion to and from the surface. Considering the extremely laminar flow regimes in most biosensors, this may not be possible. However, due to the fractal nature of the surface that involves, for example, grooves and ridges, the surface morphology may contribute substantially toward the presence of eddy diffusion. This eddy diffusion enhances mixing and helps extend the characteristic length of the boundary layer to affect the bulk diffusion to and from the surface.

The characterization of the surface by a fractal dimension provides extra flexibility and suggest an avenue whereby the nature of the surface may be modulated in desired directions, and thereby simultaneously affecting or changing the dissociation and binding rate coefficients in desired directions. Experimentalists may find it worth their effort to pay a little more attention to the nature of the surface, and how it may be manipulated to control relevant parameters and biosensor performance in desired directions.

Finally, the fractal analysis presented together in this chapter for the binding (and dissociation) of different heart-related analytes in solution to different and appropriate receptors (or receptorless) present on the sensor chip surface hopefully provides much needed physical insights into these reactions occurring *in vivo*. Any additional physical insights, no matter how small, should prove to be beneficial in helping to manage, control, and minimize the potentially devastating effects of heart-related ailments that are caused by these types of interactions – financially, emotionally, and health-wise.

REFERENCES

Adimoolam, S., and Jonas, A. (1997). Identification of a domain of lecithin-cholesterol acyltransferase that is involved in interfacial recognition. *Biochemical and Biophysical Research Communication*, **232**, 783–787.

Apple, F.S., Christenson, R.H., and Valdes, R. (1999). Simultaneous rapid measurement of whole blood myoglobin, creatine kinase MB, and cardiac troponin I by a cardiac panel for detection of myocardial infarction. *Clinical Chemistry*, **45**, 199–205.

Aviram, M. (1993). Modified forms of low-density lipoprotein and atherosclerosis. *Atherosclerosis*, **98**, 1–9.

Bailey, R.C., and Hupp, J.T. (2002a). Large-scale resonance amplification of optical sensing of volatile compounds with chemoresponsive visible-region diffraction gratings. *Journal of the American Chemical Society*, **124**, 6767–6774.

Bailey, R.C. and Hupp, J.T. (2002b). Micropatterned polymeric gratings as chemoresponsive volatile compound sensors: implications for analyte detection and identification via diffraction-based sensor arrays. *Analytical Chemistry*, **75**, 2392–2398.

Brown, M.S., and Goldstein, J.L. (1983). Lipoprotein metabolism in the microphage: implications for cholesterol deposition in atherosclerosis. *Annual Reviews in Biochemistry*, **52**, 223–261.

Corel Corporation. (1997). *Corel Quattro Pro 8.0*, Ottawa, Canada.

Delrey, J.M., Madrid, A.H., and Valino, J.M. (1998). Cardiac troponin I and minor cardiac damage: biochemical markers in a clinical model of myocardial lesions. *Clinical Chemistry*, **44**, 2270–2276.

Fielding, C.J., and Fielding, P.E. (1995). Molecular physiology of reverse cholesterol transport. *Journal of Lipid Research*, **36**, 211.

Fogelman, A.M., Schecter, I., and Seager, J. (1980). Malondialdehyde alteration of low density lipoproteins leads to cholesteryl ester accumulation in human monocyte-macrophages. *Proceedings of the National Academy of Sciences USA*, **77**, 2214–2218.

Gaus, K., and Hall, E.A.H. (1999). Detection of oxidized low-density lipoproteins using surface plasmon resonance. *Analytical Chemistry*, **71**, 2459–2467.

Glomset, J. (1972). In: *Blood Lipids and Lipoproteins* (ed. G. Nelson), Wiley, New York, pp. 745–787.

Goh, J.B., Loo, R.W., McAloney, R.A., and Goh, M.C. (2002). Diffraction-based assay for detecting multiple analytes. *Analytical and Bioanalytical Chemistry*, **374**, 54–56.

Havlin, S. (1989). Molecular diffusion and reaction. In: *The Fractal Approach to Heterogeneous Chemistry: Surfaces, Colloids, Polymers* (ed. D. Avnir), Wiley, New York, pp. 251–269.

Hirose, N., Blankenship, D.T., Krivanek, M.A., Jacksson, R.L., and Cardin, A.D. (1987). Isolation and characterization of four heparin-binding cyanogen bromide peptides of human plasma apolipoprotein B. *Biochemistry*, **26**, 5505–5512.

Jin, L., Shieh, J.J., Grabbe, E., Adimoolam, S., Durbin, D., and Jonas, A. (1999). Surface plasmon resonance biosensor studies of human wild-type and mutant lecithin cholesterol acyltransferase interactions with lipoproteins. *Biochemistry*, **38**, 15659–15665.

Jonas, A. (1991). *Biochimica Biophysica Acta*, **1084**, 205–220.

Jonas, A. (1998). Regulation of lecithin cholesterol acyltransferase activity. *Progress in Lipid Research*, **37**, 209–234.

Karlsson, O.P., and Lofas, S. (2002). Flow-mediated on-surface reconstitution of G-protein coupled receptors for applications in surface plasmon biosensors. *Analytical Biochemistry*, **300**, 132–138.

Lawn, R.M. (1992). Lipoprotein (A) in heart disease. *Scientific American*, (June), 54–60.

Loo, R.W., Tam, P.L., Goh, J.B., and Goh, M.C. (2005). An enzyme-amplified diffraction-based immunoassay. *Analytical Biochemistry*, **337**, 338–342.

Mair, J. (1997). Progress in myocardial damage detection: new biochemical markers for clinicians. *Critical Reviews in Clinical Laboratory Science*, **34**, 1.

Martin, S.J., Granstaff, V.E., and Frye, G.C. (1991). Effect of surface roughness on the response of thickness-shear mode resonators in liquids. *Analytical Chemistry*, **65**, 2910–2922.

Mines, G.A., Tzeng, B.C., Stevenson, K.J., Li, J., and Hupp, J.T. (2002). Microporous supramolecular coordination compounds as chemosensory photonic lattices. *Angewandte Chemistry International Edition English*, **410**, 154–157.

Morhard, F., Pipper, J., Dahint, R., and Gruze, M. (2000). Immobilization of antibodies in micropatterns for cell detection by optical diffraction. *Sensors & Actuators B: Chemical*, **70**, 232–240.

Peelman, F., Vanloo, B., Perez-Mendez, O., Decout, A., Verschelde, J.L., Labeur, C., Vinaimont, N., Verhee, A., Durverger, N., Brasseur, R., Vandekerckhove, J., Tavernier, J., and Rosseneu, M. (1999). Characterization of functional residues in the interfacial recognition domain of lecithin cholesterol acyltransferase (LCAT). *Protein Engineering*, **12**, 71–78.

Penn, M.S., and Chisolm, G. (1994). Oxidized lipoproteins, altered cell function and atherosclerosis. *Atherosclerosis*, **108**, Supplement 1, S21–S29.

Rininger, F., and Pittman, R.C. (1988). Regulation of the elective uptake of high density lipoprotein-associated cholesterol esters by human fibroblast and HEP G2 hepatoma cells. *Journal of Lipid Research*, **29**, 1179–1194.

Ramakrishnan, A., and Sadana, A. (2001). A single-fractal analysis of cellular analyte-receptor binding kinetics utilizing biosensors. *Biosystems*, **59**(1), 35–45.

Ross, R. (1993). The pathogenesis of atherosclerosis: a perspective for the 1990s. *Nature*, **362**, 801–809.

Sadana, A. (2001). A fractal analysis approach for the evaluation of hybridization kinetics in biosensors. *Journal of Colloid and Interface Science*, **234**, 9–18.

Singh, B.R., and Poirier, M.A. (1993). Interaction of high density and low density lipoproteins to solid surfaces coated with cholesterol as determined by an optical fiber-based biosensor. *SPIE Volume 1886 Fiber Optic Sensors in Medical Diagnostics*, 27–34.

Sorenson, C.M., and Roberts, G.C. (1997). The prefactor of fractal aggregates. *Journal of Colloid and Interface Science*, **186**, 447–452.

Steinbrecher, U.P. (1991). *Clinical Cardiology*, **14**, 865–867.

Steinbrecher, U.P. (1996). Effects of oxidized low density lipoprotein on endothelin secretion by cultured endothelial cells and macrophages. *Atherosclerosis*, **7**, 135–143.

Steinlund, P., Babcock, G.J., Sodroski, J., and Myszka, D.G. (2003). Capture and reconstitution of G protein-coupled receptor on a biosensor surface. *Analytical Biochemistry*, **326**, 243–250.

St. John, P.M., Davis, R., Cady, N., Czaka, J., Batt, C.A., and Craighead, H.G. (1998). Diffraction-based cell detection using a microcontact printed antibody grating. *Analytical Chemistry*, **70**, 1108–1111.

Tsay, Y.G., Lin, C.I., Lee, J., Gustafson, E.K., Appelqvist, R., Magnetti, P., Norton, R., Teng, N., and Charlton, D. (1991). Optical biosensor assay (Oba). *Clinical Chemistry*, **37**, 1502–1505.

Weisgraber, K.H., and Rall, S.C. (1987). Human apolipoprotein B-100 heparin binding sites. *Journal of Biological Chemistry*, **262**, 11097–11103.

– 5 –

Fractal Analysis of Binding and Dissociation Kinetics and Interactions of Cancer Markers on Biosensor Surfaces

5.1 INTRODUCTION

The early detection of the different types of cancer and cancer biomarkers or tumor biomarkers has very understandably gained considerable interest recently. Biosensors are playing an increasingly important role in the early detection of these cancer markers. For example, Okumura *et al.* (2005) indicate that gene diagnosis is playing an increasingly important role in medical therapy, and point mutation in the K-*ras* gene is frequently found in pancreatic, colonic, uterine, and lung cancers. All of these are human-based cancers. These authors have developed a highly sensitive and precise K-*ras* point detection method by using hydrogel nanospheres to enhance the detection by DNA arrays on an SPR (surface plasmon resonance) biosensor. Miller *et al.* (2003) have used antibody microarray profiling of human prostate cancer sera. They have developed a strategy for serum protein profiling, and have identified different proteins that have significantly different levels between the prostate cancer sample and control samples. These proteins are the von Willebrand Factor, immunoglobulin M, alpha-1-antichymoytrypsin, Villin, and immunoglobulin G.

Wilson (2005) has recently developed an electrochemical immunosensor for the simultaneous detection of two tumor markers: carcinoembryonic antigen (CEA) and α-fetoprotein (AFP). Their sensors were two iridium oxide electrodes of 1 mm diameter each that were patterned on a glass substrate. (3-aminopropyl) triethoxysilane and glutaraldehyde were used by the covalent method to attach the capture antibodies on the glass substrate surface. Rochette *et al.* (2005) have recently used a mediatorless biosensor for putrescene detection using multiwalled carbon nanotubes in mouse plasma. These authors used two types of genetically modified mice: (a) a mouse mammary tumor virus (MMTV) type that had no sign of tumor, and (b) an NDL1-2 type mouse with an Erb B-2 mutated gene. This type of mouse, these authors indicate, is predisposed to rapidly developing mammary tumors.

Von Tiedemann and Bilitewski (2002) indicate the strong interest in characterizing and analyzing the vascular endothelial growth factor (VEGF)-receptor interactions, since it plays a significant role in angiogenesis (which is the formation of new capillaries from pre-existing blood cells). Cao (2001) and Mallonne *et al.* (1999) indicate that if the formation of new capillaries can be blocked, then presumably the growth of the tumor can be stopped. Von Tiedemann and Bilitewski (2002) indicate that VEGF is a homodimeric glycoprotein that belongs to the cysteine-knot family. These authors used the Biacore SPR biosensor to analyze the binding and the dissociation kinetics of VEGF in solution to its receptor, Flt-1 (VEGR-1) immobilized on a sensor chip surface. The actual receptor is a soluble protein, which is referred to as sFlt-1. Hennequin *et al.* (1999) and Wood *et al.* (2000) initially emphasized the importance of analyzing the kinetics of VEGF and its receptor interactions in order to develop therapeutic agents to help influence these types of interactions.

In this chapter, we use the fractal analysis method to analyze the binding and dissociation kinetics on a biosensor surface of (a) putrescene in solution to a PDDA/APTES/MWCNT/PuO-modified GCE (Rochette *et al.*, 2005), and (b) different concentrations of VEGF in solution to sFlt-1 immobilized on a sensor chip surface (von Tiedemann and Bilitewski, 2002). Values of the binding and the dissociation rate coefficients are provided. Fractal dimension values in the binding and in the dissociation phase are also provided. As indicated elsewhere in this book, the fractal analysis may be considered as an alternative analysis to the SPR Biacore software analysis. It is appropriate, however, to indicate that the Biacore SPR kinetic analysis ignores the presence of heterogeneity of receptors on the sensing surface, and presumes that diffusional limitations are not present if the SPR biosensor is run properly.

5.2 THEORY

Havlin (1989) has reviewed and analyzed the diffusion of reactants toward fractal surfaces. The details of the theory and the equations involved for the binding and the dissociation phases for analyte–receptor binding are available (Sadana, 2001). The details are not repeated here, except that just the equations are given to permit an easier reading. These equations have been applied to other biosensor systems (Ramakrishnan and Sadana, 2001; Sadana, 2001). For most applications, a single- or a dual-fractal analysis is often adequate to describe the binding and the dissociation kinetics. Peculiarities in the values of the binding and the dissociation rate coefficients as well as in the values of the fractal dimensions with regard to the dilute analyte systems being analyzed will be carefully noted, if applicable.

5.2.1 Single-fractal analysis

Binding rate coefficient

Havlin (1989) indicates that the diffusion of a particle (analyte [Ag]) from a homogeneous solution to a solid surface (e.g., receptor [Ab]-coated surface) on which it reacts to form a product (analyte–receptor complex; (Ab·Ag)) is given by

$$(\text{Analyte}\cdot\text{Receptor}) \sim \begin{cases} t^{(3-D_{\text{f,bind}})/2} = t^{p} & (t < t_{\text{c}}) \\ t^{1/2} & (t > t_{\text{c}}) \end{cases} \tag{5.1a}$$

Here $D_{f,bind}$ or D_f (used later on in the chapter) is the fractal dimension of the surface during the binding step; t_c is the cross-over value. Havlin (1989) indicates that the cross-over value may be determined by using $r_c^2 \sim t_c$ Above the characteristic length r_c, the self-similarity of the surface is lost and the surface may be considered homogeneous. Above time t_c, the surface may be considered homogeneous, since the self-similarity property disappears and 'regular' diffusion is now present. For a homogeneous surface where $D_f = 2$ and when only diffusional limitations are present, $p = ½$ as it should be. Another way of looking at the $p = ½$ case (where $D_{f,bind} = 2$) is that the analyte in solution views the fractal object; in our case, the receptor-coated biosensor surface from a 'large distance.' In essence, in the association process, the diffusion of the analyte from the solution to the receptor surface creates a depletion layer of width $(Ɖt)^{½}$ where $Ɖ$ is the diffusion constant. This gives rise to the fractal power law

$$(\text{Analyte}\cdot\text{Receptor}) \sim t^{(3-D_{f,bind})/2}.$$

For the present analysis, t_c is arbitrarily chosen and we assume that the value of the t_c is not reached. One may consider the approach as an intermediate 'heuristic' approach that may be used in the future to develop an autonomous (and not time-dependent) model for diffusion-controlled kinetics.

Dissociation rate coefficient

The diffusion of the dissociated particle (receptor [Ab] or analyte [Ag]) from the solid surface (e.g., analyte [Ag]–receptor [Ab]) complex coated surface) into solution may be given, as a first approximation by

$$\begin{aligned}(\text{Analyte}\cdot\text{Receptor}) &\sim -t^{(3-D_{f,diss})/2} \qquad (t > t_{diss}) \\ &= -k_{diss}t^{(3-D_{f,diss})/2}\end{aligned} \tag{5.1b}$$

Here $D_{f,diss}$ is the fractal dimension of the surface for the dissociation step. The start of the dissociation step corresponds to the highest concentration of the analyte–receptor complex on the surface. Henceforth, its concentration only decreases. The dissociation kinetics may be analyzed in a manner 'similar' to the binding kinetics.

5.2.2 Dual-fractal analysis

Binding rate coefficient

Sometimes, the binding curve exhibits complexities and two parameters (k, D_f) are not sufficient to adequately describe the binding kinetics. This is further corroborated by low values of r^2 factor (goodness-of-fit). In that case, one resorts to a dual-fractal analysis (four parameters: k_1, k_2, D_{f1}, and D_{f2}) to adequately describe the binding kinetics. The single-fractal analysis presented above is thus extended to include two fractal dimensions. At present, the time ($t = t_1$) at which the 'first' fractal dimension 'changes' to the 'second' fractal dimension is arbitrary and empirical. For the most part, it is dictated by the data analyzed and experience gained by handling a single-fractal analysis. A smoother curve is obtained in the 'transition' region, if care is taken to select the correct number of points for the two regions. In this case, the product (antibody–antigen; or analyte–receptor complex, Ab·Ag or analyte·receptor) is given by

$$(\mathrm{Ab}\cdot\mathrm{Ag}) \approx \begin{cases} t^{(3-D_{\mathrm{f1,bind}})/2} = t^{p_1} & (t < t_1) \\ t^{(3-D_{\mathrm{f2,bind}})/2} = t^{p_2} & t_1 < t < t_2 = t_{\mathrm{c}} \\ t^{1/2} & (t > t_{\mathrm{c}}) \end{cases} \tag{5.1c}$$

In some cases, as mentioned above, a triple-fractal analysis with six parameters (k_1, k_2, k_3, D_{f1}, D_{f2}, and D_{f3}) may be required to adequately model the binding kinetics. This is when the binding curve exhibits convolutions and complexities in its shape perhaps due to the very dilute nature of the analyte (in some of the cases to be presented) or for some other reasons. Also, in some cases, a dual-fractal analysis may be required to describe the dissociation kinetics.

5.3 RESULTS

A fractal analysis will be applied to the binding and dissociation kinetics of putrescine in solution to a PDDA/APTES/MWCNT/PuO-modified GCE (Rochette *et al.*, 2005), and to (b) different concentrations of VEGF in solution to sFLt-1 immobilized on a sensor chip surface (von Tiedemann and Bilitewski, 2002). This is one possible way to analyze diffusion-limited binding kinetics assumed to be present in the systems to be analyzed. Alternate expressions as mentioned elsewhere in other chapters in the book, involving saturation, first-order reaction, and no diffusion limitations are possible, but they are apparently deficient in describing the heterogeneity that inherently exists on the surface. Hardly anybody else in the open literature mentions or talks about heterogeneities on the surface, and if they do so they do it in a cursory manner. No attempt is made by others in the literature to quantitatively assess the impact of heterogeneity on the binding and the dissociation reaction occurring on the biosensor surface. The binding (and dissociation) on the biosensor surface (SPR or other biosensor) is a complex reaction, and the fractal dimension and the binding and the dissociation rate coefficient provide a useful lumped parameter(s) analysis of the diffusion-limited situation.

Rochette *et al.* (2005) have analyzed the binding of different concentrations of putrescine in solution on a PDDA poly (diallyldimethylammonium) chloride/APTES (3-aminopropyltrietoxy-silane)/MWCNT (multiwalled carbon nanotube/PuO (putrescine oxidase)-modified glassy carbon electrode (GCE) in 0.05 M phosphate buffer. Figures 5.1a–c show the binding of 25, 50, and 100 μM putrescine in solution to the above-mentioned modified GCE. In Figures 5.1a and c, a single-fractal analysis is required to describe the binding kinetics and in Figure 5.1b a dual-fractal analysis is required to describe the binding kinetics. Table 5.1 shows the values of (a) the binding rate coefficient k and the fractal dimension

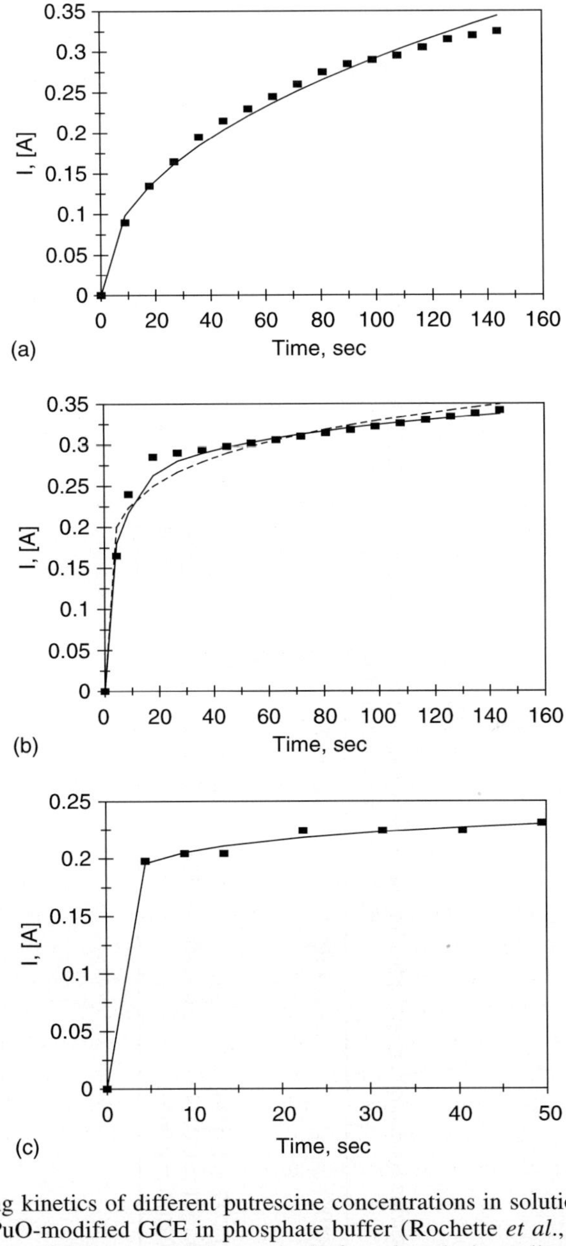

Figure 5.1 Binding kinetics of different putrescine concentrations in solution (in μM) to PDDA/APTES/MWCNT/PuO-modified GCE in phosphate buffer (Rochette *et al.*, 2005):(a) 25, (b) 50, (c) 100. When only a solid line is used then a single-fractal analysis applies. When both a dashed (----) and a solid (____) line are used, then the dashed line is for a single-fractal analysis and the solid line is for dual-fractal analysis. In this case, the solid line provides the better fit.

Table 5.1

Influence of putrescine concentration on the binding rate coefficients and fractal dimensions on a PDDA/APTES/MWCNTPuO-modified GCE in 0.05 M phosphate buffer (Rochette *et al.*, 2005)

Putrescine concentration, (μM)	$k \times 10^6$	$k_1 \times 10^6$	$k_2 \times 10^6$	D_f	D_{f1}	D_{f2}
25	0.17712 ± 0.0036	na	na	2.8657 ± 0.0189	na	na
50	0.1571 ± 0.0114	0.1200 ± 0.0126	0.1948 ± 0.0017	2.6784 ± 0.0176	2.4578 ± 0.1082	2.7788 ± 0.0113
100	0.03676 ± 0.00167	na	na	2.0998 ± 0.02990	na	na

D_f for a single-fractal analysis, and (b) the binding rate coefficients, k_1 and k_2, and the fractal dimensions, D_{f1} and D_{f2}, for a dual-fractal analysis.

Corel Quattro Pro (1997) was used to fit the data. The Corel Quattro Pro program provided the parameter values presented in Table 5.1. The values of the parameters presented in Table 5.1 are within 95% confidence limits. For example, the value of the binding rate coefficient k provided for a single-fractal analysis is $(0.17712 \pm 0.0036) \times 10^{-6}$. The 95% confidence limit indicates that 95% of the k values will lie between 0.14112 and 0.18072. This indicates that the values are precise and significant.

Apparently, the binding mechanism at the lowest (25 μM) and at the highest (100 μM) putrescine concentration in solution show similarities since a single-fractal analysis is adequate to describe the binding kinetics. At the intermediate putrescine concentration (50 μM) in solution, since a dual-fractal mechanism is required to describe the binding kinetics, the binding mechanism is apparently different from that observed at the lowest (25 μM) and at the highest (100 μM) putrescine concentrations in solution.

Figures 5.2a–c show the binding kinetics of 65, 55, and 50 μM putrescine in solution for a plasma sample from tumor-predisposed mice (A–C). Figure 5.2d shows the binding kinetics of putrescine (concentration not given) in solution for a plasma sample from a healthy mouse (D). In Figures 5.2a–c, the putrescine concentrations were based on calibration results

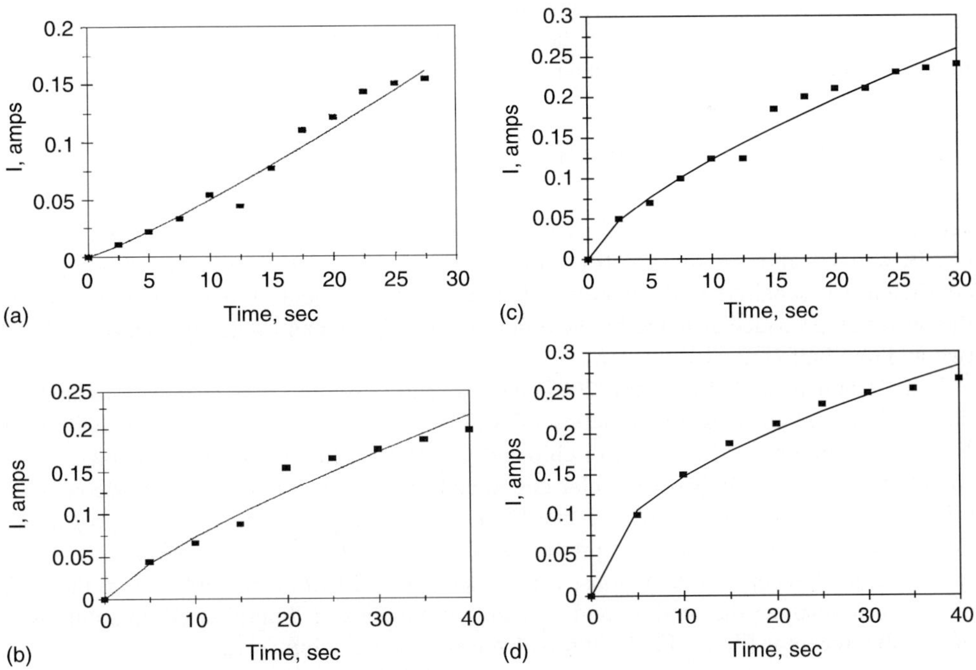

Figure 5.2 Binding kinetics for putrescine in solution for three plasma samples (A, B, and C) from three tumor-predisposed mice and a healthy mouse (Rochette *et al.*, 2005): A – putrescine concentration 65 μm, B – putrescine concentration 55 μM, C – putrescine concentration 50 μM, D – putrescine concentration not given.

Table 5.2

Binding rate coefficients and fractal dimensions for (a) putrescine in three plasma samples (A, B, and C) from three tumor-predisposed mice and in a healthy mouse (D) (Rochette *et al.*, 2005)

Analyte in solution, plasma sample	$k \times 10^6$	D_f
A	0.003422 ± 0.00057	0.6740 ± 0.1308
B	0.02561 ± 0.00226	1.6392 ± 0.00676
C	0.011995 ± 0.00156	1.4320 ± 0.1312
D	0.04951 ± 0.0255	2.0542 ± 0.0538

and on dilution factors. The binding kinetics is adequately described by a single-fractal analysis. The values of the binding rate coefficient k and the fractal dimension D_f are given in Table 5.2.

Note that the fractal dimension D_f and the binding rate coefficient k values for the healthy mouse (D) were the highest when compared with their respective values obtained for the tumor-predisposed mice (A, B, and C). For the tumor-predisposed mice (A, B, and C) not only were the values of the binding rate coefficient k lower but also so were values of the fractal dimension D_f.

Table 5.2 and Figure 5.3 show for a single-fractal analysis the increase in the binding rate coefficient k with an increase in the fractal dimension D_f. For the data obtained for the mice (A–D) the binding rate coefficient k is given by

$$k = (0.00764 \pm 0.00285)\, D_f^{2.3131 \pm 0.3798} \tag{5.2}$$

The fit is good. Only four data points are available. The data for the tumor-disposed mice (A, B, and C) and for the healthy mouse (D) are plotted together. The binding rate coefficient k is sensitive to the degree of heterogeneity or fractal dimension D_f present on the sensor chip surface as noted by the order of dependence exhibited between second and two and one-half (=2.3131) order.

Von Tiedemann and Bilitewski (2002) have recently used the SPR biosensor to analyze the binding and dissociation kinetics of VEGF in solution to its receptor (sFlt-1 or VEGFR-1) immobilized on a sensor chip surface. These authors indicate that sFlt-1 is the soluble form of the VEGFR (vascular endothelial growth factor receptor). Barleon *et al.* (1996) and Shibuya (2001) indicate that Flt-1 is apparently responsible for the organization of blood vessels. In order to provide a better understanding of the kinetics of these types of ligand–receptor interactions, von Tiedemann and Bilitewski (2002) used the SPR biosensor to analyze the binding and dissociation kinetics of ligand VEGF in solution to the soluble receptor Flt-1 (sFlt-1) immobilized on a sensor chip surface.

Figure 5.4a shows the binding and dissociation of 2250 ng/ml VEGF in solution to sFlt-1 immobilized on a sensor chip surface (von Tiedemann and Bilitewski, 2002). 1.03 ng/mm^2 or 1030 RU of sFlt-1 was immobilized on the sensor chip surface. A dual-fractal analysis is required to adequately describe the binding kinetics. The dissociation kinetics is adequately described by a single-fractal analysis. The values of (a) the binding rate coefficient k and the

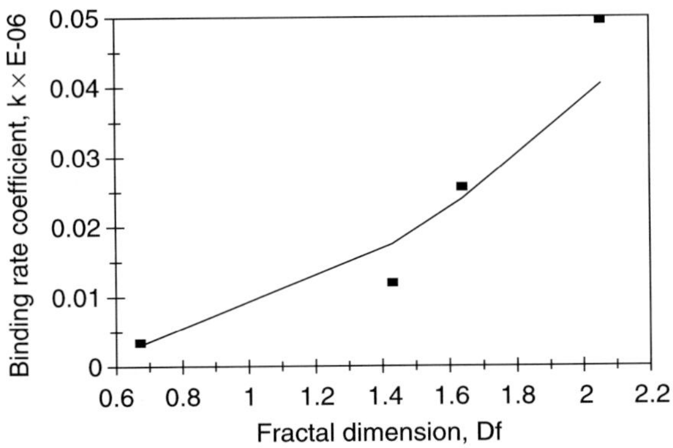

Figure 5.3 Increase in the binding rate coefficient k with an increase in the fractal dimension D_f.

fractal dimension D_f for a single-fractal analysis, and (b) the binding rate coefficient, k_1 and k_2, and the fractal dimensions, D_{f1} and D_{f2} for a dual-fractal analysis are given in Tables 5.3a and b.

Figure 5.4b shows the binding and dissociation of 1125 ng/ml VEGF in solution to sFlt-1 immobilized on a sensor chip surface (von Tiedemann and Bilitewski, 2002). A single-fractal analysis is adequate to describe the binding and the dissociation kinetics. The values of the binding rate coefficient k and the fractal dimension D_f for a single-fractal analysis are given in Tables 5.3a and b. It is of interest to note that there is a change in the binding mechansim as one goes from the lower (1125 ng/ml VEGF) to the higher (2250 ng/ml VEGF) concentration in solution. At the lower concentration, a single-fractal analysis is adequate to describe the binding kinetics, whereas at the higher concentration a dual-fractal analysis is required to describe the binding kinetics. In both cases, a single-fractal analysis is adequate to describe the dissociation kinetics.

Figures 5.5a–c show the binding and dissociation of 20, 10, and 5 nM VEGF in solution to 0.34 ng/mm^2 sFlt-1 immobilized on a sensor chip surface. A dual-fractal analysis is required to adequately describe the binding kinetics. A single-fractal analysis is adequate to describe the dissociation kinetics. Note that a decrease in the analyte (VEGF) concentration in solution by a factor of 2 from 20 to 10 nM leads to (a) a decrease in the binding rate coefficient k_1 by a factor of 1.83 from a value of 18.0077 to 9.8213, and (b) to an increase in the binding rate coefficient k_2 by a factor of 1.072 from a value of 30.2177 to 32.4065.

Also note that a decrease in the analyte (VEGF) concentration in solution by a factor of 2 from 10 to 5 nM leads to (a) a decrease in the binding rate coefficient k_1 by a factor of 6.395 from a value of 9.8213 to 1.5358, and (b) to a decrease in the binding rate coefficient k_2 by a factor of 1.47 from a value of 32.4065 to 22.0158.

(1) Table 5.4a and Figure 5.6a show for a dual-fractal analysis the increase in the binding rate coefficient k_1 with an increase in the VEGF concentration in solution. In the 5–20 nM VEGF concentration in solution, the binding rate coefficient k_1 is given by

$$k_1 = (0.10859 \pm 0.0723)[\mathrm{VEGF}]^{1.775 \pm 0.5204} \tag{5.3a}$$

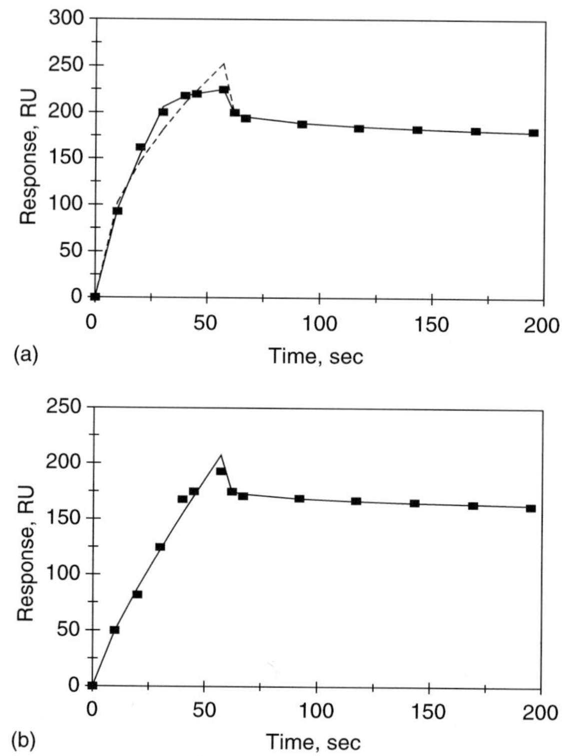

Figure 5.4 Binding and dissociation kinetics for the interaction of vascular endothelial growth factor (VEGF) in solution to the soluble form of the VEGF receptor (sFlt-1 or VEGF-1) immobilized on a sensor chip surface (von Tiedemann and Bilitewski, 2002): (a) 2250 ng/ml, (b) 1125 ng/ml. When only a solid line is used then a single-fractal analysis applies. When both a dashed (----) and a solid (____) line are used, then the dashed line is for a single-fractal analysis and the solid line is for dual-fractal analysis. In this case, the solid line provides the better fit.

(2) Table 5.4a and Figure 5.6b show for a dual-fractal analysis the increase in the fractal dimension D_{f1} with an increase in the VEGF concentration in solution. In the 5–20 nM VEGF concentration in solution, the fractal dimension, D_f is given by

$$D_{f1} = (0.7453 \pm 0.0723)[\mathrm{VEGF}]^{0.4268 \pm 0.1479} \tag{5.3b}$$

The fit is good. Only three data points are available. The availability of more data points would lead to a more reliable fit. In case (1) the binding rate coefficient k_1 exhibits an order of dependence between one and one-half and second (=1.775) order on the VEGF concentration in solution. The non-integer order of dependence exhibited by the binding rate coefficient k_1 on the VEGF concentration in solution lends support to the fractal nature of the system. The binding rate coefficient k_1 is sensitive to the

Table 5.3a

Binding and dissociation rate coefficients for the interaction of vascular endothelial growth factor (VEGF) in solution to the soluble form of the VEGF receptor[a] (sFlt-1 or VEGF-1) immobilized on a sensor chip surface (von Tiedemann and Bilitewski, 2002)

Analyte concentration in solution (ng/mL) /receptor on surface	k	k_1	k_2	k_d
2250	31.272 ± 3.455	18.528 ± 1.107	156.277 ± 0.173	20.289 ± 0.836
1125	7.4038 ± 0.5024	na	na	14.315 ± 0.449

[a]Amount of sFlt-1 immobilized is 1030 RU = 1.03 ng/mm^2.

Table 5.3b

Fractal dimensions for the binding and the dissociation phases for the interaction of vascular endothelial growth factor (VEGF) in solution to the soluble form of the VEGF receptor[a] (sFlt-1 or VEGF-1) immobilized on a sensor chip surface (von Tiedemann and Bilitewski, 2002)

Analyte concentration concentration in solution (ng/mL) /receptor on surface	D_f	D_{f1}	D_{f2}	D_{fd}
2250	1.9668 ± 0.1472	1.5836 ± 0.1477	2.8199 ± 0.00866	2.6734 ± 0.02604
1125	1.3510 ± 0.0924	na	na	2.7010 ± 0.01994

[a]Amount of sFlt-1 immobilized is 1030 RU = 1.03 ng/mm^2.

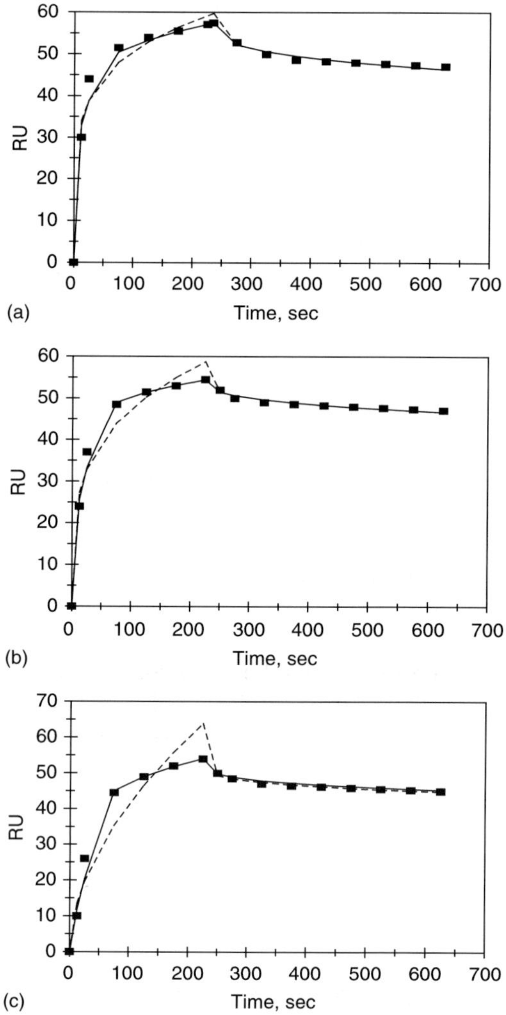

Figure 5.5 Binding and dissociation kinetics for the interaction of different concentrations (in nM) of the vascular endothelial growth factor (VEGF) in solution to sFlt-1 immobilized on a sensor chip surface (von Tiedemann and Bilitewski, 2002): (a) 20, (b) 10, (c) 5. When only a solid line is used then a single-fractal analysis applies. When both a dashed (----) and a solid (____) line are used, then the dashed line is for a single-fractal analysis and the solid line is for dual-fractal analysis. In this case, the solid line provides the better fit.

VEGF concentration in solution. In case (2) the fractal dimension D_{f1} exhibits an order of dependence less than one-half (=0.4268) order on the VEGF concentration in solution.

Tables 5.4a and b and Figures 5.6c and d show the increase in the binding rate coefficients k_1 and k_2 with an increase in the fractal dimensions D_{f1} and D_{f2}. In the

Table 5.4a

Binding and dissociation rate coefficients for the interaction of different concentrations (in nM) of VEGF in solution to sFlt-1[a] immobilized on a sensor chip surface (von Tiedemann and Bilitewski, 2002)

Analyte in solution/ receptor on surface	k	k_1	k_2	k_d
20 nM VEGF/sFlt-1	20.9335±1.9527	18.0077±2.2426	30.2177±0.0402	1.5937±0.1417
10 nM VEGF/sFlt-1	13.9999±1.6297	9.8213±2.3647	32.4065±0.0978	1.0165±0.1204
5 nM VEGF/sFlt-1	3.4615±1.0013	1.5358±0.5499	22.0158±0.0681	2.1433±0.0762

[a]340 RU = 0.34 ng/mm².

5–20 nM VEGF concentration range in solution, the binding rate coefficients, k_1 and k_2, is given by

$$k_1 = (0.3866 \pm 0.0356) D_{f1}^{4.0921 \pm 0.1989} \tag{5.3c}$$

$$k_2 = (0.01082 \pm 0.00043) D_{f2}^{7.7724 \pm 1.0538} \tag{5.3d}$$

The fit is good. Only three data points are available. The availability of more data points would lead to a more reliable fit. The binding rate coefficient k_1 is very sensitive to the degree of heterogeneity or fractal dimension D_{f1} that exists on the surface as noted by the higher than fourth (=4.0921) order of dependence exhibited on the fractal dimension D_{f1}. The binding rate coefficient k_2 is extremely sensitive to the degree of heterogeneity or fractal dimension D_{f2} that exists on the surface as noted by the order of dependence between seven and eight (=7.7724) exhibited on the fractal dimension D_{f2}.

Table 5.4a and Figure 5.6e show the increase in the dissociation rate coefficient k_d with an increase in the fractal dimension D_{fd}. In the 5–20 nM VEGF concentration in solution, the dissociation rate coefficient, k_d is given by

$$k_d = (0.001919 \pm 0.00055) D_{fd}^{7.6425 \pm 4.1229} \tag{5.3e}$$

Table 5.4b

Fractal dimensions for the binding and the dissociation phase for the interaction of different concentrations (in nM) of VEGF in solution to sFlt-1[a] immobilized on a sensor chip surface (von Tiedemann and Bilitewski, 2002)

Analyte in solution/receptor on surface	D_f	D_{f1}	D_{f2}	D_{fd}
20 nM VEGF/sFlt-1	2.6180±0.0638	2.5228±0.1299	2.7646±0.1182	2.3502±0.0839
10 nM VEGF/sFlt-1	2.4704±0.08592	2.2414±0.2036	2.8086±0.01444	2.3206±0.0841
5 nM VEGF/sFlt-1	1.9214±0.1982	1.3962±0.4790	2.6682±0.01484	2.5150±0.0263

[a]340 RU = 0.34 ng/mm².

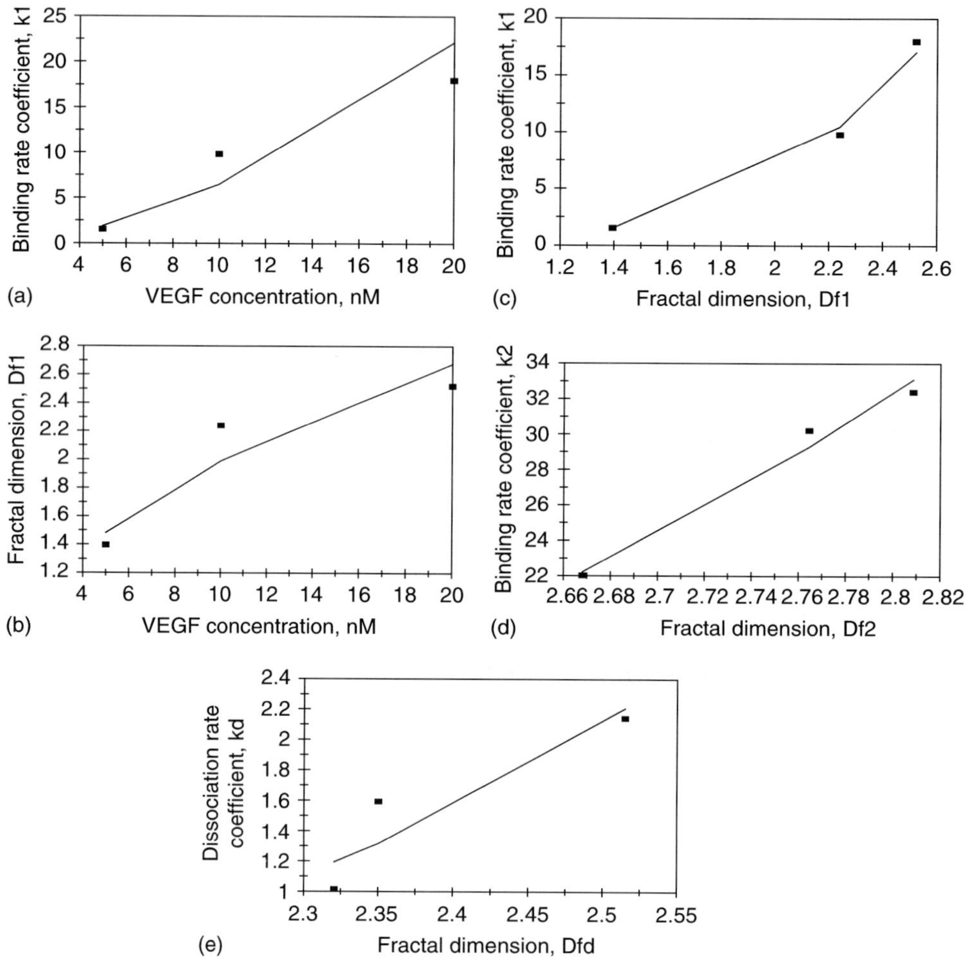

Figure 5.6 (a) Increase in the binding rate coefficient k_1 with an increase in the VEGF concentration (in nM) in solution, (b) Increase in the fractal dimension D_f with an increase in the VEGF concentration (in nM) in solution, (c) Increase in the binding rate coefficient k_1 with an increase in the fractal dimension D_{f1}, (d) Increase in the binding rate coefficient k_2 with an increase in the fractal dimension D_{f2}, (e) Increase in the dissociation rate coefficient k_{d2} with an increase in the fractal dimension in the dissociation phase D_{fd2}.

The fit is reasonable. Only three data points are available. The availability of more data points would lead to a more reliable fit. The dissociation rate coefficient k_d is extremely sensitive to the degree of heterogeneity that exists on the sensor chip surface in the dissociation phase, or the fractal dimension D_{fd} as noted by the order of dependence between seven and eight (=7.6425) exhibited.

Affinity values are of interest to practicing biosensorists as indicated elsewhere in different chapters in this book. Figure 5.7a shows the increase in the affinity, K_1 (= k_1/k_d) with

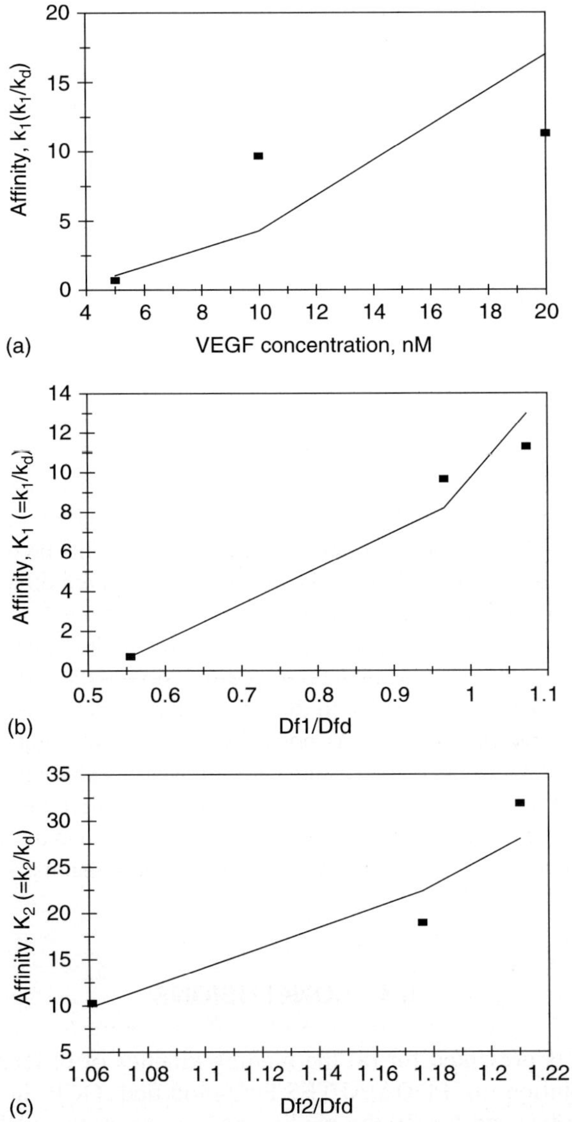

Figure 5.7 (a) Increase in the affinity K_1 (= k_2/k_d) with an increase in the VEGF concentration in solution, (b) Increase in the affinity K_1 (= k_1/k_d) with an increase in the ratio of the fractal dimensions D_{f1}/D_{fd}, (c) Increase in the affinity K_2 (= k_2/k_d) with an increase in the ratio of the fractal dimensions D_{f2}/D_{fd}.

an increase in the VEGF concentration (in nM) in solution. For the data presented in Table 5.4a and in Figure 5.7a, the affinity K_1 is given by

$$K_1(= k_1/k_d) = (0.04328 + 0.075)[\text{VEGF}]^{1.9894 \pm 1.0182} \tag{5.4}$$

There is scatter in the fit of the data. Only three data points are available. The availability of more data points would lead to a more reliable fit. The scatter in the fit of the data is reflected in the error in the exponent (where only the positive error is given since a negative affinity value is not possible), and in the order of dependence close to second (=1.9894) order on the VEGF concentration in solution.

Figures 5.7b and c show the increase in the affinities K_1 $(= k_1/k_d)$ and K_2 $(= k_2/k_d)$ with an increase in the ratio of the fractal dimensions D_{f1}/D_{fd} and D_{f2}/D_{fd}. For the data presented in Table 5.4a and in Figures 5.7b and c, the affinities K_1 and K_2 are given by

$$K_1 = (9.5311 \pm 2.3047)(D_{f1}/D_{fd})^{4.3517 \pm 0.4323} \tag{5.4b}$$

$$K_2(k_2/k_d) = (6.2188 \pm 1.4771)(D_{f1}/D_{fd})^{7.884 \pm 2.174} \tag{5.4c}$$

The fit is good. Once again, only three data points are available. The availability of more data points would lead to a more reliable fit. The affinity K_1 is very sensitive to the ratio of the fractal dimensions present in the binding and in the dissociation phase as noted by the higher than fourth (=4.3517) order of dependence exhibited. This predictive equation is of particular use if one wants to manipulate the affinity K_1 in desired directions. For example, if one is interested in obtaining higher affinity values, then one needs to selectively increase the degree of heterogeneity or the fractal dimension in the binding phase (D_{f1}) while simultaneously getting a smaller increase in the fractal dimension in the dissociation phase (D_{fd}).

The affinity K_2 is extremely sensitive to the ratio of the fractal dimensions present in the binding and in the dissociation phase as noted by the close to eighth (=7.884) order of dependence exhibited. Once again, and as indicated above, this predictive equation is of particular use if one wants to manipulate the affinity K_2 in desired directions. Note that the affinity K_2 $(= k_2/k_d)$ is more sensitive (order of dependence=7.884) than the affinity, K_1 $(= k_1/k_d)$ (order of dependence=4.3517) on the ratio of fractal dimensions, D_{f2}/D_{fd} and D_{f1}/D_{fd}, respectively.

5.4 CONCLUSIONS

A fractal analysis is presented for (a) the binding kinetics of different putrescine concentrations in solution to PDDA/APTES/PuO-modified GCE in phosphate buffer (Rochette *et al.*, 2005) and for (b) the binding and dissociation of VEGF in solution to the soluble form of the VEGF receptor (sFlt-1 or VEGF-1) immobilized on the sensor chip surface (von Tiedemann and Bilitewski, 2002).The kinetic analysis of these types of interactions is important, for example, since VEGF plays an important role in angiogenesis, and if the formation of the new capillaries can be blocked then presumably the growth of the tumor can be stopped (Cao, 2001; Mallonne *et al.*, 1999). The fractal analysis is used to provide a better understanding of the kinetics of reactions (involving cancer biomarkers or cancer-related analytes), and to relate the binding and the dissociation rate coefficients with the degree of heterogeneity or fractal dimension present on the sensor chip surface.

The fractal analysis provides a quantitative indication of the state of disorder (fractal dimension) and the binding (and dissociation) rate coefficient values on the sensor chip surface. Both types of examples are given wherein either a single- or a dual-fractal analysis was used. The dual-fractal analysis was used only when the single-fractal analysis did not provide an adequate fit. This was done by the regression analysis provided by Corel Quattro Pro 8.0 (1997). In accord with the prefactor analysis for fractal aggregates (Sorenson and Roberts, 1997), quantitative predictive expressions are developed for (a) the binding rate coefficient k as a function of the fractal dimension D_f for a single-fractal analysis for the detection of putrescine in tumor-predisposed mice (A, B, and C) and for a healthy mouse (D) (Rochette *et al.*, 2005); (b) the binding rate coefficients, k_1 and k_2, as a function of the fractal dimensions, D_{f1} and D_{f2}, respectively for the interaction of different VEGF concentrations (in nM) in solution to its receptor, sFlt-1 (or VEGF-1) immobilized on a sensor chip surface (von Tiedemann and Bilitewski, 2002); (c) the dissociation rate coefficient k_d as a function of the fractal dimension in the dissociation phase, D_{fd}; and (d) the affinity K_1 or K_2 as a function of the ratio of fractal dimensions present in the binding and in the dissociation phases, respectively.

The fractal dimension is not a classical independent variable such as analyte concentration in solution. Nevertheless, the expressions obtained for the binding (and the dissociation) rate coefficients for a single- and dual-fractal analysis as a function of the fractal dimension indicate a high sensitivity of these rate coefficients on their respective fractal dimensions on the SPR sensor chip surface. For example, in the 5–20 nM VEGF concentration range in solution (von Tiedemann, 2002) for a dual-fractal analysis, the binding rate coefficient k_1 exhibits an order of dependence slightly higher than fourth (=4.0921) on the fractal dimension D_{f1} or the degree of heterogeneity that exists on the sensor chip surface.

As indicated elsewhere in different chapters throughout the book, the analysis is extendable to different analyte–receptor reactions occurring on different types of biosensor and non-biosensor surfaces (for example, cell surfaces). Here it is hoped that the kinetic analysis (albeit using fractals) would help shed some physical insights into the interactions between cancer markers or biomarkers or analytes that are cancer-related (such as VEGF or putrescine) with their corresponding receptors. Any insights that one may obtain to better understand these types of cancer-related reactions (be it thermodynamic or kinetic) would significantly assist in the management and control of the different types of cancers. Needless to say, as in heart-related and other auto-immune diseases (such as cancer, arthritis, diabetes, etc.) the earlier one is able to detect or diagnose these types of life-threatening or quality of life-impacting diseases, and put one on a treatment protocol, the better it is for the individual(s) involved. For example, it is common knowledge, that if one is able to detect cancer in stage I, rather than in stage II or III, the prognosis is significantly better.

REFERENCES

Barleon, B., Sozzani, S., Zhou, D., Welch, H.A., Mantovani, A., and Marme, D. (1996). Migration of human monocytes in response to vascular growth factor (VEGF) is mediated via the VEGF receptor, flt-1. *Blood*, **87**(8), 3336–3343.

Cao, Y. (2001). Endogeneous angiogenesis inhibitors and their therapeutic implications. *International Journal of Biochemistry and Cell Biology*, **33**, 357–369.

Corel Quattro Pro 8.0 (1997). Corel Corporation, Ottawa, Canada.

Havlin, S. (1989). Molecular diffusion and reaction. In: *The Fractal Approach to Heterogeneous Chemistry: Surfaces, Colloids, Polymers* (ed. D. Avnir), Wiley, New York, pp. 251–269.

Hennequin, L.F., Thomas, A.P., Johnstone, C., Stokes, E.S.E., Ple, P.A., Lohmann, J.J.M., Ogilve, D.J., Dukes, M., Wedge, S.R., Curwen, J.O., Kendrew, J., and Lambert-van der Brempt, C. (1999). Design and structure-activity relationship of a new class of potent VEGF receptor tyrosine kinase inhibitors. *Journal of Molecular Chemistry*, **42**, 5369–5389.

Mallonne, H., Langer, I., Kiss, R., and Atassi, G. (1999). Mechanisms of tumor angiogenesis and therapeutic implications: angiogenesis inhibitors. *Clinical Experiments in Metastasis*, **17**, 1–14.

Miller, J.C., Zhou, H., Kwekel, J., Cavallo, R., Burke, J., Butler, E.B., The, B.S., and Haab, B.B. (2003). Antibody microarray profiling of human prostate cancer sera: antibody screening and identification of potential biomarkers. *Proteomics*, **3**(1), 56–63.

Okumura, A., Sato, Y., Kyo, M., and Kawaguchi, H. (2005). Point mutation detection with the sandwich method employing hydrogel nanospheres by the surface plasmon resonance imaging technique. *Analytical Biochemistry*, **339**, 328–337.

Ramakrishnan, A., and Sadana, A. (2001). A single-fractal analysis of cellular analyte-receptor binding kinetics utilizing biosensors. *Biosystems*, **59**(1), 35–45.

Rochette, J.F., Sacher, E., Meunier, M., and Luong, J.H.T. (2005). A mediatorless biosensor for putrescene using multiwalled carbon nanotubes. *Analytical Biochemistry*, **336**, 305–311.

Sadana, A. (2001). A fractal analysis approach for the evaluation of hybridization kinetics in biosensors. *Journal of Colloid and Interface Science*, **234**, 9–18.

Shibuya, M. (2001). Structure and dual function of vascular endothelial growth factor receptor-1 (Flt-1). *International Journal of Biochemistry and Cell Biology*, **33**, 409–420.

Sorenson, C.M., and Roberts, G.C. (1997). The prefactor of fractal aggregates. *Journal of Colloid and Interface Science*, **186**, 447–452.

Von Tiedemann, B., and Bilitewski, U. (2002). Characterization of the vascular endothelial growth factor–receptor interaction and determination of the recombinant protein by an optical receptor sensor. *Biosensors and Bioelectronics*, **17**, 983–991.

Wilson, M.S. (2005). Electrochemical immunosensors for the simultaneous detection of two tumor markers. *Analytical Chemistry*, **77**, 1496–1502.

Wood, J.M., Bold, G., Buchdunger, E., Cozens, R., Ferrari, S., Frei, J., Hofman, F., Mestan, J., Mett, H., O'Reilly, T., Persohn, E., Rossel, J., Schnell, C., Stover, D., Theuer, A., Towbin, H., Wenger, F., Woods-Cook, K., Menrad, A., Siemeister, G., Schirner, M.R., Drevs, J., Martiny-Bron, G., Totzke, F., and Marme, D. (2000). PTK787/ZK 222584, a novel and potent inhibitor of vascular endothelial growth factor receptor tyrosine kinases, impairs vascular endothelial growth factor-induced responses and tumor growth after oral administration. *Cancer Research*, **60**, 2178–2189.

– 6 –

Fractal Analysis of Binding and Dissociation of Autoimmune Disease Markers on Biosensor Surfaces

6.1 INTRODUCTION

Feng *et al.* (2004) recently indicate that autoimmune diseases affect 1–3% of the population (Jacobsen *et al.*, 1997). Elderly people are most affected by these diseases (Badley and Tennant, 1993; CDC Report, 2005). Autoantibodies are detected in these types of diseases, and von Muhlen and Tan (1995) indicate that the pattern of these autoantibodies is used to distinguish among the disorders in this group. Feng *et al.* (2004) emphasize that sensitive testing is required for these types of diseases, especially for rheumatoid diseases, which are difficult to recognize in their early stages. During the latter stages of these diseases multiple organs may be affected and the damage may be irreversible. These authors have recently developed a method for the parallel detection of autoantibodies with microarrays for rheumatoid diseases.

Feng *et al.* (2004) further emphasize the development of mathematical tests that may be used in conjunction with their microarray method since these mathematical tests minimize *matrix effects* (Wood, 1991). Feng *et al.* (2004) indicate that the data generated by their microarrays may be analyzed by artificial intelligence software that assists in differential disease diagnosis.

Researchers have also developed diagnostics to detect trace amounts of proteins called amyloid β-derived diffusible ligands (ADDLs) in body fluids such as cerebrospinal fluid (Analytical Currents, 2005). These ADDLs are biomarkers for the onset of Alzheimer's disease and occur at elevated levels in the brain of Alzheimer's patients. Mirkin *et al.* (2005) used an ultrasensitive nanoparticle-based bio-barcode. They used their method in Alzheimer's patients (postmortem analysis) and in normal subjects. These authors noted that the ADDLs were consistently higher for the patients than for the control subjects. Van-Duyne (2005) used localized surface plasmon resonance (LSPR) nanosensor technology to detect ADDLs by analyzing the interaction between ADDLs and specific anti-ADDL antibodies.

Hueber *et al.* (2002) recently reviewed autoantibody profiling for the study and treatment of autoantibody diseases. These authors indicate that recent advances in miniaturized proteonomics has permitted the detection of autoantibodies in submicroliter quantities of biological fluids. Geysen *et al.* (1984) define proteonomics as a large-scale study of expression, function, and interaction of proteins. Hueber *et al.* (2002) emphasize that these proteonomic technologies permit the characterization of autoreactive B-cell responses in diseases such as rheumatoid arthritis, multiple sclerosis, autoimmune diabetes, and systemic lupus erythematosus (SLE).

Welschof *et al.* (1997) indicate that the immune response is tightly regulated by stimulatory and suppressive mechanisms. The efficiency of the immune responses against foreign antigens is affected by the excessive or deficient suppression of the immune response and may trigger the initiation of an autoimmune reaction. These authors emphasize the immunoregulatory role of natural IgG-anti-F$(ab')_2$ antibodies in both healthy individuals and in patients with certain diseases. These authors analyzed the antigen-binding domain of a human IgG-anti-F$(ab')_2$ autoantibody. They used the Biacore surface plasmon resonace (SPR) biosensor to analyze the binding and dissociation kinetics of recombinant antibody fragments and immobilized antigens.

In this chapter we use fractal analysis to analyze (a) the binding and dissociation kinetics of different concentrations of scFv2 and scFv6 (single chain Fv) to F$(ab')_2$ antibody immobilized on a sensor chip surface (Welschof *et al.*, 1997). We emphasize, however, that the SPR Biacore analysis (and as abundantly indicated elsewhere in this book) does not take into consideration the heterogeneity of receptors on the sensor chip surface, and the influence of external diffusional limitations in its software analysis.

6.2 THEORY

Havlin (1989) has reviewed and analyzed the diffusion of reactants toward fractal surfaces. The details of the theory and the equations involved for the binding and dissociation phases for analyte–receptor binding are available (Sadana, 2001). The details are not repeated here, except that just the equations are given to permit an easier reading. These equations have been applied to other biosensor systems (Ramakrishnan and Sadana, 2001; Sadana, 2001). For most applications, a single- or a dual-fractal analysis is often adequate to describe the binding and the dissociation kinetics. Peculiarities in the values of the binding and dissociation rate coefficients, as well as in the values of the fractal dimensions with regard to the dilute analyte systems being analyzed will be carefully noted, if applicable.

6.2.1 Single-fractal analysis

Binding rate coefficient

Havlin (1989) indicates that the diffusion of a particle (analyte [Ag]) from a homogeneous solution to a solid surface (e.g. receptor [Ab]-coated surface) on which it reacts to form a product (analyte–receptor complex (Ab.Ag)) is given by

$$(\text{Analyte.Receptor}) \sim \begin{cases} t^{(3-D_{\text{f,bind}})/2} = t^{p}\,(t < t_{\text{c}}) \\ t^{1/2}\,(t > t_{\text{c}}) \end{cases} \tag{6.1a}$$

Here $D_{\text{f,bind}}$ or D_{f} (used later on in the chapter) is the fractal dimension of the surface during the binding step and t_{c} the cross-over value. Havlin (1989) indicates that the crossover value may be determined by $r_{\text{c}}^{2} \sim t_{\text{c}}$. Above the characteristic length r_{c}, the self-similarity of the surface is lost and the surface may be considered homogeneous. Above time t_{c} the surface may be considered homogeneous, since the self-similarity property disappears, and 'regular' diffusion is now present. For a homogeneous surface where $D_{\text{f}} = 2$, and when only diffusional limitations are present, $p = ½$ as it should be. Another way of looking at the p = ½ (where $D_{\text{f,bind}} = 2$) is that the analyte in solution views the fractal object, in our case, the receptor-coated biosensor surface, from a 'large distance.' In essence, in the association process, the diffusion of the analyte from the solution to the receptor surface creates a depletion layer of width $(Ðt)^{½}$, where $Ð$ is the diffusion constant. This gives rise to the fractal power law, (Analyte·Receptor) $\sim t^{(3-D_{\text{f,bind}})/2}$. For the present analysis, t_{c} is arbitrarily chosen and we assume that the value of the t_{c} is not reached. One may consider the approach as an intermediate 'heuristic' approach that may be used in the future to develop an autonomous (and not time-dependent) model for diffusion-controlled kinetics.

Dissociation rate coefficient

The diffusion of the dissociated particle (receptor [Ab] or analyte [Ag]) from the solid surface (e.g., analyte [Ag]–receptor [Ab]) complex coated surface) into solution may be given, as a first approximation, by

$$(\text{Analyte.Receptor}) \sim -k' t^{(3-D_{\text{f,diss}})/2}\,(t > t_{\text{diss}}) \tag{6.1b}$$

Henceforth, its concentration only decreases. The dissociation kinetics may be analyzed in a manner 'similar' to the binding kinetics.

6.2.2 Dual-fractal analysis

Binding rate coefficient

Sometimes, the binding curve exhibits complexities and two parameters (k, D_{f}), are not sufficient to adequately describe the binding kinetics. This is further corroborated by low values of r^2 factor (goodness-of-fit). In that case, one resorts to a dual-fractal analysis (four parameters: k_1, k_2, D_{f1}, and D_{f2}) to adequately describe the binding kinetics. The single-fractal analysis presented above is thus extended to include two fractal dimensions. At present, the time ($t = t_1$) at which the 'first' fractal dimension 'changes' to the 'second' fractal dimension is arbitrary and empirical. For the most part, it is dictated by the data analyzed and experience gained by handling a single-fractal analysis. A smoother curve is obtained in the 'transition' region, if care is taken to select the correct number of points for the two regions. In this case, the product (antibody–antigen or analyte–receptor complex, Ab.Ag or analyte.receptor) is given by

$$(\text{Ab.Ag}) \approx \begin{cases} t^{(3-D_{\text{f1,bind}})/2} = t^{p1} & (t < t_1) \\ t^{(3-D_{\text{f2,bind}})/2} = t^{p2} & t_1 < t < t_2 = t_c \\ t^{1/2} & (t > t_c) \end{cases} \tag{6.1c}$$

In some cases, as mentioned above, a triple-fractal analysis with six parameters (k_1, k_2, k_3, D_{f1}, D_{f2}, and D_{f3}) may be required to adequately model the binding kinetics. This is when the binding curve exhibits convolutions and complexities in its shape due to the very dilute nature of the analyte (in some of the cases to be presented) or for some other reasons. Also, in some cases, a dual-fractal analysis may be required to describe the dissociation kinetics.

6.3 RESULTS

At the outset it is appropriate to indicate that a fractal analysis will be applied to the data obtained for the binding and dissociation of recombinant antibody fragment ScFv2 and ScFv6 in solution to antigen $F(ab')_2$ immobilized on a sensor chip surface (Welschof *et al.*, 1997). This is one possible explanation for analyzing the diffusion-limited binding and dissociation kinetics assumed to be present in the system being analyzed. Alternate expressions involving saturation, first-order reaction, and diffusion limitations are possible, but they are apparently deficient in describing the heterogeneity that inherently exists on the surface. The analyte–receptor binding as well as dissociation is a complex reaction, and the fractal dimension (either $D_{\text{f,bind}}$ or $D_{\text{f,diss}}$) and the rate coefficient for binding (k_{bind}) or dissociation (k_{diss}) provide a useful lumped parameter analysis of the diffusion-limited situation.

Welschof *et al.* (1997) indicate that recent studies show that the autoantibody, IgG-anti-$F(ab')_2$, is involved in the natural pathogenesis of certain autoimmune diseases. These authors indicate that the IgG-anti-$F(ab')_2$ autoantibodies belong to the large family of anti-IgG antibodies generally termed as rheumatoid factors (RFs). These authors emphasize that the classical RFs bind to the Fc region and are an IgM isotype (Carson *et al.*, 1987). The antibodies analyzed by Welschof *et al.* (1997) bind to the $F(ab')_2$ region and belong to the IgG isotype. These authors indicate that Rose *et al.* (1949) were the first to describe the classical RF in patients with rheumatoid arthritis. In order to obtain better physical insights into the interaction of these autoantibodies, these authors analyzed the gene segment structure of their antigen-binding domains and their binding (and dissociation) kinetics.

Figure 6.1a shows the binding and dissociation of 400 nM recombinant antibody fragment, ScFv2 (single chain Fv2) in solution to antigen $F(ab')_2$ immobilized on a sensor chip surface. A dual-fractal analysis is required to adequately describe the binding kinetics. A single-fractal analysis is required to describe the dissociation kinetics. The values of (a) the binding rate coefficient k and the fractal dimension D_f for a single-fractal analysis, (b) the binding rate coefficients, k_1 and k_2, and the fractal dimensions, D_{f1} and D_{f2}, and (c)

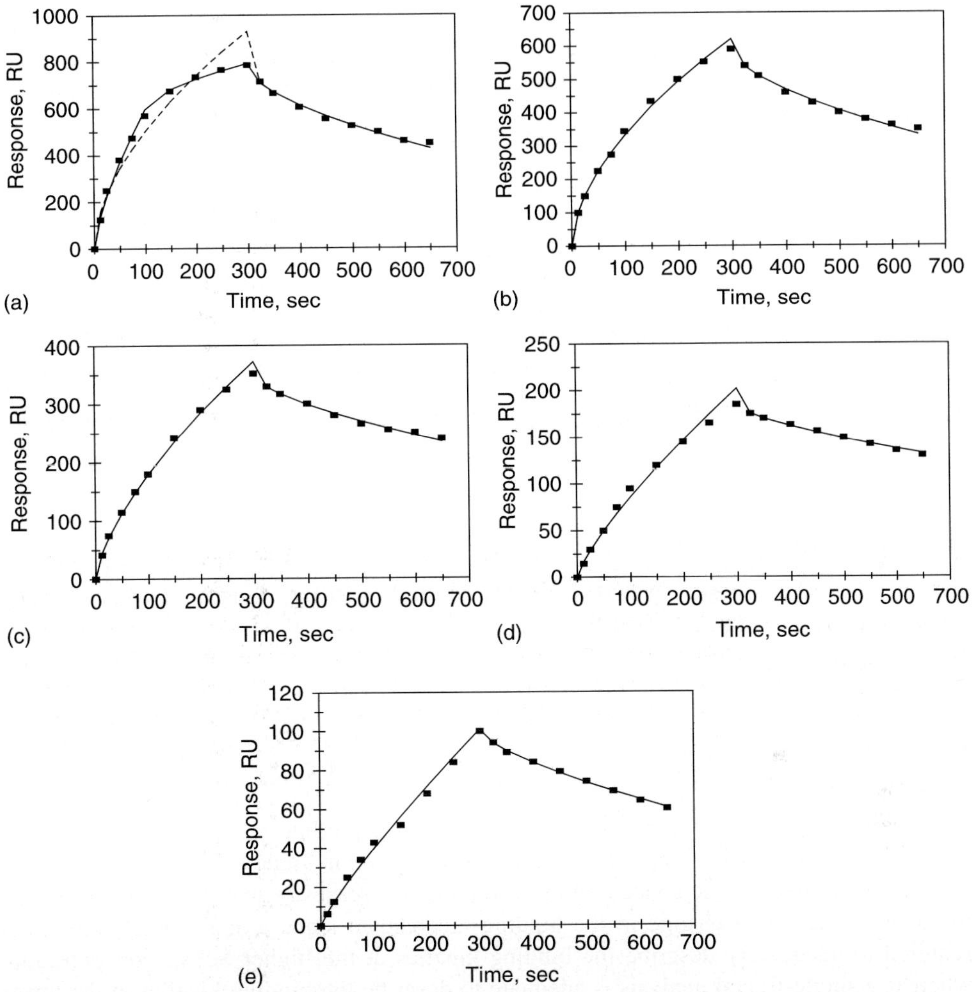

Figure 6.1 Binding and dissociation of different concentrations (in nM) of recombinant antibody fragment, ScFv2 (single-chain Fv2) in solution to antigen $F(ab')_2$ immobilized on a sensor chip surface (Welschof *et al.*, 1997): (a) 400 (b) 200 (c) 100 (d) 50 (e) 25. When only a solid line is used then a single-fractal analysis applies. When both a dashed (-----) line and solid (_____) lines are used, then the dashed line is for a single-fractal analysis, and the solid line is for a dual-fractal analysis. In this case the solid line provides the better fit.

the dissociation rate coefficient k_d and the fractal dimension D_{fd} for a single-fractal analysis are given in Table 6.1a and b.

Figures 6.1b–e show the binding and dissociation of (1) 200, (2) 100, (3) 50, and (4) 25 nM recombinant antibody fragment, ScFv2 (single chain Fv2) in solution to antigen $F(ab')_2$ immobilized on a sensor chip surface. A single-fractal analysis is adequate to describe the

Table 6.1

(a) Binding and dissociation rate coefficients and (b) fractal dimensions for the binding and dissociation phase for the interaction of different concentrations of recombinant antibody fragment scFv2 (single chain Fv2) in solution to immobilized antigen $F(ab')_2$ on a sensor chip surface (Welschof *et al.*, 1997)

(a)

ScFv2 concentration, nM	k	k_1	k_2	k_d
400	37.939 ± 5.722	22.701 ± 2.166	228.72 ± 3.12	11.437 ± 0.682
200	24.085 ± 0.7668	NA	NA	7.645 ± 0.399
100	8.351 ± 0.3581	NA	NA	3.0026 ± 0.1266
50	2.3264 ± 0.1968	NA	NA	1.1796 ± 0.0562
25	0.8253 ± 0.0694	NA	NA	0.6562 ± 0.0295

(b)

ScFv2 concentration, nM	D_f	D_{f1}	D_{f2}	D_{fd}
400	1.8786 ± 0.8921	1.5804 ± 0.1079	2.5644 ± 0.0524	1.8266 ± 0.0471
200	1.8612 ± 0.0205	NA	NA	1.8004 ± 0.0415
100	1.6684 ± 0.02752	NA	NA	1.2499 ± 0.377
50	1.4352 ± 0.05324	NA	NA	1.7012 ± 0.03794
25	1.3148 ± 0.05294	NA	NA	1.6024 ± 0.03596

binding and the dissociation kinetics. The values of the binding rate coefficient k and the fractal dimension D_f for a single-fractal analysis are given in Table 6.1a and b.

Note that as the analyte concentration in solution decreases from 400 to 200 nM ScFv2 in solution, there is a change in the binding mechanism since a dual-fractal analysis is required to adequately describe the binding kinetics at the higher ScFv2 concentration, whereas a single-fractal analysis is adequate to describe the binding kinetics at the lower ScFv2 concentration.

It is of interest to note that as the ScFv2 concentration decreases by a factor of one-half from 200 to 100 nM in solution, the binding rate coefficient, k, for a single-fractal analysis decreases by a factor of 2.884. Similarly, the fractal dimension, D_f, exhibits a decrease by 10.4% from a value of 1.8612–1.6684.

The trend of the binding rate coefficient k with a change in ScFv2 concentration in solution is similar, and thus it is not repeated here to avoid repetition.

Table 6.1a and Figure 6.2a indicate for a single-fractal analysis the increase in the binding rate coefficient k with an increase in the ScFv2 concentration in solution. In the 25–400 nM ScFv2 concentration range in solution, the binding rate coefficient k is given by

$$k = (0.00891 \pm 0.002683)[\text{ScFv2}]^{1.4417 \pm 0.1202} \tag{6.2a}$$

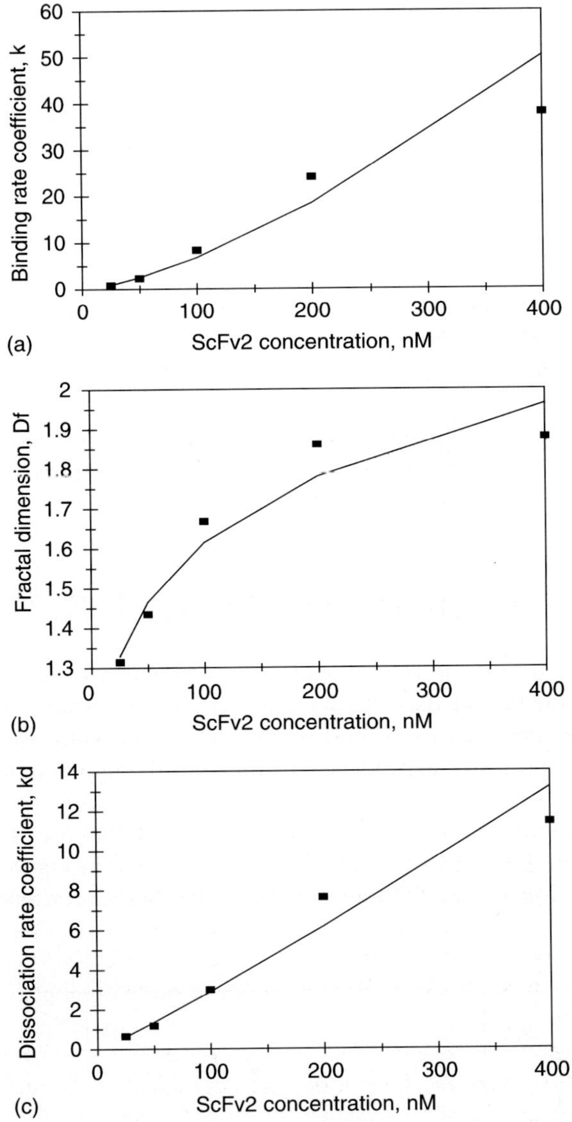

Figure 6.2 Increase in (a) the binding rate coefficient k, for a single-fractal analysis with an increase in the ScFv2 concentration in solution; (b) the fractal dimension D_f for a single-fractal analysis with an increase in the ScFv2 concentration in solution; (c) the dissociation rate coefficient k_d with an increase in the ScFv2 concentration in solution.

The fit is good. Five data points are available. The availability of more data points would lead to a more reliable fit. The binding rate coefficient k exhibits close to a one and one-half order of dependence on the ScFv2 concentration in solution. The non-integer order of dependence exhibited lends support to the fractal nature of the system.

Table 6.1a and Figure 6.2b indicate for a single-fractal analysis the increase in the fractal dimension D_f with an increase in the ScFv2 concentration in solution. In the 25–400 nM ScFv2 concentration range in solution, the fractal dimension D_f is given by

$$D_f = (0.8461 \pm 0.0369)[\text{ScFv2}]^{0.1405 \pm 0.0195} \tag{6.2b}$$

The fit is good. There is a little scatter in the data. Five data points are available. The availability of more data points particularly at the higher ScFv2 concentration would lead to a more reliable fit. The fractal dimension D_f is only mildly sensitive to the ScFv2 concentration in solution as noted by the low (= 0.1405) order of dependence exhibited. It should be noted that the fractal dimension is based on a log scale, and even small changes in the fractal dimension value indicate significant changes in the degree of heterogeneity on the sensor chip surface.

Table 6.1a and Figure 6.2c indicate for a single-fractal analysis the increase in the dissociation rate coefficient k_d with an increase in the ScFv2 concentration in solution. In the 25–400 nM ScFv2 concentration range in solution, the dissociation rate coefficient k_d is given by

$$k_d = (0.01875 \pm 0.00349)[\text{ScFv2}]^{1.094 \pm 0.0779} \tag{6.2c}$$

The fit is good. Five data points are available. The availability of more data points would lead to a more reliable fit. The dissociation rate coefficient k_d exhibits an order of dependence slightly higher than one (= 1.094) on the ScFv2 concentration in solution. Once again, the non-integer order of dependence exhibited lends support to the fractal nature of the system.

As indicated elsewhere in the book, affinity K (= k/k_d) values are of interest to practicing biosensorists. From the data presented in Table 6.1a and in Figure 6.3 one notes the increase in the affinity K value with an increase in the ScFv2 concentration in solution.

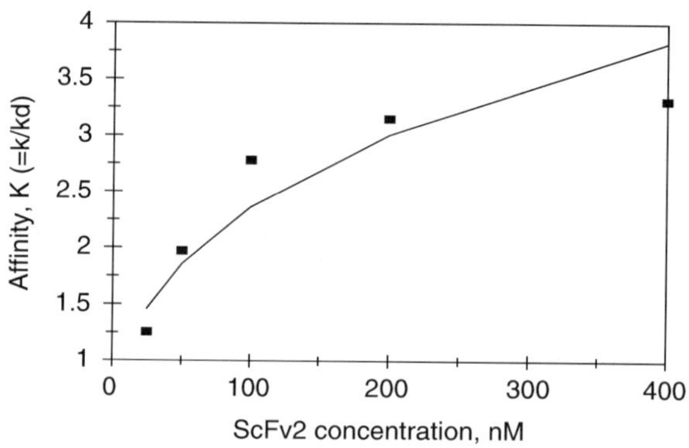

Figure 6.3 Increase in the affinity K (= k/k_d) with an increase in the ScFv2 concentration in solution.

In the 25–400 nM ScFv2 concentration range, the affinity K is given by

$$K = (0.4751 \pm 0.0814)[\text{ScFv2}]^{0.3474 \pm 0.0721} \tag{6.3}$$

The fit is good. Five data points are available. The availability of more data points would lead to a more reliable fit. The affinity K is only mildly sensitive to the ScFv2 concentration in solution as noted by the low (= 0.3474) order of dependence exhibited. Once again, the non-integer order of dependence exhibited lends support to the fractal nature of the system.

Figures 6.4a–e show the binding and dissociation of (1) 400, (2) 200, (3) 100, (4) 50, and (5) 25 nM recombinant antibody fragment, ScFv6 (single chain Fv6) in solution to antigen $F(ab')_2$ immobilized on a sensor chip surface. In cases (1) and (5), dual-fractal analysis is required to adequately describe the binding kinetics. A single-fractal analysis is required to describe the dissociation kinetics. The values of (a) the binding rate coefficient k and the fractal dimension D_f for a single-fractal analysis, (b) the binding rate coefficients, k_1 and k_2, and the fractal dimensions, D_{f1} and D_{f2} for a dual-fractal analysis, and (c) the dissociation rate coefficient k_d and the fractal dimension D_{fd} for a single-fractal analysis are given in Table 6.2a and b.

In cases (2), (3), and (4), a single-fractal analysis is adequate to describe the binding and the dissociation kinetics. The values of (a) the binding rate coefficient k and the fractal dimension D_f for a single-fractal analysis are given in Table 6.2a and b. Note that as the analyte concentration in solution decreases from 400 to 200 nM ScFv6 in solution, there is a change in the binding mechanism since a dual-fractal analysis is required to adequately describe the binding kinetics at the higher ScFv6 concentration, whereas a single-fractal analysis is adequate to describe the binding kinetics at the lower ScFv6 concentration.

In case (3), it is of interest to note that as the ScFv6 concentration decreases by a factor of one-half from 200 to 100 nM in solution, the binding rate coefficient k for a single-fractal analysis decreases by a factor of 2.569. Similarly, the fractal dimension D_f exhibits a decrease of 5.2% from a value of 1.9874 to 1.8842.

In case (4), the trend of the binding rate coefficient k with a change in ScFv6 concentration in solution is similar, and thus it is not repeated here to avoid repetition.

The binding mechanism is similar at the lowest (25 nM) and at the highest (400 nM) ScFv6 concentration in solution used since a dual-fractal analysis is required to adequately describe the binding mechanism. In the intermediate ScFv6 concentration range (50–200 nM) in solution the binding mechanism is similar since a single-fractal analysis is adequate to describe the binding mechanism.

Table 6.2a and Figure 6.5a indicate for a single-fractal analysis the increase in the binding rate coefficient k with an increase in the ScFv6 concentration in solution. In the 50–200 nM ScFv6 concentration range in solution, the binding rate coefficient k is given by

$$k = (0.0632 \pm 0.0110)[\text{ScFv6}]^{1.079 \pm 0.1625} \tag{6.4a}$$

The fit is good. Only three data points are available. The availability of more data points would lead to a more reliable fit. The binding rate coefficient k exhibits close to a first (= 1.079) order of dependence on the ScFv6 concentration in solution. The non-integer order of dependence exhibited lends support to the fractal nature of the system.

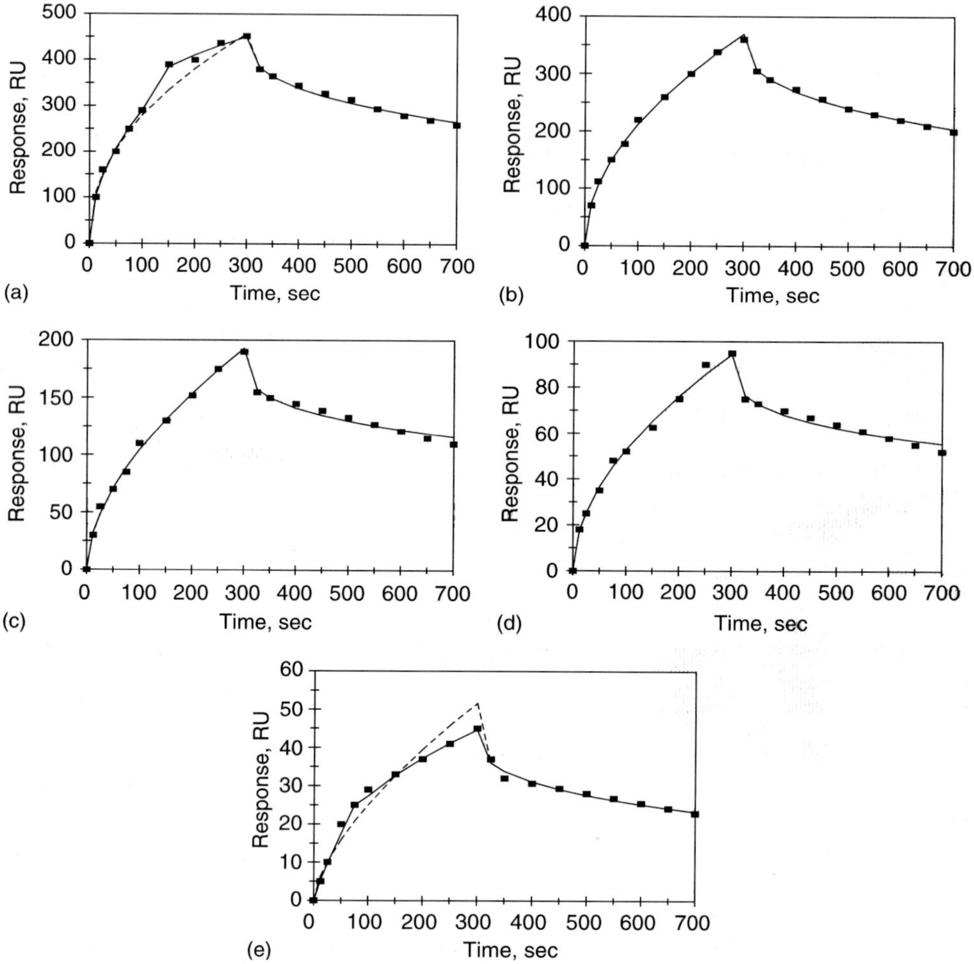

Figure 6.4 Binding and dissociation of different concentrations (in nM) of recombinant antibody fragment ScFv2 (single-chain Fv2) in solution to antigen F(ab′)$_2$ immobilized on a sensor chip surface (Welschof *et al.*, 1997): (a) 400 (b) 200 (c) 100 (d) 50 (e) 25. When only a solid line is used then a single-fractal analysis applies. When both a dashed (-----) line and solid (____) lines are used, then the dashed line is for a single-fractal analysis, and the solid line is for a dual-fractal analysis. In this case the solid line provides the better fit.

Table 6.2a and Figure 6.5b indicate that for a single-fractal analysis the increase in the dissociation rate coefficient k_d with an increase in the ScFv6 concentration in solution. In the 25–400 nM ScFv6 concentration range in solution, the dissociation rate coefficient k_d is given by

$$k_d = (0.4931 \pm 0.1262)[\text{ScFv6}]^{0.6523 \pm 0.1039} \qquad (6.4b)$$

Table 6.2

(a) Binding and dissociation rate coefficients and (b) fractal dimensions for the binding and dissociation phase for the interaction of different concentrations of recombinant antibody fragment scFv6 (single chain Fv6) in solution to immobilized antigen F $(ab')_2$ on a sensor chip surface (Welschof *et al.*, 1997)

(a)

ScFv6 concentration, nM	k	k_1	k_2	k_d
400	35.825 ± 4.032	30.295 ± 1.831	123.52 ± 0.960	21.401 ± 0.960
200	20.574 ± 0.797	NA	NA	15.276 ± 0.472
100	8.009 ± 0.546	NA	NA	12.427 ± 0.870
50	4.606 ± 0.1859	NA	NA	7.5552 ± 0.521
25	1.2069 ± 0.225	0.6053 ± 0.0647	3.5091 ± 0.0325	3.1773 ± 0.245

(b)

ScFv6 concentration, nM	D_f	D_{f1}	D_{f2}	D_{fd}
400	2.1078 ± 0.0694	2.0190 ± 0.0694	2.5462 ± 0.0874	2.2796 ± 0.0329
200	1.9874 ± 0.0249	NA	NA	2.2246 ± 0.0229
100	1.8842 ± 0.0432	NA	NA	2.4050 ± 0.0508
50	1.9408 ± 0.0259	NA	NA	2.4778 ± 0.05
25	1.6824 ± 0.1120	1.2772 ± 0.1204	2.1078 ± 0.03564	2.3584 ± 0.0558

The fit is reasonable. There is some scatter in the data. Only five data points are available. The availability of more data points would lead to a more reliable fit. The dissociation rate coefficient k_d exhibits an order of dependence between 1/2 and 1 (= 0.6523) on the ScFv6 concentration in solution. The non-integer order of dependence again lends support to the fractal nature of the system.

Table 6.2b and Figure 6.5c indicate for a single-fractal analysis the decrease in the fractal dimension in the dissociation phase D_{fd} with an increase in the ScFv6 concentration in solution. In the 50–200 nM ScFv6 concentration range in solution, the fractal dimension D_{fd} is given by

$$D_{fd} = (3.2382 \pm 0.0808)[\text{ScFv6}]^{-0.06 \pm 0.0251} \tag{6.4c}$$

The fit is reasonable. Only three data points are available. The availability of more data points would lead to a more reliable fit. The fractal dimension exhibits a very slight negative (=–0.06) order of dependence on the ScFv6 concentration in solution.

Table 6.2a and b and Figure 6.5d indicate that for a single-fractal analysis the decrease in the dissociation rate coefficient k_d with an increase in the fractal dimension, D_{fd}. In the

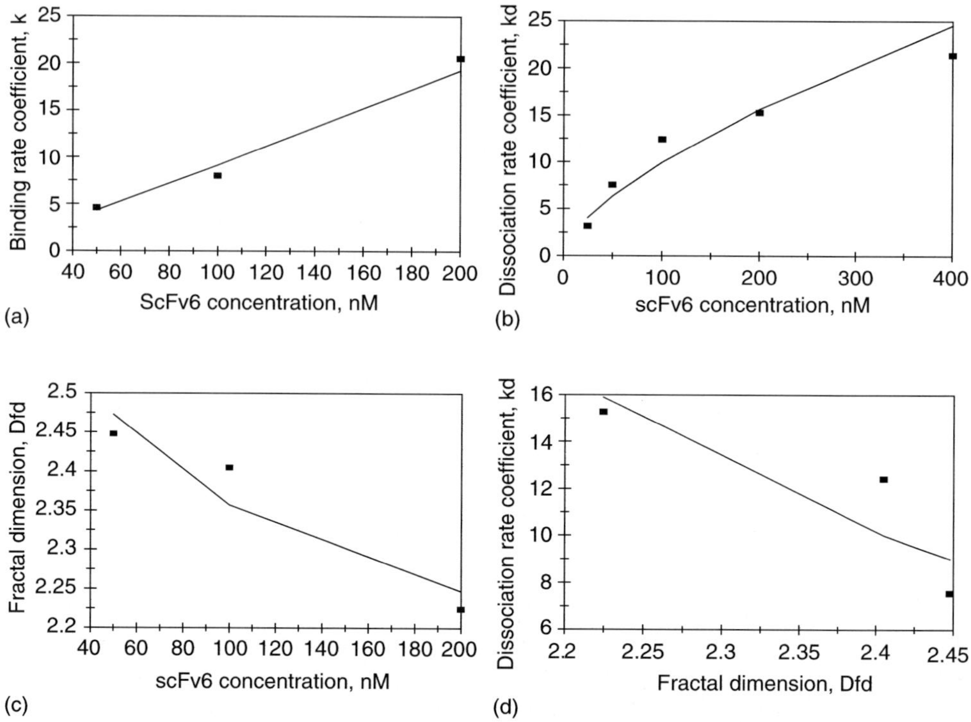

Figure 6.5 Increase in (a) the binding rate coefficient k for a single-fractal analysis with an increase in the ScFv6 concentration in solution; (b) the dissociation rate coefficient k_d with an increase in the ScFv6 concentration in solution; (c) decrease in the fractal dimension D_{fd} for a single-fractal analysis with an increase in the ScFv6 concentration in solution; and (d) decrease in the dissociation rate coefficient k_d with an increase in the fractal dimension for dissociation D_{fd}.

25–400 nM ScFv6 concentration range in solution, the dissociation rate coefficient k_d is given by

$$k_d = (1829.51 \pm 596.32)[D_{fd}]^{-5.935 \pm 3.921} \tag{6.4d}$$

The fit is reasonable. There is some scatter in the data. Only five data points are available. The availability of more data points would lead to a more reliable fit. The dissociation rate coefficient k_d is very sensitive to the degree of heterogeneity present on the surface during the dissociation phase D_{fd} as noted by the close to negative sixth (=–5.935) order of dependence exhibited.

Figure 6.6 shows for a single fractal analysis the increase in the affinity, K (=k/k_d) with an increase in the fractal dimension ratio, D_f/D_{fd}. In the 50–200 nM ScFv6 concentration

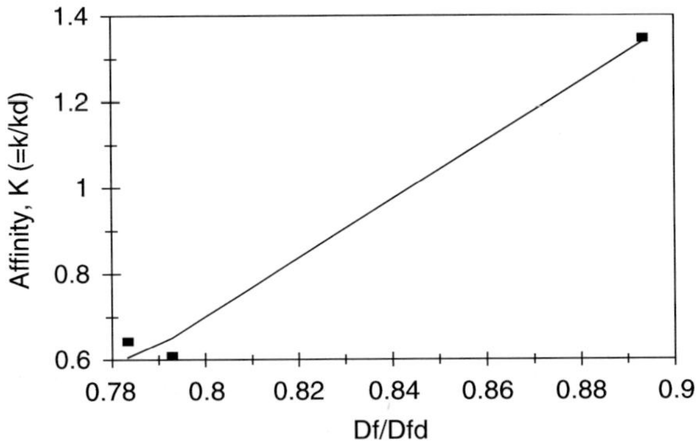

Figure 6.6 Increase in the affinity, K (= k/k_d) for a single-fractal analysis with an increase in the fractal dimension ratio, D_f/D_{fd}.

range in solution, the affinity K is given by

$$K(= k/k_d) = (2.6407 \pm 0.2514)[D_f/D_{fd}]^{6.0279 \pm 0.8854} \quad (6.4e)$$

The fit is good. Only three data points are available. The availability of more data points would lead to a more reliable fit. The affinity K is very sensitive to the ratio of fractal dimensions, D_f/D_{fd}, as noted by the slightly higher than sixth (= 6.0279) order of dependence exhibited. In other words, very slight changes in the ratio of fractal dimensions (binding and dissociation) on the surface could lead to significant changes in the affinity values. This is of significance since practicing biosensorists as well as chemists and biochemists and others are particularly interested in affinity values.

Figure 6.7a shows the binding and the dissociation of CCR5 in solution to gp120/CD4 immobilized on a sensor chip surface (Navratilova *et al.*, 2005). A single-fractal analysis is adequate to describe the binding and the dissociation kinetics. The values of (a) the binding rate coefficient k and the fractal dimension D_f for a single-fractal analysis, and (b) the dissociation rate coefficient k_d and the fractal dimension for the dissociation phase D_{fd} are given in Table 6.3a and b.

Navratilova *et al.* (2005) indicate that the small molecule TAK-779 inhibits HIV-1 replication since it blocks the interaction between the viral surface protein gp120 and CCR5 (Dragic *et al.*, 2000; Zhao *et al.*, 2004). Figure 6.7b shows the binding and dissociation of CCR5 in solution to gp120/CD4 immobilized on the sensor chip when the inhibitor TAK-779 is not present. A dual-fractal analysis is required to describe the binding and the dissociation kinetics. The values of (a) the binding rate coefficient k and the fractal dimension D_f for a single-fractal analysis, (b) the binding rate coefficients, k_1 and k_2, and the fractal dimensions, D_{f1} and D_{f2}, for a dual-fractal analysis, (c) the dissociation rate coefficient k_d and the fractal dimension D_{fd} for a single-fractal analysis, and (d) the dissociation rate coefficients, k_{d1} and k_{d2}, and the fractal dimensions, D_{fd1} and D_{fd2}, for a dual-fractal analysis are given in Table 6.3a and b.

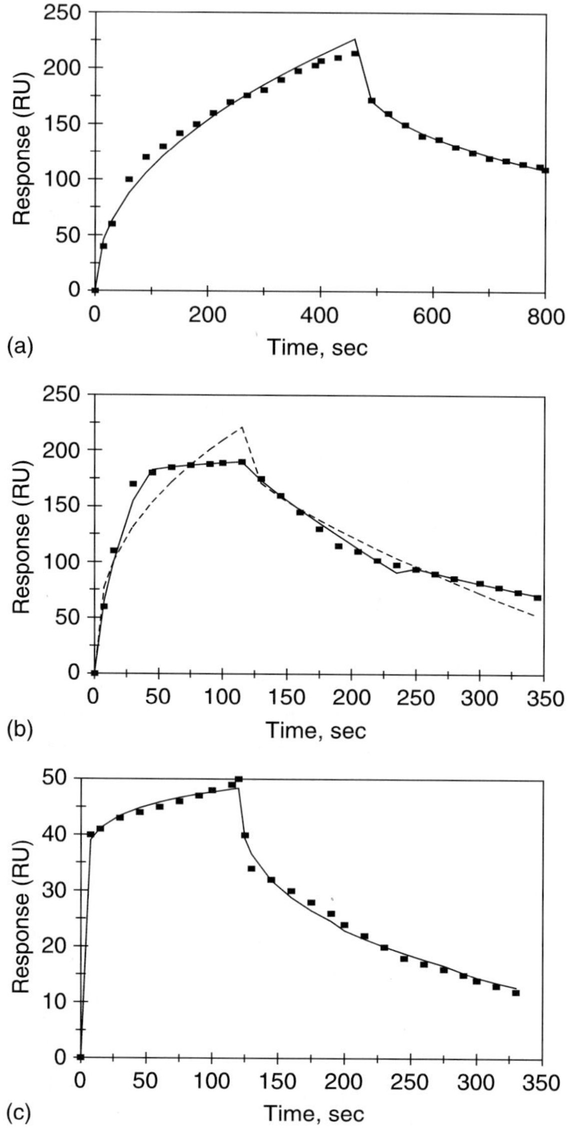

Figure 6.7 (a) Binding and dissociation of CCR5 in solution to gp120:CD4 immobilized on a sensor chip surface (Navratilova *et al.*, 2005); (b) Binding and dissociation of CCR5 in solution to gp120:CD4 immobilized on a sensor chip surface in the absence of the inhibitor TAK-779 (Navratilova *et al.*, 2005); and (c) Binding and dissociation of CCR5 in solution to gp120: CD4 immobilized on a sensor chip surface in the presence of the inhibitor TAK-779 (Navratilova *et al.*, 2005). When only a solid line is used then a single-fractal analysis applies. When both dashed (-------) line and solid (_____) lines are used, then the dashed line is for a single-fractal analysis, and the solid line is for a dual-fractal analysis. In this case the solid line provides the better fit.

Table 6.3

(a) Binding and dissociation rate coefficients and (b) fractal dimensions for the binding and the dissociation phase for CCR5 in solution to (i) gp120/CD4 immobilized on a sensor chip surface, and (ii) inhibition of binding of CCR5 in solution to CD4/gp120 in the absence and in the presence of inhibitor TAK779 (Navratilova *et al.*, 2005)

(a)

Analyte in solution/Receptor on surface	K	k_1	k_2	k_d	k_{d1}	k_{d2}
(i) CCR5//gp120/CD4	12.999 ± 0.939	NA	NA	12.384 ± 0.198	NA	NA
(ii) CCR5//gp120/CD4 no inhibitor TAK present	35.470 ± 6.601	18.238 ± 2.550	156.93 ± 0.143	2.6993 ± 0.373	1.6054 ± 0.1584	12.681 ± 0.071
CCR5//gp120/CD4 10 μM inhibitor TAK present	33.416 ± 0.66	NA	NA	6.3218 ± 0.3916	NA	NA

(b)

Analyte in solution/Receptor on surface	D_f	D_{f1}	D_{f2}	D_{fd}	D_{fd1}	D_{fd2}
(i) CCR5//gp120/CD4	2.0674 ± 0.0354	NA	NA	2.2678 ± 0.119	NA	NA
(ii) CCR5//gp120/CD4 no inhibitor TAK present	2.2290 ± 0.1297	1.7412 ± 0.1916	2.9194 ± 0.00359	1.5552 ± 0.8778	1.2778 ± 0.0924	2.176 ± 0.0235
CCR5//gp120/CD4 10 μM inhibitor TAK present	2.8452 ± 0.014	NA	NA	2.3452 ± 0.0279	NA	NA

Figure 6.7c shows the binding and dissociation of CCR5 in solution to gp120/CD4 immobilized on the sensor chip surface in the presence of 10 μM inhibitor, TAK-779. A single-fractal analysis is adequate to describe the binding and the dissociation kinetics. It is of interest to note that there is a change in the binding and in the dissociation mechanism when the inhibitor, TAK–779, is present and when it is absent. When the inhibitor, TAK-779, is used a single-fractal analysis is adequate to describe the binding and dissociation kinetics. When the inhibitor, TAK-779 is absent, then a dual-fractal analysis is required to adequately describe the binding and the dissociation kinetics.

Navratilova *et al.* (2005) analyzed the influence of different CD4:gp120 ratios on the binding and dissociation of gp120. Figures 6.8a–e show the binding and dissociation of CCR5 in solution to a (1) 1:100, (2) 1:33, (3) 1:10, (4) 1:4, and (5) 1:2 gp120:CD4 ratio immobilized on a sensor chip surface. In case (1), dual-fractal analysis is required to adequately describe the binding kinetics. A single-fractal analysis is adequate to describe the dissociation kinetics. The values of (a) the binding rate coefficient k and the fractal dimension D_f for a single-fractal analysis, (b) the binding rate coefficients, k_1 and k_2, and the fractal dimensions, D_{f1} and D_{f2}, for a dual-fractal analysis, and the (c) the dissociation rate coefficient k_d and the fractal dimension D_{fd} for a single-fractal analysis are given in Table 6.3a and b.

In case (2), single-fractal analysis is required to adequately describe the binding and the dissociation kinetics. A dual-fractal analysis could very well have been used to describe the binding kinetics as may be noted from Figure 6.8b. However, the coefficient of regression (R^2 value) was high (= 0.9614) and thus a single-fractal analysis is used for the binding phase. Needless to say, if one wants a higher coefficient of regression or R^2 value, then a dual-fractal analysis could be used. In cases (3), (4), and (5), the values of (a) the binding rate coefficient k and the fractal dimension D_f for a single-fractal analysis, and (b) the dissociation rate coefficient k_d and the fractal dimension D_{fd} for a single-fractal analysis are given in Table 6.3a and b.

In case (4), it is of interest to note that for a dual-fractal analysis as the gp120:CD4 ratio changes from 1:10 to 1:4, (a) the binding rate coefficient k_1 increases by a factor of 1.80 from a value of 7.6087 to 13.666, and (b) the binding rate coefficient k_2 increases by a factor of 2.04 from a value of 27.395 to 55.9602. Similarly, the dissociation rate coefficient k_d increases by a factor of 2.018 from a value of 4.9017 to 9.8903.

In case (5), it is of interest to note that for a dual-fractal analysis as the gp120:CD4 ratio changes from 1:4 to 1:2 (a) the binding rate coefficient k_1 increases by a factor of 1.18 from a value of 13.666 to 16.129, and (b) the binding rate coefficient k_2 increases by a factor of 1.727 from a value of 55.9602 to 96.674. In Figure 6.8e the single-fractal analysis binding curve is not shown (----- dotted line) to show the clarity in the plot. Only the dual-fractal analysis is shown in this case for the binding phase.

Table 6.4a and Figures 6.9a and b indicate for a dual-fractal analysis the increase in the binding rate coefficients, (1) k_1 and (2) k_2, with an increase in the relative concentration of gp120 and CD4. In the original paper (Navratilova *et al.*, 2005) the ratios were given such as gp120:CD4 as 1:10 (see Table 6.4a). One is unable to use ratios in the Corel Quattro Pro program (Corel Corporation, 1997) to plot the ratios in the x-axis. These ratios have been converted to relative concentration values. For example, the gp120:CD4 1:10 ratio is converted to 1/10 = 0.1, etc. Similarly, the gp120:CD4 1:4 ratio is 1/4 = 0.25.

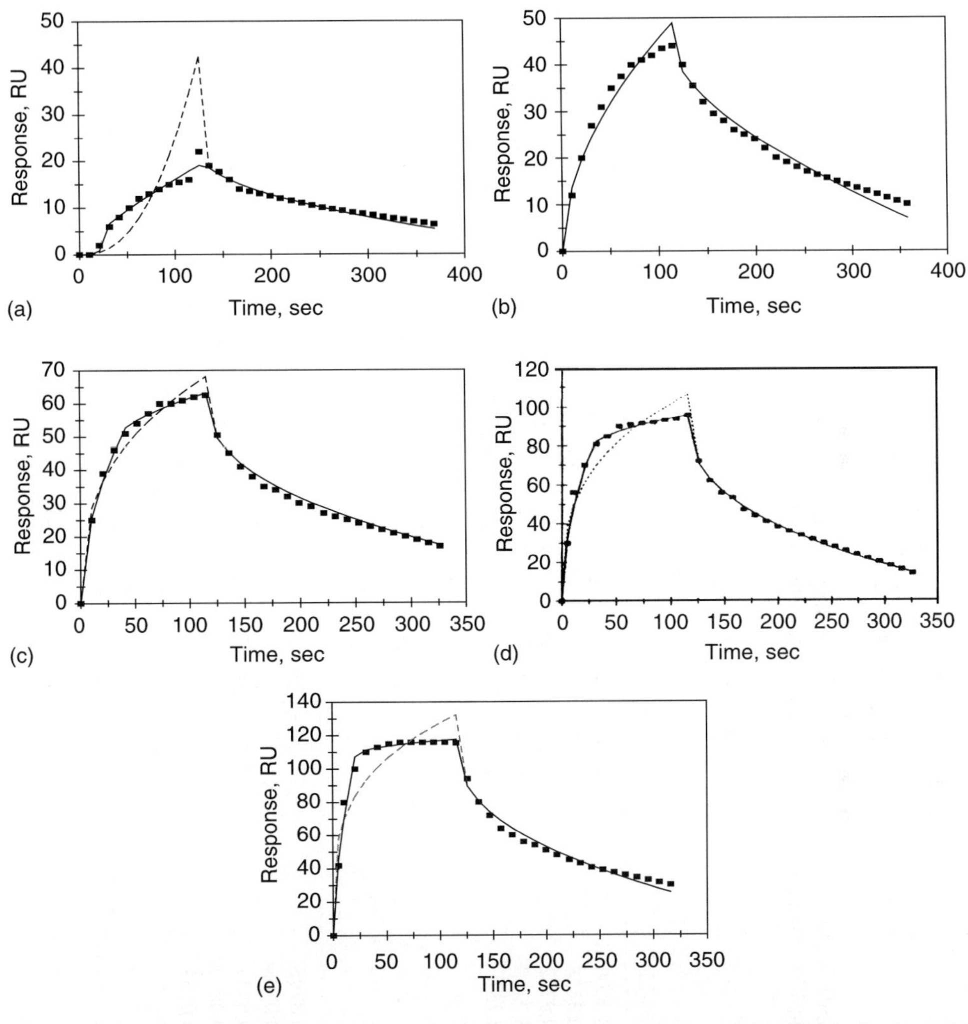

Figure 6.8 Binding and dissociation of CCR5 in solution to different gp120:CD4 ratios immobilized on a sensor chip surface (Navratilova *et al.*, 2005): (a) 1:100 (b) 1:33 (c) 1:10 (d) 1:4 (e) 1:2. When only a solid line is used then a single-fractal analysis applies. When both dashed(-----) line and solid (_____) lines are used, then the dashed line is for a single-fractal analysis, and the dashed line is for a dual-fractal analysis. In this case the solid line provides the better fit.

In the relative concentration range of 0.1–0.5 for gp120:CD4, the binding rate coefficients, k_1 and k_2, are given by

$$k_1 = (23.809 \pm 3.269)\,[\text{relative concentration, gp 120 : CD4}]^{0.4757 \pm 0.1127} \tag{6.5a}$$

$$k_2 = (166.153 \pm 0.491)\,[\text{relative concentration, gp 120 : CD4}]^{0.7833 \pm 0.002586} \tag{6.5b}$$

Table 6.4

(a) Binding and dissociation rate coefficients and (b) fractal dimensions for the binding and dissociation phase for the interaction of CCR5 in solution to different gp120:CD4 ratios immobilized on a sensor chip surface (Navratilova *et al.*, 2005)

(a)

Analyte in solution/gp120:CD4 ratio	k	k_1	k_2	k_d
CCR5/1:100	0.000368 ± 0.000074	9.7 E-09 + 17.3 E-09	0.4869 ± 0.0800	1.0695 ± 0.1033
CCR5/1:33	3.9352 ± 0.3428	NA	NA	1.3147 ± 0.1342
CCR5/1:10	12.327 ± 0.9722	7.6087 ± 0.4036	27.395 ± 0.4388	4.9017 ± 0.1399
CCR5/1:4	23.2031 ± 3.2717	13.666 ± 1.986	55.9602 ± 0.8933	9.8903 ± 0.2205
CCR5/1:2	37.182 ± 6.153	16.129 ± 2.98	96.674 ± 0.9373	9.6747 ± 0.6054

(b)

Analyte in solution/gp120:CD4 ratio	D_f	D_{f1}	D_{f2}	D_{fd}
CCR5/1:100	0 + 0.8798	0 + 2.5942	1.4826 ± 0.1508	2.0018 ± 0.0484
CCR5/1:33	1.9386 ± 0.07086	NA	NA	1.7836 ± 0.0502
CCR5/1:10	2.2802 ± 0.6444	1.9628 ± 0.0993	2.6478 ± 0.0459	2.1706 ± 0.01589
CCR5/1:4	2.3576 ± 0.0819	1.9310 ± 0.1976	2.7738 ± 0.0256	2.2132 ± 0.01245
CCR5/1:2	2.4658 ± 0.0951	1.7558 ± 0.3434	2.9175 ± 0.01756	2.1576 ± 0.03674

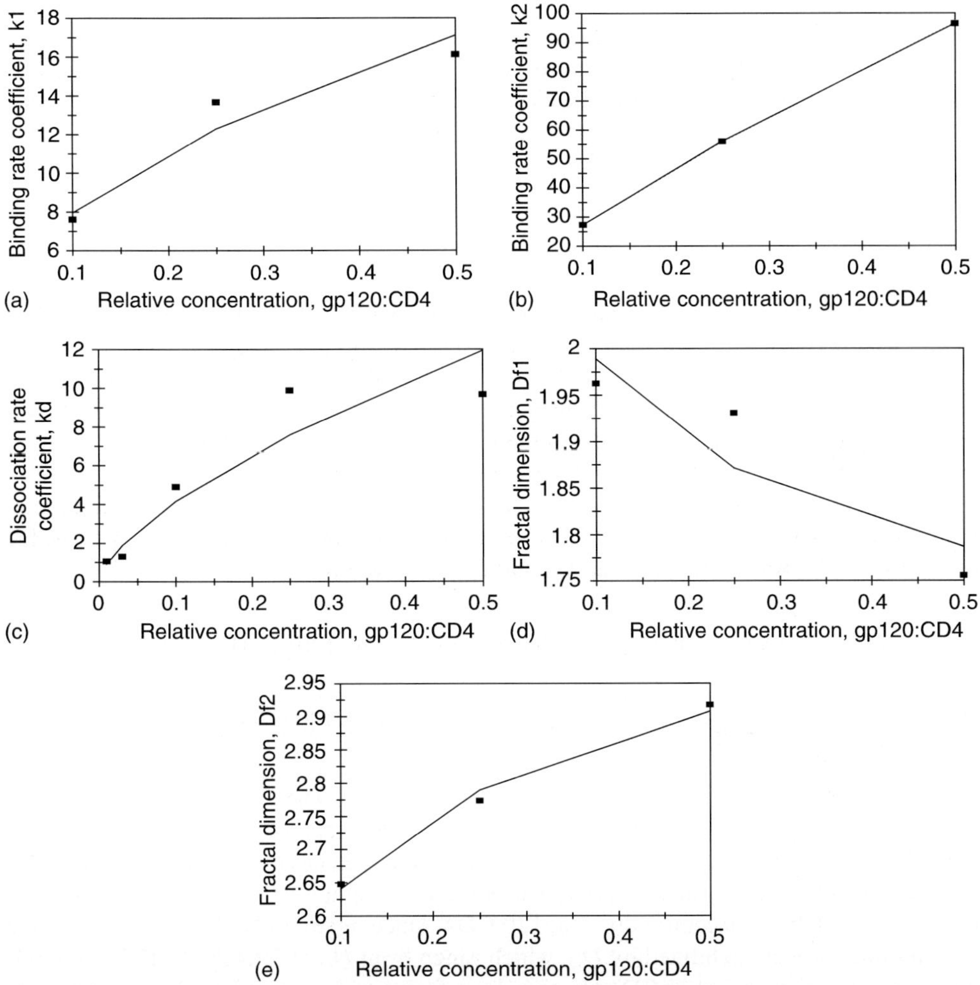

Figure 6.9 (a) Increase in the binding rate coefficient k_1 for a dual-fractal analysis with an increase in the relative concentration gp120:CD4 in solution. (b) Increase in the binding rate coefficient k_2 for a dual-fractal analysis with an increase in the relative concentration gp120:CD4 in solution. (c) Increase in the dissociation rate coefficient k_d for a single-fractal analysis with an increase in the relative concentration gp120:CD4 in solution. (d) Decrease in the fractal dimension D_{f1} with an increase in the relative concentration gp120:CD4. (e) Decrease in the fractal dimension D_{f2} with an increase in the relative concentration gp120:CD4.

The fit is good. Only three data points are available. The availability of more data points would lead to a more reliable fit. The binding rate coefficient k_1 exhibits an order of dependence close to one-half (= 0.4757) and k_2 exhibits an order of dependence between one-half and one (= 0.7833) on the relative concentration gp120:CD4 in solution. The non-integer order of dependence once again lends support to the fractal nature of the system.

Note that the binding rate coefficient k_2 exhibits a higher order of dependence than k_1 on the relative concentration gp120:CD4 in solution.

Table 6.4a and Figure 6.9c indicate for a single-fractal analysis the increase in the dissociation rate coefficient k_d with an increase in the relative concentration of gp120 and CD4. In the relative concentration range of 0.01–0.5 for gp120:CD4, the dissociation rate coefficient k_d is given by

$$k_d = (18.816 \pm 6.947)\,[\text{relative concentration, gp 120 : CD4}]^{0.6549 \pm 0.0994} \tag{6.5c}$$

The fit is reasonable. Only five data points are available. The availability of more data points would lead to a more reliable fit. The dissociation rate coefficient k_d exhibits an order of dependence between one-half and one (= 0.6549) on the relative concentration gp120:CD4 in solution. The non-integer order of dependence exhibited by k_d on gp120:CD4 once again lends support to the fractal nature of the system.

Table 6.4b and Figures 6.9d and e show for a dual-fractal analysis the (1) decrease in the fractal dimension D_{f1} and (2) increase in the fractal dimension D_{f2} with an increase in the relative concentration of gp120:CD4. In the relative concentration range, gp120:CD4 of 0.1 to 0.5, the fractal dimensions, D_{f1} and D_{f2} are given by

$$D_{f1} = (1.7066 \pm 0.0677)\,[\text{relative concentration, gp 120 : CD4}]^{-0.0665 \pm 0.0336} \tag{6.6a}$$

$$D_{f2} = (3.0309 \pm 0.0216)\,[\text{relative concentration, gp 120 : CD4}]^{0.0598 \pm 0.00662} \tag{6.6b}$$

The fit is reasonable. Only three data points are available. The availability of more data points would lead to a more reliable fit. In case (1), the fractal dimension D_{f1} exhibits a very slight (=–0.0665) negative order of dependence and in case (2) the fractal dimension D_{f2}, just like the fractal dimension D_{f1}, exhibits a very slight (= 0.0598) order of dependence on the relative concentration of gp120:CD4. Once again, only slight changes in the fractal dimension are exhibited by D_{f2}, which range from D_{f2} = 2.6478–2.9175 as the relative concentration of gp120:CD4 changes by a factor of 5 from 0.1 to 0.5. As indicated elsewhere in this book, the fractal dimension is based on a log scale and even small changes in the fractal dimension indicate significant changes in the degree of heterogeneity on the sensor chip surface.

Table 6.4a and b and Figures 6.10a and b indicate for a dual-fractal analysis (1) the decrease in the binding rate coefficient k_1 and (2) the increase in the binding rate coefficient k_2, with an increase in the fractal dimension, D_{f1} and D_{f2}, respectively. For the 1:10 to 1:2 gp120:CD4 ratio (or the relative concentration, gp120:CD4 from 0.1 to 0.5), the binding rate coefficients, k_1 and k_2, are given by

$$k_1 = (4.4706 \pm 0.1103) D_{f1}^{-0.7534 \pm 0.2866} \tag{6.7a}$$

$$k_2 = (0.02386 \pm 0.00747) D_{f2}^{7.5548 \pm 0.4968} \tag{6.7b}$$

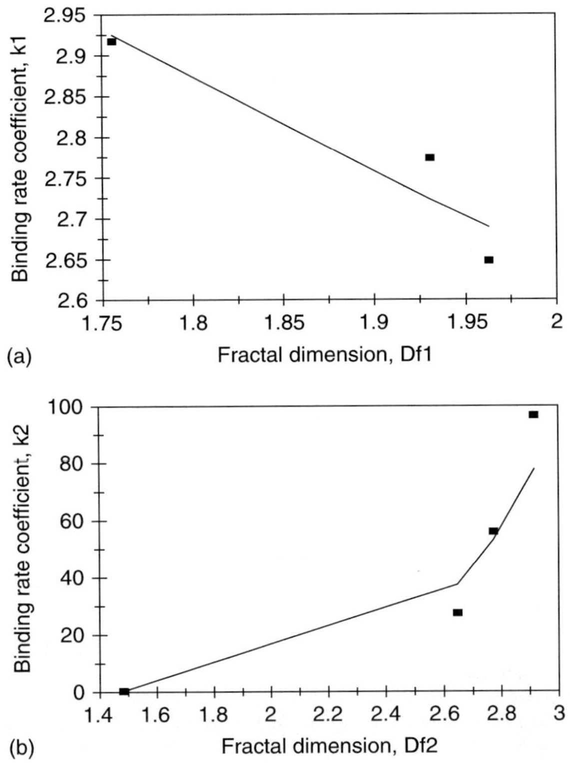

Figue 6.10 (a) Decrease in the binding rate coefficient k_1 for a dual-fractal analysis with an increase in the fractal dimension D_{f1}. (b) Increase in the binding rate coefficient k_2 for a dual-fractal analysis with an increase in the fractal dimension D_{f2}.

In case (1), the fit is reasonable. Only three data points are used though four are available. The error for one of the points (gp120:CD4 ratio 1:100) is large, and thus it is not used. In fact, in Table 6.4 the value of k_1 is presented as 9.7 E-09 + 17.3E-09. The negative error is not presented since a negative value for the binding rate coefficient k_1 is not possible. The binding rate coefficient k_1 exhibits an order of dependence between one-half and one (=–0.7534) on the fractal dimension D_{f1} or the degree of heterogeneity on the sensor chip surface.

In case (2), the fit is reasonable. Only four data points are available. The availability of more data points would lead to a more reliable fit. The binding rate coefficient k_2 is very sensitive to the fractal dimension D_{f2} or the degree of heterogeneity on the sensor chip surface as noted by the very high order of dependence between 7 and 8 (= 7.5548) exhibited.

As indicated elsewhere in this book, affinity values are of interest to practicing biosensorists. Table 6.4a and Figures 6.11a and b indicate (1) the decrease in the affinity K_1 ($=k_1/k_d$) and (2) increase in the affinity K_2 ($=k_2/k_d$) with an increase in the relative concentration, gp120:CD4. In the 0.1 to 0.5 gp120:CD4 relative concentration range, the

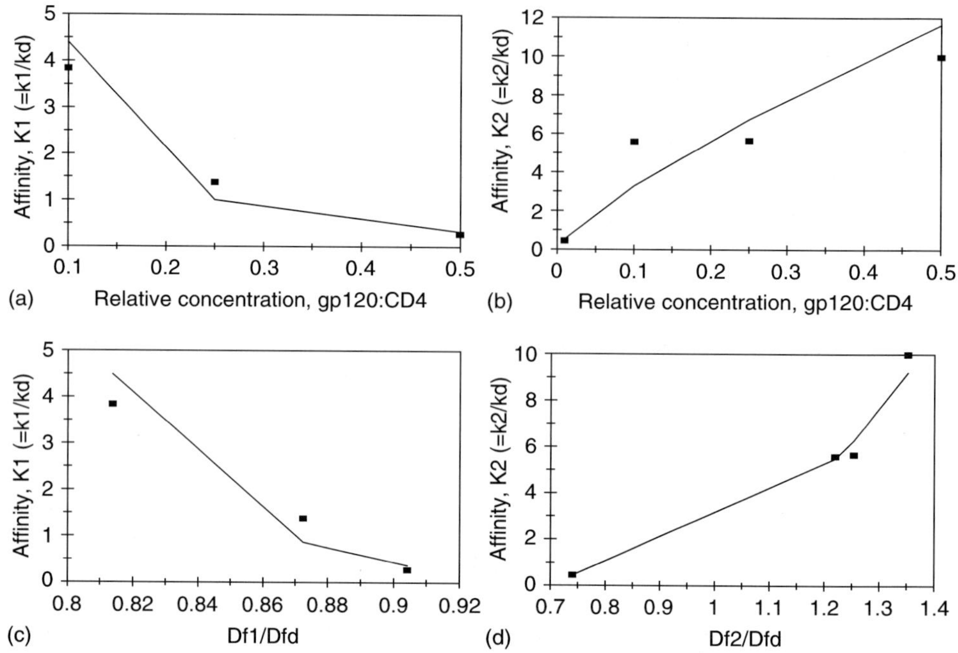

Figure 6.11 (a) Decrease in the affinity K_1 ($= k_1/k_d$) with an increase in the relative concentration gp120:CD4. (b) Increase in the affinity K_2 ($= k_2/k_d$) with an increase in the relative concentration gp120:CD4. (c) Decrease in the affinity K_1 ($= k_1/k_d$) with an increase in ratio of the fractal dimension D_{f1}/D_{fd}. (d) Increase in the affinity K_2 ($= k_2/k_d$) with an increase in the ratio of fractal dimension D_{f2}/D_{fd}.

affinities, K_1 and K_2 are given by

$$K_1 = (0.1067 \pm 0.0515)\,[\text{relative concentration, gp 120 : CD4}]^{-1.6163 \pm 0.3445} \qquad (6.8a)$$

$$K_2 = (20.114 \pm 10.703)\,[\text{relative concentration, gp 120 : CD4}]^{0.7829 \pm 0.1415} \qquad (6.8b)$$

In case (1), the fit is good. Only three data points are available. The availability of more data points would lead to a more reliable fit. The affinity K_1 exhibits an order of dependence slightly more than negative one and one-half (=–1.6163). If higher affinity K_1 values are required then it behooves one to use lower relative concentration, gp120:CD4 values, and vice versa at least in the range of 0.1–0.5 of relative concentration, gp120:CD4 analyzed.

In case (2), the fit is good. Only four data points are available. The availability of more data points would lead to a more reliable fit. The affinity K_2 exhibits an order of dependence between one-half and one (= 0.7829). If higher affinity K_2 values are required then it behooves one to use higher relative concentration, gp120:CD4 values, and vice versa at least in the range of 0.1–0.5 of relative concentration, gp120:CD4 analyzed. Note that higher affinity K_1 and K_2 values are obtained at lower and higher relative concentration, gp120:CD4 values, respectively.

Figure 6.11c shows the decrease in the affinity K_1 with an increase in the ratio of the fractal dimensions, D_{f1}/D_{fd}. In the 0.1–0.5 gp120:CD4 relative concentration range, the affinity K_1 is given by

$$K_1 = (0.03519 \pm 0.0274)(D_{f1}/D_{fd})^{-23.536 \pm 7.568} \quad (6.9a)$$

The fit is good. Only three data points are available. The availability of more data points would lead to a more reliable fit. The affinity K_1 is extremely sensitive to the ratio of the fractal dimensions, D_{f1}/D_{fd}, as noted by the extremely high (=–23.536) negative order of dependence exhibited. In this case, for higher affinity K_1 values, one needs lower values of the fractal dimension ratio, D_{f1}/D_{fd}.

Figure 6.11d shows the increase in the affinity K_2 with an increase in the ratio of the fractal dimensions, D_{f2}/D_{fd}. In the 0.01–0.5 (with the exception of 0.03) gp120:CD4 relative concentration range, the affinity K_2 is given by

$$K_2 = (2.0265 \pm 0.2034)(D_{f2}/D_{fd})^{5.0225 \pm 0.2010} \quad (6.9b)$$

The fit is good. Only four data points are available. The availability of more data points would lead to a more reliable fit. The affinity K_2 is very sensitive to the ratio of the fractal dimensions, D_{f2}/D_{fd}, as noted by the slightly higher than fifth (= 5.0225) order of dependence exhibited. Once again, this represents one method of controlling or changing the affinity K_2, value on the sensor chip surface in this 0.01–0.5 (with the exception of 0.03) relative concentration, gp120:CD4 range. Note once again, that the affinity K_1 decreases with an increase in the ratio of fractal dimensions, D_{f1}/D_{fd}, and the affinity K_2 increases with an increase in the ratio of fractal dimensions, D_{f2}/D_{fd}.

6.4 CONCLUSIONS

A fractal analysis is presented for (a) the binding and dissociation of different concentrations of the recombinant antibodies ScFv2 and ScFv6 in solution to antigen $F(ab')_2$ immobilized on a sensor chip surface (Welschof *et al.*, 1997), and (b) binding and dissociation of CCR5 in solution to gp120:CD4 immobilized on a sensor chip surface (Navratilova *et al.*, 2005). The interactions involved in both these examples presented above are involved in autoimmune reactions that lead to intractable diseases. The first one is involved in arthritis, and the second one is involved in HIV-1. It is suggested that a better understanding of the binding (and dissociation) interactions involved in the above two reaction systems would help lead to better physical insights into arthritis, HIV-1, and other similar types of autoimmune diseases. The fractal analysis is used with this purpose in mind to relate the binding and dissociation coefficients with the degree of heterogeneity (or the fractal dimension) present on the sensor chip surface.

As indicated elsewhere in the different chapters in this book the fractal analysis provides a quantitative indication of the state of disorder on the sensor chip surface, and relates this to the binding and dissociation rate coefficient as well as to the affinity values. Both types of examples are given wherein either a single- or a dual-fractal analysis was used. The dual-fractal analysis is used only when the single-fractal analysis did not provide an adequate fit. This was done by the regression analysis provided by Corel Quattro Pro 8.0 (1997).

In accordance with the prefactor analysis for fractal aggregates (Sorenson and Roberts, 1997), quantitative (predictive) expressions are developed for (a) the binding rate coefficient k, as a function of the analyte concentration in solution or the fractal dimension or the degree of heterogeneity in the binding phase D_f that exists on the sensor chip surface, (b) the dissociation rate coefficient k_d as a function of the analyte concentration in solution or the fractal dimension in the dissociation phase D_{fd}, (c) the fractal dimension in the binding phase as a function of the analyte concentration in solution, and (d) the affinity K as a function of the analyte concentration in solution and as a function of the ratio of fractal dimensions present in the binding and dissociation phases.

The fractal dimension is not a classical independent variable such as analyte (antigen, antibody, or other biological molecule) concentration in solution. Nevertheless, the expression obtained for (a) the binding and the dissociation rate coefficients for a single- and for a dual-fractal analysis as a function of the fractal dimension, and (b) the affinity K as a function of the ratio of the fractal dimensions in the binding and in the dissociation phase indicate, in general, a high sensitivity of these rate coefficients (or affinities) on their respective fractal dimensions (or ratios of fractal dimensions as the case may be) on the sensor chip surface.

For example, for the binding and dissociation of different concentrations of recombinant antibody fragment, ScFv6 (single-chain Fv6) in solution to antigen $F(ab')_2$ immobilized on a sensor chip surface (Welschof *et al.*, 1997), the dissociation rate coefficient k_d exhibits close to a negative sixth (=–5.935) order of dependence on the fractal dimension in the dissociation phase D_{fd}. Similarly, the affinity K (=k/k_d) for this same interaction exhibits a slightly higher than sixth (=6.279) order of dependence on the ratio of fractal dimensions present in the binding and in the dissociation phases, D_f/D_{fd}. Both these predictive relations indicate the very high sensitivity that is involved. Also, for the binding of CCR5 in solution to different gp120:CD4 ratios immobilized on a sensor chip surface, the binding rate coefficient k_2, for a dual-fractal analysis exhibits a 7.5548 order of dependence on the fractal dimension D_{f2}. This again indicates the very high sensitivity involved of the surface in determining the binding rate coefficient. This once again reinforces one of the major points of this book, that is to pay more and more attention to the nature (or the degree of heterogeneity) of the sensor chip surface, and how it influences the binding and dissociation rate coefficients, as well as the affinity values.

Note that the data analysis in itself does not provide any evidence for surface roughness or heterogeneity, and the existence of surface roughness or heterogeneity assumed may not be correct. Considering the complexity involved on the SPR chip surface, this is not an unreasonable assumption. Furthermore, there is deviation in the data that may be minimized by providing a correction for the depletion of the analyte in the vicinity of the surface (imperfect mixing).

An increase in the binding rate coefficient value should lead to enhanced sensitivity and to a decrease in the response time of the SPR biosensor. Both these aspects would be beneficial to SPR biosensor construction. For a selective (or multiple) reaction system, if an increase in the fractal dimension leads to an increase in the binding rate coefficient (of interest), then this would enhance selectivity. Stability is a more complex issue, and one might intuitively anticipate that a distribution or heterogeneity of the receptor on the biosensor surface would lead to a more stable biosensor. In this case, one type of distribution may be better than the other depending on the biosensor performance of interest. Considering the complexity that is involved on the SPR biosensor chip surface, and if we are looking at stability, then this aspect is particularly valuable here. Similar behavior has been observed for the deactivation of enzymes wherein a distribution of activation energies for deactivation (as compared to a single activation energy for deactivation) leads to a more stable enzyme (Malhotra and Sadana, 1987).

For the most part, and as indicated throughout in different chapters in this book, the analysis is extendable to analyte–receptor reactions occurring on (a) other types of biosensors and on (b) nonbiosensor surfaces. Here it is hoped that the biosensor-based analysis would lead to a better understanding of the binding and dissociation interactions of (a) binding and dissociation of the different concentrations of the recombinant antibodies ScFv2 and ScFv6 in solution to antigen $F(ab')_2$ immobilized on a sensor chip surface (Welschof *et al.*, 1997), and (b) the binding and dissociation of CCR5 in solution to gp120:CD4 immobilized on a sensor chip surface (Navratilova *et al.*, 2005).

Autoimmune diseases are particularly difficult to treat, and at best one 'manages' them. For some autoimmune diseases, such as systemic lupus erythematosus (SLE) the etiology is not well known, and thus the treatment is often complicated such as the use of steroids and chemotherapy drugs, which have significant side effects (such as hair loss). Even the treatment of arthritis has come into scrutiny thanks to the turmoil raised by treatment of drugs such as vioxx and celebrex. It would be excellent if one is able to 'catch' these insidious diseases at an early stage, just as in the case of different forms of cancer where the prognosis is much better. Biosensors do exhibit the potential to assist in this effort. The aim of this chapter was to assist in this effort of being able to detect early autoimmune diseases, by providing a couple of examples on arthritis and HIV-1.

REFERENCES

Analytical Currents (2005). Protein biomarkers for Alzheimer's. *Analytical Chemistry*, **77**(April 1), 128A.

Badley, E.M., and Tennant, A. (1993). Impact of disablement due to rheumatic disorders in a British population; estimates of severity and prevalence from the Calderdale Rheumatic Disablement Survey. *Annals of Rheumatic Disease*, **52**, 6–13.

Carson, D.A., Chen, P., Fox, R.I., Kippi, T.J., Jirik, F., Goldfen, R.D., Silverman, G., Radoux, V., and Feng, S. (1987). *Annual Reviews in Immunology*, **5**, 109–126.

CDC report (2005)-2030 Public health and aging: projected prevalence of self-reported arthritis or chronic joint symptoms among persons aged ≤65 years. *Morbid Mortal Weekly Report*, 2003, **52**, 489–491.

Corel Corporation (1997). *Corel Quattro Pro*, Ottawa, Canada.

Dragic, T., Trkola, A., Thompson, D.A., Cormier, E.G., Kajumo, F.A., Maxwell, E., Lin, S.W., Ying, W., Smith, S.O., Sakmar, T.P., and Moore, J.P. (2000). A binding pocket for a small molecule inhibitor of HIV-1 entry within the transmembrane helices of CCR5. *Proceedings of the National Academy of Sciences USA*, **97**, 5639–5644.

Feng, Y., Ke, X., Ma, R., Chen, Y., Hu, G., and Liu, F. (2004). Parallel detection of autoantibodies with microarrays in rheumatoid diseases. *Clinical Immunology*, **50**, 416–422.

Geysen, H.M., Meloen, R.H., and Bartling, S.J. (1984). Use of peptide synthesis to probe viral antigens for epitopes to a solution of a single amino acid. *Proceedings of the National Academy of Sciences USA*, **81**, 3998–4002.

Havlin, S. (1989). Molecular diffusion and reaction. In: *The Fractal Approach to Heterogeneous Chemistry*: *Surfaces, Colloids, Polymers* (ed. D. Avnir), Wiley, New York, pp. 251–269.

Hueber, W., Utz, P.J., Steeinman, L., and Robinson, W.H. (2002). Autoantibody profiling for the study and treatment of autoimmune disease. *Arthritis Research*, **4**(5), 290–295.

Jacobson, D.L., Gange, S.J., Rose, N.R., and Graham, N.M. (1997). Epidemiology and estimated population burden of selected autoimmune diseases in the United States. *Immunology and Immunopathology*, **84**, 223-243.

Malhotra, A., and Sadana, A. (1987). Microheterogeneity of enzymes and deactivation. *Biotechnology and Bioengineering*, **30**(9), 1041–1056.

Mirkin, C., Klein, W., Chang, L., and Nam, J.M. (2005). Nanoparticle-based detection in cerebral spinalfluid of a soluble pathogenic biomarker for Alzheimer's disease. *Proceedings of the National Academy of Sciences*, **102**, 2273–2276.

Navratilova, I., Sodroski, J., and Myszka, D.G. (2005). Solubilization, stabilization, and purification of chemokine receptors using biosensor technology. *Analytical Biochemistry*, **339**, 271–281.

Ramakrishnan, A., and Sadana, A. (2001). A single-fractal analysis of cellular analyte-receptor binding kinetics utilizing biosensors. *Biosystems*, **59**(1) 35–45.

Rose, H.M., Ragan, C., Pearce, E., and Lipman, M.O. (1949). *Proceedings of the Society of Experimental Biology and Medicine*, **68**, 1–5.

Sadana, A. (2001). A fractal analysis approach for the evaluation of hybridization kinetics in biosensors. *Journal of Colloid and Interface Science*, **234**, 9–18.

Sorenson, C.M., and Roberts, G.C. (1997). The prefactor of fractal aggregates. *Journal of Colloid and Interface Science*, **186**, 447–452.

Van Duyne, R. (2005). Detection of a biomarker for Alzheimer's disease from synthetic and clinical samples using a nanoscale optical biosensor. *Journal of the American Chemical Society*, **127**, 2264–2271.

Von Muhlen, C.A., and Tan, E.M. (1995). Autoantibodies in the diagnosis of systemic rheumatic disease. *Seminars in Arthritis and Rheumatology*, **24**, 323–358.

Welschof, M., Terness, P., Kirpiyanov, S.M., Stanescu, D., Boeitling, F., Dorsan, H., Dubel, S., Little, M., and Opelz, G. (1997). The antigen-binding domain of a human IgG-anti-F$(ab')_2$ autoantibody. *Proceedings of the National Academy of Sciences USA*, **94**, 1902–1907.

Wood, W. (1991). Matrix Effects in immunoassays. *Scandinavian Journal of Clinical Laboratory Investigations*, **51,** 105–112.

Zhao, Q., He, Y., Alespeti, G., and Debnath, A.K. (2004). A novel assay to identify entry inhibitors that block binding of HIV-1 gp120 to CCR5. *Virology*, **326**, 299–309.

– 7 –

Fractal Binding and Dissociation Kinetics of Prion Proteins on Biosensor Surfaces

7.1 INTRODUCTION

Misfolded proteins lead to a variety of diseases. One disease that has been in the forefront for quite a while and has gained quite a bit of publicity (notoriety) is mad-cow disease, or otherwise medically known as bovine spongioform encephalopathy (BSE). The fading of the epidemic of the human form of the disease (Crutzfeld-Jacob disease, vCJD) in the United Kingdom has helped minimize the fear generated in the general population. A few cases of mad-cow disease were also reported in the North Western part of the United States. However, the authorities quickly stepped in to stop the spread of the disease. Carrell (2004) has recently emphasized prion dormancy and disease. The author indicates the presence of thousands of dormant carriers of vCJD, and the possibility of infecting others. There is also emphasis on the quandary faced by the public health officials in the United Kingdom, and the necessity of political will to take corrective action. Marx (2005) recently indicates that play and exercise prevents amyloid build up in the mouse brain. Research apparently suggests that in an enriched environment in cages the β-amyloid build up is reduced compared to when the mice are kept in standard laboratory cages. Thus, exercise is apparently good for the mind as well as for the body. Furthermore, Marx (2005) adds that research at the University of Chicago at Illinois in the U.S. indicates the increased activity of neprilysin in the brain. This is a β-amyloid-degrading enzyme. Kagan (2005) in a very recent letter to the magazine *Science* indicates that the therapeutic approach to Alzheimer's is to inhibit or reverse protein aggregation or to remove the aggregated proteins. The author emphasizes that how the aggregated proteins kill or damage cells is not yet clear. Apparently, research suggests that the aggregates damage the cells by forming ion-permeable channels in cellular membranes (Arispe *et al.*, 1993; Mirzabekov *et al.*, 1996; Lin *et al.*, 1997; Bhatia *et al.*, 2000; Hirakura and Kagan, 2001; Hirakura *et al.*, 2002). Apparently, cellular energy is drained through these 'leakage' pathways. Cellular signaling is inhibited, and apoptosis results. Dobson (2005) commenting on Kagan's letter (2005) emphasizes that the onset of diseases due to misfolding of protein have similar underlying origins (Dobson, 2003). Dobson (2005) concludes his letter by indicating that there are fatal forms of proteins that can have beneficial effects. He provides the example of the misfolded forms of α-lactalbumin found in milk that have the ability to kill cancer cells in a selective manner.

Aguzzi and Polymenidou (2004) recently indicate that transmissible spongioform encephalopathy (TSE) is an infectious neurodegenerative disease. It is characterized by the aggregation and accumulation of prion protein, PrP^{Sc} in the central nervous system.

This is an abnormal structural form of the PrP^{C}. PrP^{C} is the normal form, and is expressed at the surface of neurons and other cells. Blanquet-Grossard *et al.* (2005) very recently indicate that the present hypothesis in the spread of TSE is that, the PrP^{Sc} form is the infecting agent, and propagates the disease by imparting its abnormal conformation to the normal (non-pathogenic) PrP^{C} form.

Ishii *et al.* (1984) and Kovacs *et al.* (2004) indicate the presence of amyloid plaques as characteristic of vCJD. Dandoy-Dron *et al.* (1988) and Riemer *et al.* (2000) indicate that C1q is one of the 19 proteins that exhibit an increased synthesis in scrapie-infected brain tissues. Blanquet-Grossard *et al.* (2005) hypothesize that the abnormal form, PrP^{Sc}, activates the C1q protein. These authors wanted to test the hypothesis of the activation (or implication) of the classical complement pathway during the early stages of the pathogenesis of this disease. These authors used the surface plasmon resonance (SPR) biosensor to test if the human C1q complement protein can bind to mouse PrP immobilized on a sensor chip. Blanquet-Grossard *et al.* (2005) emphasize that the immobilized PrP undergoes structural modification that may mimic the conformational changes undergone by this protein during pathogenesis. They emphasize that C1q is a natural sensor for these changes.

In this chapter we re-analyze the binding and dissociation kinetics data presented by Blanquet-Grossard *et al.* (2005) using fractal kinetics. The complement protein, C1q is in solution and the PrP form is immobilized on the sensor chip surface. In no way are we indicating that the present analysis is in any way better than the original Biacore (Biacore X apparatus) presented by Blanquet-Grossard *et al.* (2005). One may consider the analysis to be presented as an alternate analysis that provides physical insights into the interactions occurring on the sensor chip surface.

7.2 THEORY

Havlin (1989) has reviewed and analyzed the diffusion of reactants towards fractal surfaces. The details of the theory and the equations involved for the binding and dissociation phases for analyte–receptor binding are available (Sadana, 2001). The details are not repeated here; except that just the equations are given to permit an easier reading. These equations have been applied to other biosensor systems (Ramakrishnan and Sadana, 2001; Sadana, 2001). For most applications, a single- or a dual-fractal analysis is often adequate to describe the binding and the dissociation kinetics. Peculiarities in the values of the binding and dissociation rate coefficients, as well as in the values of the fractal dimensions with regard to the dilute analyte systems being analyzed will be carefully noted, if applicable.

7.2.1 Single-fractal analysis

Binding rate coefficient

Havlin (1989) indicates that the diffusion of a particle (analyte [Ag]) from a homogeneous solution to a solid surface (e.g. receptor [Ab]-coated surface) on which it reacts to form a product (analyte–receptor complex, Ab.Ag) is given by:

$$\text{(Analyte.Receptor)} \sim \begin{cases} t^{(3-D_{\mathrm{f,bind}})/2} = t^{\mathrm{p}} & (t < t_{\mathrm{c}}) \\ t^{1/2} & (t > t_{\mathrm{c}}) \end{cases} \tag{7.1a}$$

Here $D_{\mathrm{f,bind}}$ or D_{f} (used later on in the chapter) is the fractal dimension of the surface during the binding step. t_{c} is the cross-over value. Havlin (1989) indicates that the cross-over value may be determined by $r_{\mathrm{c}}^2 \sim t_{\mathrm{c}}$. Above the characteristic length, r_{c}, the self-similarity of the surface is lost and the surface may be considered homogeneous. Above time t_{c}, the surface may be considered homogeneous, since the self-similarity property disappears and 'regular' diffusion is now present. For a homogeneous surface where $D_{\mathrm{f}}=2$, and when only diffusional limitations are present, $p=½$ as it should be. Another way of looking at the $p=½$ case (where $D_{\mathrm{f,bind}}=2$) is that the analyte in solution views the fractal object, in our case, the receptor-coated biosensor surface, from a 'large distance.' In essence, in the association process, the diffusion of the analyte from the solution to the receptor surface creates a depletion layer of width $(Ð)^{½}$ where $Ð$ is the diffusion constant. This gives rise to the fractal power law, $\text{(Analyte.Receptor)} \sim t^{(3-D_{\mathrm{f,bind}})/2}$. For the present analysis, t_{c} is arbitrarily chosen and we assume that the value of the t_{c} is not reached. One may consider the approach as an intermediate 'heuristic' approach that may be used in future to develop an autonomous (and not time-dependent) model for diffusion-controlled kinetics.

Dissociation rate coefficient

The diffusion of the dissociated particle (receptor [Ab] or analyte [Ag]) from the solid surface (e.g., analyte [Ag]–receptor [Ab]) complex coated surface) into solution may be given, as a first approximation, by

$$\text{(Analyte.Receptor)} \sim -\mathrm{k}' t^{(3-D_{\mathrm{f,diss}})/2} \quad (t > t_{\mathrm{diss}}) \tag{7.1b}$$

Here $D_{\mathrm{f,diss}}$ is the fractal dimension of the surface for the dissociation step. This corresponds to the highest concentration of the analyte–receptor complex on the surface. Henceforth, its concentration only decreases. The dissociation kinetics may be analyzed in a manner 'similar' to the binding kinetics.

7.2.2 Dual-fractal analysis

Binding rate coefficient

Sometimes, the binding curve exhibits complexities and two parameters (k, D_f) are not sufficient to adequately describe the binding kinetics. This is further corroborated by low values of r^2 factor (goodness-of-fit). In that case, one resorts to a dual-fractal analysis (four parameters: k_1, k_2, D_{f1}, and D_{f2}) to adequately describe the binding kinetics. The single-fractal analysis presented above is thus extended to include two fractal dimensions. At present, the time ($t=t_1$) at which the 'first' fractal dimension 'changes' to the 'second' fractal dimension is arbitrary and empirical. For the most part, it is dictated by the data analyzed and experience gained by handling a single-fractal analysis. A smoother curve is obtained in the 'transition' region, if care is taken to select the correct number of points for the two regions. In this case, the product (antibody–antigen; or analyte–receptor complex, Ab.Ag or analyte.receptor) is given by

$$(\text{Ab.Ag}) \approx \begin{cases} t^{(3-D_{f1,\text{bind}})/2} = t^{p_1} & (t < t_1) \\ t^{(3-D_{f2,\text{bind}})/2} = t^{p_2} & t_1 < t < t_2 = t_c \\ t^{1/2} & (t > t_c) \end{cases} \tag{7.1c}$$

In some cases, as mentioned above, a triple-fractal analysis with six parameters (k_1, k_2, k_3, D_{f1}, D_{f2}, and D_{f3}) may be required to adequately model the binding kinetics. This is when the binding curve exhibits convolutions and complexities in its shape due to perhaps the very dilute nature of the analyte (in some of the cases to be presented) or for some other reasons. Also, in some cases, a dual-fractal analysis may be required to describe the dissociation kinetics.

7.3 RESULTS

In this chapter we analyze the binding and dissociation kinetics of human C1q in solution to PrP immobilized on a sensor chip surface (Blanquet-Grossard *et al.*, 2005) using fractal analysis. As indicated earlier in this book this is just one method of analyzing the binding and dissociation kinetic data that takes into consideration the different morphologies that may exist on the biosensor surface.

Alternate expressions for fitting the data are available that include saturation, first-order reaction, and no diffusion limitations, but these expressions are apparently deficient in describing the heterogeneity that inherently exists on the surface. One might justifiably argue that appropriate modeling may be achieved by using a Langmuirian or other approach. The Langmuirian approach may be used to model the data presented if one presumes the presence of discrete classes of sites. However, that being said, we will as indicated above use fractal analysis (specifically eqs. 7.1a–c) to analyze the binding and dissociation data of the interactions between C1q and PrP (Blanquet-Grossard *et al.*, 2005).

Figures 7.1a–e show the binding and dissociation of (1) 5, (2) 10, (3) 15, (4) 20, and (5) 30 nM human C1q in solution to PrP immobilized on a sensor chip surface. A single-fractal analysis is adequate to describe the binding and dissocition kinetics. The values of (a) the binding rate coefficient k and the fractal dimension D_f for a single-fractal analysis, and (b) the dissociation rate coefficient k_d and the fractal dimension for the dissociation phase D_{fd} are given in Table 7.1. Corel Quattro-Pro 8.0 (Corel Corporation, 1997) is used to fit the data. Eq. (7.1a) is used to obtain the binding rate coefficient k and the fractal dimension D_f for a single-fractal analysis. Similarly, eq. (7.1b) is used to obtain the value

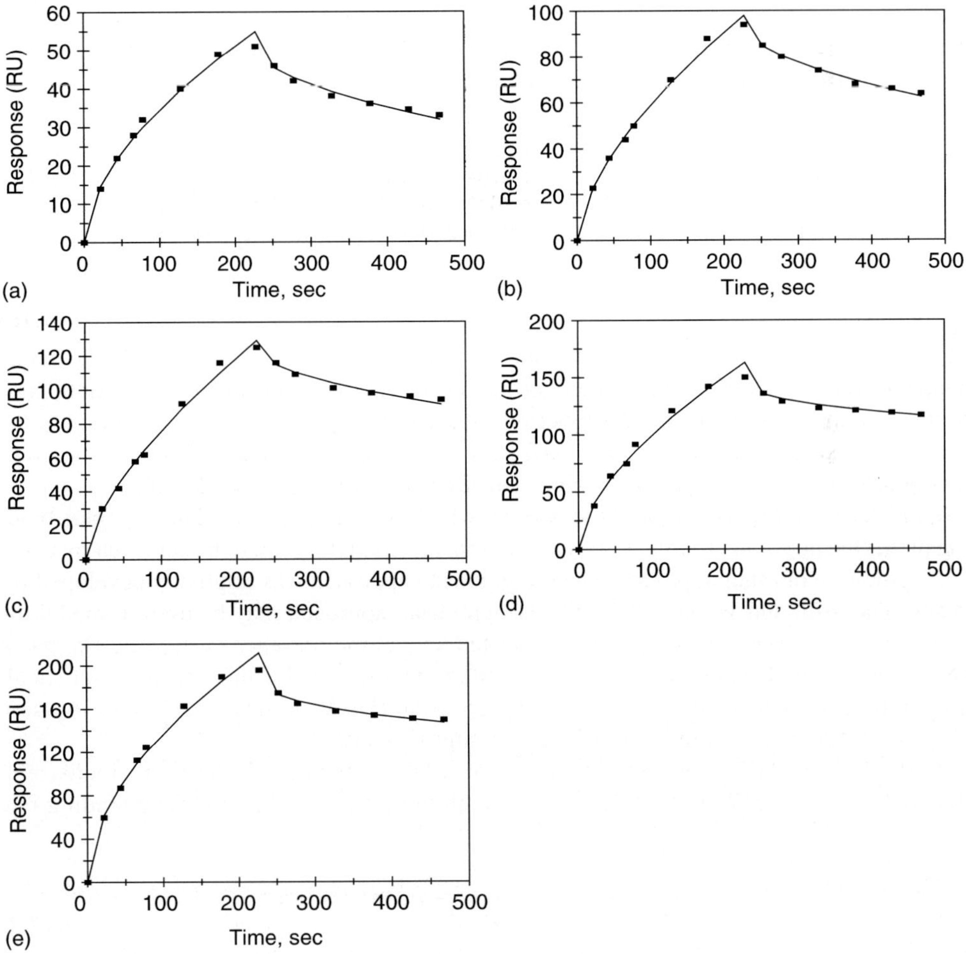

Figure 7.1 Binding and dissociation curves for different concentrations (in nM) of human C1q in solution to PRP immobilized on a sensor chip surface (Blanquet-Grossard *et al.*, 2005): (a) 5, (b) 10, (c) 15, (d) 20, (e) 30.

Table 7.1

(a) Binding and dissociation rate coefficients and (b) fractal dimensions for the binding and dissociation phase for the interaction of different concentrations of C1q in solution to mouse prion protein (PrP) immobilized on a sensor chip surface (Blanquet-Grossard *et al.*, 2005)

(a)

C1q concentration, nM	k	k_d
5	2.591 ± 0.1392	0.9633 ± 0.0966
10	3.3928 ± 0.1069	1.6619 ± 0.0813
15	3.893 ± 0.203	1.8193 ± 0.2112
20	6.385 ± 0.427	4.6320 ± 0.3172
30	12.090 ± 0.599	7.675 ± 0.532

(b)

C1q concentration, nM	D_f	D_{fd}
5	1.8750 ± 0.0518	1.9104 ± 0.0978
10	1.7612 ± 0.03382	1.9266 ± 0.0488
15	1.7104 ± 0.0506	1.9354 ± 0.1124
20	1.8072 ± 0.0646	2.2794 ± 0.0678
30	1.9454 ± 0.0482	2.3272 ± 0.0686

of the dissociation rate coefficient k_d and the fractal dimension D_{fd} for a single-fractal analysis. The values of the parameters presented in Table 7.1 are within 95% confidence limits. For example, the value of k reported for the binding of 5 nM C1q in solution to the PrP immobilized on the sensor chip surface is 2.591 ± 0.1392. The 95% confidence limits indicate that that 95% on the k values will lie between 2.4518 and 2.7302.

In all fairness, this is one possible way by which to analyze the analyte–receptor binding data. One might justifiably argue that appropriate modeling might be achieved by using a Langmuirian or other approach. The Langmuirian approach was originally developed for gases (Thomson and Webb, 1968). The Langmuirian approach may be used to model the data presented if one assumes, and as indicated above, the presence of discrete classes of sites. One might also attempt to model the data to be presented using a fractal (statistical) growth process such as diffusion-limited aggregation (DLA), kinetic gelation, or invasion percolation. However, at present, no such attempt is made.

In the 5–30 nM C1q concentration range in solution, Figure 7.2 shows the increase in the binding rate coefficient k with an increase in the C1q concentration. The binding rate coefficient k is given by

$$k = (0.5699 \pm 0.1728)\,[\mathrm{C1q}]^{0.8213 \pm 0.1932} \tag{7.2}$$

The fit is quite good. There is some scatter in the data. The binding rate coefficient k exhibits less than first-order dependence on the C1q concentration in solution. The

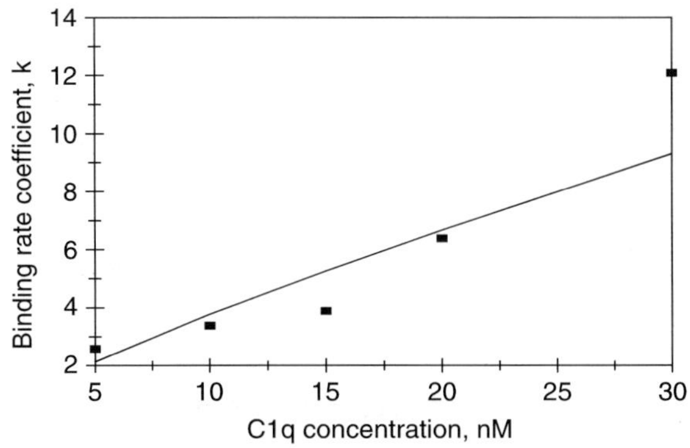

Figure 7.2 Increase in the binding rate coefficient k with an increase in the C1q concentration (in nM) in solution.

non-integral dependence exhibited by the binding rate coefficient k on the C1q concentration in solution lends support to the fractal nature of the system.

In the 5–30 nM C1q concentration range in solution, Figure 7.3a shows the increase in the dissociation rate coefficient k_d with an increase in the C1q concentration. The dissociation rate coefficient k_d is given by

$$k_d = (0.1245 \pm 0.0463)[\mathrm{C1q}]^{1.1566 \pm 0.2307} \tag{7.3a}$$

The fit is good. There is some scatter in the data. The dissociation rate coefficient k_d exhibits slightly higher than first-order (=1.156) dependence on the concentration in solution. The non-integral dependence exhibited by the dissociation rate coefficient k_d on the C1q concentration in solution lends support to the fractal nature of the system.

In the 5–30 nM C1q concentration range in solution, Figure 7.3b shows the increase in the fractal dimension in the dissociation phase D_{fd} with an increase in the C1q concentration in solution. The fractal dimension in the dissociation phase D_{fd} is given by

$$D_{fd} = (1.513 \pm 0.0988)[\mathrm{C1q}]^{0.1199 \pm 0.0462} \tag{7.3b}$$

The fit is good. There is some scatter in the data. The fractal dimension, D_{fd} exhibits a very low order (=0.1199) of dependence on the C1q concentration in solution.

In the 5–30 nM C1q concentration range in solution, Figure 7.3c shows the increase in the dissociation rate coefficient k_d with an increase in the fractal dimension in the dissociation phase D_{fd}. The dissociation rate coefficient k_d is given by

$$k_d = (0.00687 \pm 0.0021)[D_{fd}]^{8.133 \pm 1.3946} \tag{7.3c}$$

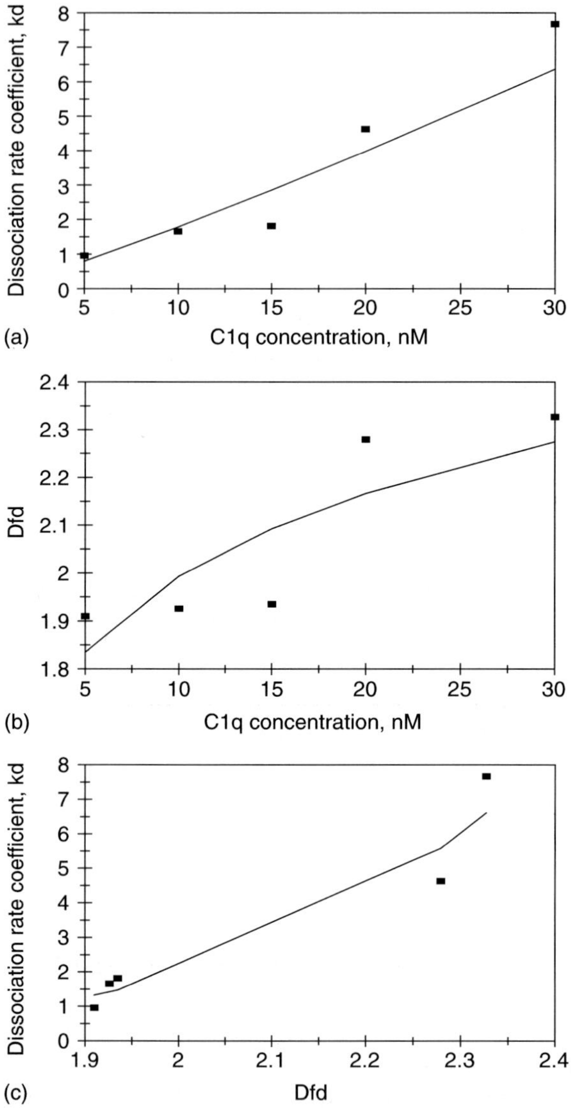

Figure 7.3 (a) Increase in the dissociation rate coefficient k_d with an increase in the C1q concentration (in nM) in solution, (b) increase in the dissociation rate coefficient k_d with an increase in the C1q concentration (in nM) in solution, (c) increase in the dissociation rate coefficient k_d with an increase in the fractal dimension in the dissociation phase D_{fd}.

The fit is good. There is some scatter in the data. The dissociation rate coefficient k_d is very sensitive to the degree of heterogeneity present in the dissociation phase D_{fd} as noted by the very high order exhibited (=8.133).

In the 5–30 nM C1q concentration range in solution, Figure 7.4a shows the increase in the affinity $K(=k/k_d)$ with an increase in the ratio of the fractal dimensions, D_f/D_{fd}.

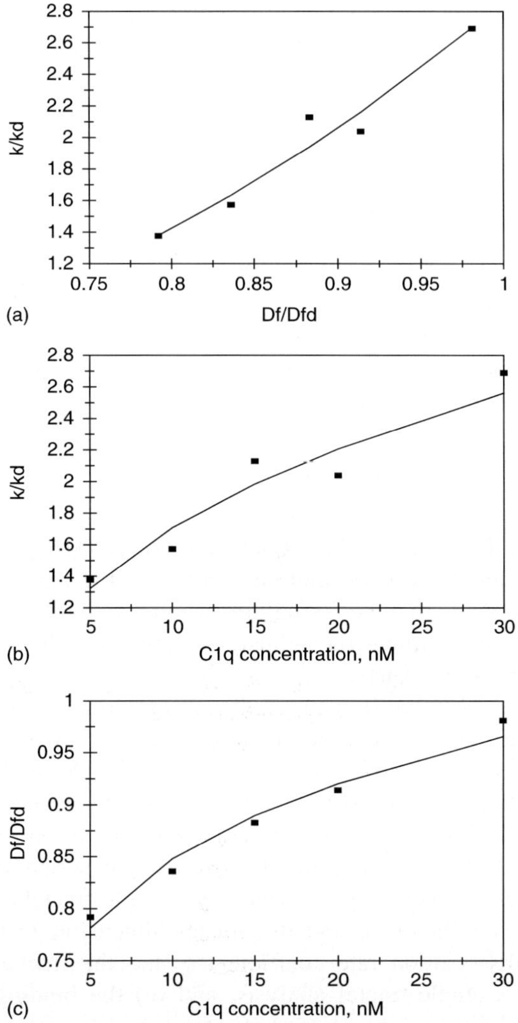

Figure 7.4 (a) Increase in the affinity $K(=k/k_d)$ with an increase in the ratio of fractal dimensions, in the binding and dissociation phase D_f/D_{fd}, (b) increase in the affinity $K(=k/k_d)$ with an increase in the C1q concentration (in nM) in solution, (c) increase in the fractal dimension ratio D_f/D_{fd} with an increase in the C1q concentration (in nM) in solution.

The affinity K is given by

$$K = k/k_d = (2.8804 \pm 0.1976)\,(D_f/D_{fd})^{3.129 \pm 0.4077} \tag{7.4a}$$

The fit is very good. The affinity K is very sensitive to the ratio of the fractal dimensions D_f/D_{fd} as noted by the greater than third-order dependence (=3.129) exhibited.

In the 5–30 nM C1q concentration range in solution, Figure 7.4b shows the increase in the affinity $K(=k/k_d)$ with an increase in the C1q concentration in solution. The affinity K is given by

$$K = k/k_d = (0.7310 \pm 0.0651)[\mathrm{C1q}]^{0.3687 \pm 0.06255} \tag{7.4b}$$

The fit is good. The affinity K is mildly sensitive to the C1q concentration in solution as noted by the low order of dependence (=0.3687) exhibited. The fractional order of dependence exhibited by the affinity on the C1q concentration in solution lends support to the fractal nature of the system.

In the 5–30 nM C1q concentration range in solution, Figure 7.4c shows the increase in the ratio of the fractal dimensions D_f/D_{fd} with an increase in the C1q concentration in solution. The ratio of fractional dimensions is given by

$$D_f/D_{fd} = (0.6461 \pm 0.0103)[\mathrm{C1q}]^{0.1181 \pm 0.01148} \tag{7.4c}$$

The fit is very good. The ratio of the fractal dimensions D_f/D_{fd} is only very slightly dependent on the C1q concentration in solution as noted by the very low order (=0.1181) of dependence exhibited. Note that the fractal dimension is based on a log scale, and even small changes in the fractal dimension indicate a significant change in the degree of heterogeneity on the sensor chip surface.

Blanquet-Grossard *et al.* (2005) also analyzed the binding and dissociation of 1–15 nM C1q concentration in solution in the presence of 10 μM $CuSO_4$ to mouse PrP protein immobilized on a sensor chip surface. Figures 7.5a–e show the binding of (1) 15, (2) 5, (3) 3, (4) 2, and (5) 1 nM C1q in solution in the presence of 10 μM $CuSO_4$ in solution to mouse PrP prion protein immobilized on a sensor chip surface. In these cases a dual-fractal analysis is required to adequately describe the binding kinetics. The dissociation kinetics is adequately described by a single-fractal analysis. The values of (a) the binding rate coefficient k and the fractal dimension D_f for a single-fractal analysis, and (b) the dissociation rate coefficient k_d and the fractal dimension in the dissociation phase for a single-fractal analysis, and (c) the binding rate coefficients, k_1 and k_2, and the fractal dimensions, D_{f1} and D_{f2}, for a dual-fractal analysis are given in Table 7.2a and b.

For the data presented in Table 7.2a and b, Figure 7.6 shows the increase in the binding rate coefficient k_2 with an increase in the fractal dimension D_{f2}. The binding rate coefficient k_2 is given by

$$k_2 = (4.718 \pm 3.481)(D_{f2})^{3.2725 \pm 0.3025} \tag{7.5a}$$

The fit is quite good. The binding rate coefficient k_2 is sensitive to the degree of heterogeneity on the sensor chip surface D_{f2} as noted by the greater than third-order dependence (=3.275) exhibited.

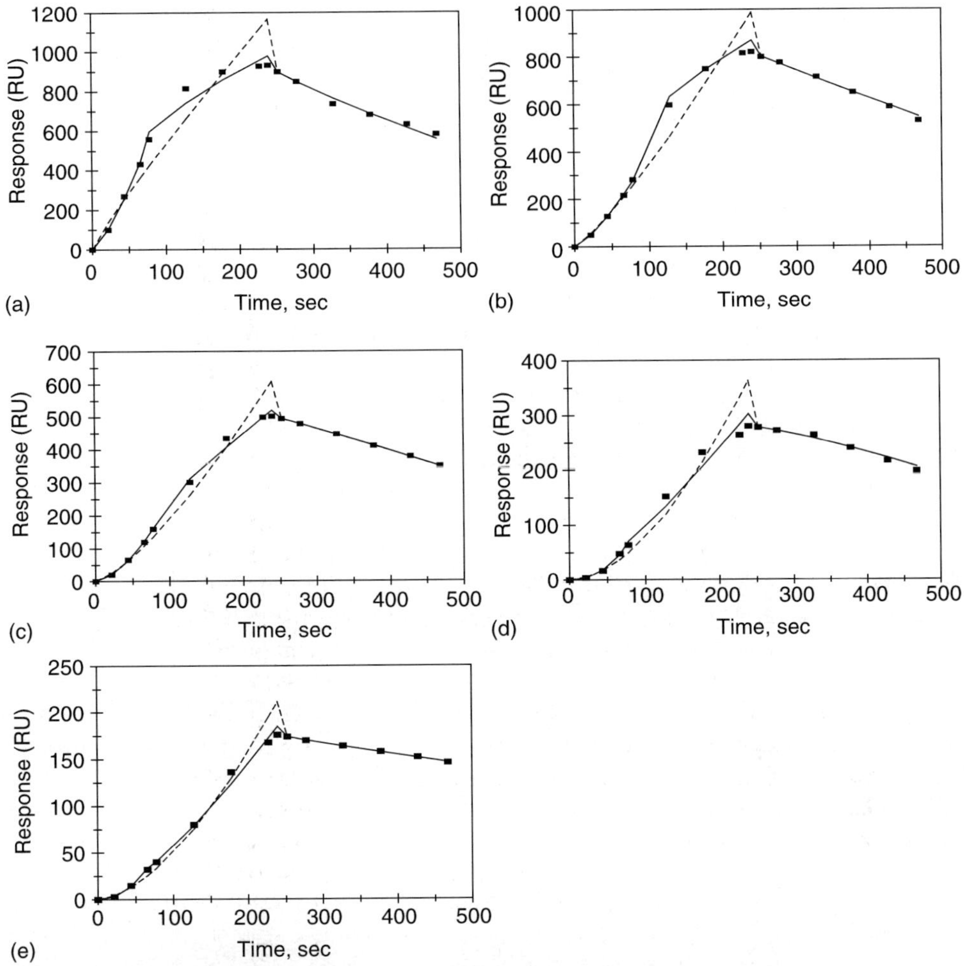

Figure 7.5 Binding and dissociation curves for different concentrations (in nM) of human C1q in solution in the presence of 10 μM $CuSO_4$ to PRP immobilized on a sensor chip surface (Blanquet-Grossard *et al.*, 2005): (a) 15, (b) 5, (c) 3, (d) 2, (e) 1.

For the data presented in Table 7.2a and b, Figure 7.7 shows the increase in the affinity $K_2(=k_2/k_d)$ with an increase in the ratio of fractal dimensions D_{f2}/D_{fd}. The affinity $K_2(=k_2/k_d)$ is given by

$$K_2(=k_2/k_d) = (6.8743 \pm 4.128)(D_{f2}/D_{fd})^{1.9738 \pm 0.2928} \tag{7.5b}$$

The fit is reasonable. There is scatter in the data. The affinity K_2 is quite sensitive to the ratio of the fractal dimensions, D_{f2}/D_{fd} as noted by the close to second-order dependence (=1.9738) exhibited.

Table 7.2

(a) Binding and dissociation rate coefficients and (b) fractal dimensions for the binding and dissociation phase for the interaction of different concentrations of C1q in the presence of 10 μM $CUSO_4$ in solution to mouse prion protein (PrP) immobilized on a sensor chip surface (Blanquet-Grossard *et al.*, 2005)

(a)

C1q concentration, nM	k	k_1	k_2	k_d
15	8.554 ± 2.262	1.6022 ± 0.0392	88.241 ± 7.239	3.8632 ± 0.3787
5	1.3939 ± 0.2348	0.7599 ± 0.0158	52.535 ± 1.828	1.6053 ± 0.1532
3	0.3955 ± 0.0714	0.1302 ± 0.0048	5.8799 ± 0.3397	0.4531 ± 0.0082
2	0.02246 ± 0.00660	0.003787 ± 0.000388	0.2517 ± 0.0350	0.0700 ± 0.0137
1	0.02590 ± 0.00620	0.003738 ± 0.000366	0.1136 ± 0.00661	0.1927 ± 0.0088

(b)

C1q concentration, nM	D_f	D_{f1}	D_{f2}	D_{fd}
15	1.2068 ± 0.2096	0.3146 ± 0.1438	2.122 ± 0.1694	1.3152 ± 0.0764
5	0.6048 ± 0.1390	0.2924 ± 0.0420	1.9896 ± 0.1410	1.1088 ± 0.0744
3	0.3224 ± 0.1482	0 + 0.752	1.3638 ± 0.2258	0.8540 ± 0.0146
2	1.8072 + 0.2232	1.8072 + 0.2484	0.4122 ± 0.2802	0.4308 ± 0.1454
1	1.9454 + 0.1918	1.9454 + 0.2382	0.3012 ± 0.0924	1.1490 ± 0.0364

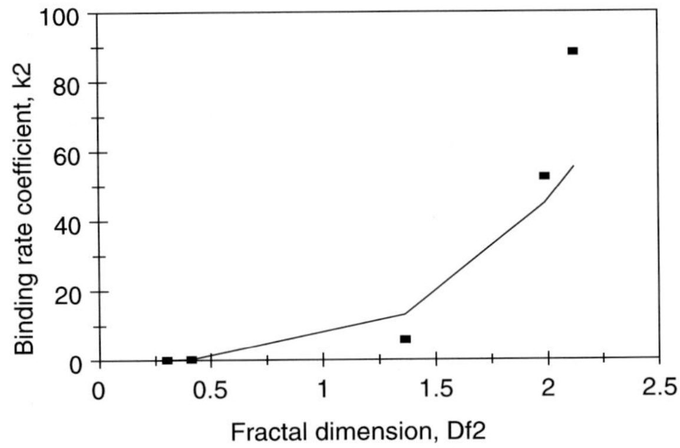

Figure 7.6 Increase in the binding rate coefficient k_2 with an increase in the fractal dimension in the binding phase D_{f2}.

Figures 7.8a and b show the binding and dissociation of C1q in solution in the (a) absence and (b) presence of EDTA in solution to recombinant murine PrP protein immobilized on a sensor chip surface. In this case, a single-fractal analysis is adequate to describe the binding and the dissociation kinetics. The values of (a) the binding rate coefficient k and the fractal dimension D_f for a single-fractal analysis, and (b) the dissociation rate coefficient k_d and the fractal dimension D_{fd} in the dissociation phase for a single-fractal analysis are given in Table 7.3.

It is of interest to note that when the two cases are compared, the fractal dimension in the binding phase D_f increases by 13% from 2.2418 to 2.5332, and the binding rate coefficient k decreases by 28.9% from a value of 57.077 to 40.557. The presence of EDTA in solution leads to an increase in the degree of heterogeneity on the sensor chip surface. Changes in the degree of heterogeneity on the sensor chip surface and in the binding rate

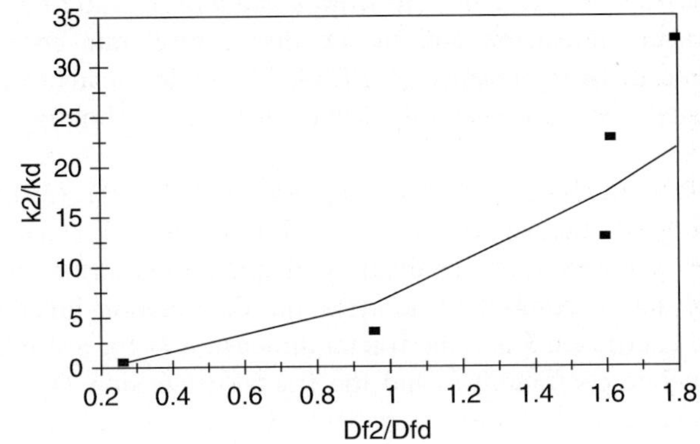

Figure 7.7 Increase in the affinity K_2 $(=k_2/k_d)$ with an increase in the ratio of fractal dimensions D_{f2}/D_{fd}.

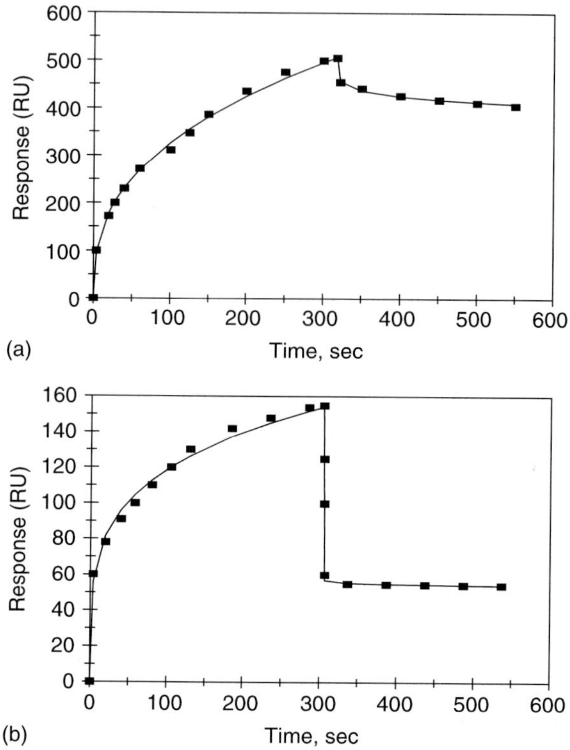

Figure 7.8 Binding and dissociation of C1q in solution in the (a) absence and in the (b) presence of EDTA in solution to recombinant murine PrP protein immobilized on a sensor chip surface.

coefficient are in opposite directions. Similarly, the fractal dimension in the dissociation phase D_{fd} increases by 1.96% from a value of 2.664 to 2.7186, and the dissociation rate coefficient k_d increases by a factor of 1.36 from a value of 38.786 to 52.674. In this case, changes in the fractal dimension and in the dissociation rate coefficient are in the same direction. Note that the presence of EDTA in solution leads to an increase in the degree of heterogeneity on the sensor chip surface for both the binding phase and the dissociation phase.

Figures 7.9a–d show the binding and dissociation of (1) 2, (2)1, (3)0.5, and (4) 0.2 μM C1q domain in solution to mouse prion protein PrP immobilized on a sensor chip surface. In cases (1) and (2), a single-fractal analysis is adequate to describe the binding kinetics. A dual-fractal analysis is required to describe the dissociation kinetics. The values of (a) the binding rate coefficient k and the fractal dimension D_f for a single-fractal analysis, (b) the dissociation rate coefficient k_d and the fractal dimension D_{fd} for a single-fractal analysis, and (c) the dissociation rate coefficients, k_{d1} and k_{d2}, and the fractal dimensions, D_{fd1} and D_{fd2}, for a dual-fractal analysis are given in Table 7.4.

In cases (3) and (4), a single-fractal analysis is adequate to describe the binding and dissociation kinetics. The values of (a) the binding rate coefficient k and the fractal dimension

Table 7.3

Binding and dissociation rate coefficients and fractal dimensions for the binding and dissociation phase for the interaction of C1q in solution to mouse prion protein (PrP) immobilized on a sensor chip surface in the absence and presence of EDTA (Blanquet-Grossard *et al.*, 2005).

Analyte in Solution	k	k_d	D_f	D_{fd}
C1q	57.077±1.519	38.786±2.003	2.2418±0.0120	2.6664±0.0296
C1q+EDTA	40.557±1.681	52.674±17.781	2.5332±0.0196	2.7186±0.0872

D_f for a single-fractal analysis, and (b) the dissociation rate coefficient k_d and the fractal dimension D_{fd} for a single-fractal analysis are given in Table 7.4. Note the change in the dissociation mechanism as the concentration of the C1q domain in solution decreases from 1 to 0.5 nM in solution.

In the 0.2–2 μM C1q domain concentration in solution, Table 7.4 and Figure 7.10a show the increase in the binding rate coefficient k with an increase in the C1q concentration in solution. In this concentration range, the binding rate coefficient k is given by

$$k = (1.9212 \pm 0.0295)[\mathrm{C1q}]^{1.3010 \pm 0.0089} \tag{7.6a}$$

The fit is very good. The availability of more data points would more firmly establish the above relationship. The binding rate coefficient k exhibits an order of dependence between 1 and one and one-half (=1.301) on the C1q concentration in solution. The non-integral order of dependence exhibited lends support to the fractal nature of the system.

In the 0.2–2 μM C1q domain concentration in solution, Table 7.4 and Figure 7.10b shows the increase in the binding rate coefficient k with an increase in the fractal dimension D_f. In this concentration range, the binding rate coefficient k is given by

$$k = (0.00247 \pm 0.00009)[D_f]^{10.70 \pm 0.1739} \tag{7.6b}$$

The fit is very good. The availability of more data points would more firmly establish the above relationship. The binding rate coefficient k is very sensitive to the fractal dimension D_f or the degree of heterogeneity on the sensor chip as noted by the greater than tenth order of dependence (=10.7) exhibited by k on D_f.

In the 0.2–2 μM C1q domain concentration in solution, Table 7.4 and Figure 7.10c shows the increase in the fractal dimension D_f with an increase in the C1q concentration in solution. In this concentration range, the fractal dimension D_f is given by

$$D_f = (1.8627 \pm 0.0089)[\mathrm{C1q\ domain}]^{0.1215 \pm 0.0028} \tag{7.6c}$$

The fit is very good. The availability of more data points would more firmly establish the above relationship. The fractal dimension D_f is only mildly sensitive to the C1q domain concentration in solution as noted by the very low order of dependence (=0.1215) exhibited.

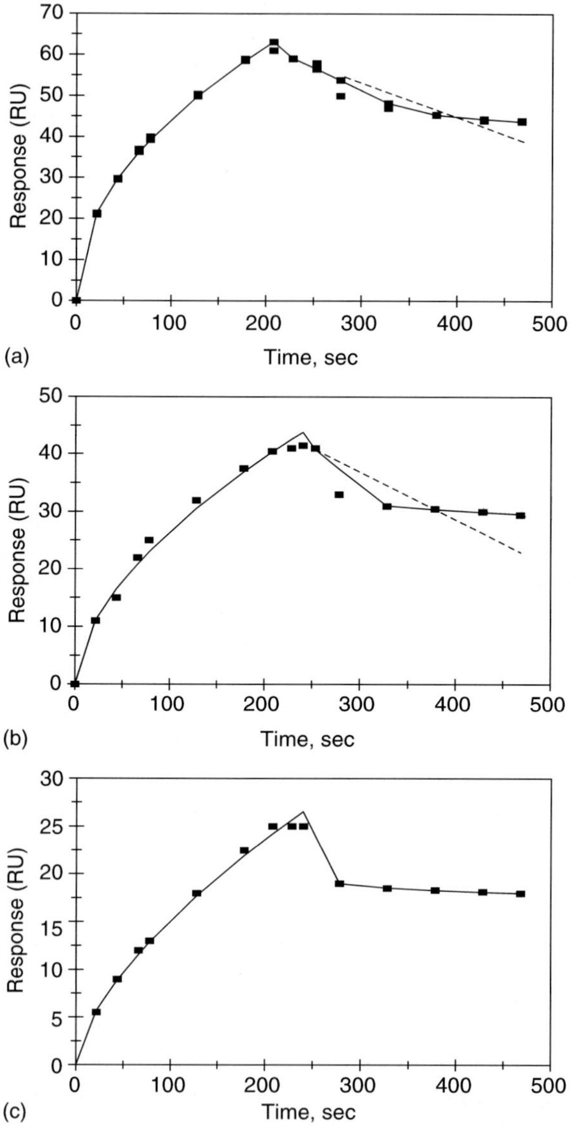

Figure 7.9 Binding and dissociation curves for different concentrations (in μM) of C1q domain in solution to mouse prion protein PrP immobilized on a sensor chip surface (Blanquet-Grossard *et al.*, 2005): (a) 2, (b) 1, (c) 0.5, (d) 0.2. When only a solid line is used then a single-fractal analysis applies. When both a dashed (----) and a solid (_____) line are used, then the dashed line is for a single-fractal analysis, and the solid line is for a dual-fractal analysis. In this case, the solid line provides the better fit.

Table 7.4

(a) Binding and dissociation rate coefficients and (b) fractal dimensions for the binding and dissociation phase for the interaction of different concentrations of C1q domain in solution to mouse prion protein (PrP) immobilized on a sensor chip surface (Blanquet-Grossard *et al.*, 2005)

(a)

C1q concentration, nM	k	k_d	k_{d1}	k_{d2}
2	4.7860 ± 0.1065	0.1420 ± 0.0561	0.0775 ± 0.0335	4.8165 ± 0.1439
1	1.8872 ± 0.1192	0.0840 + 0.0864	0.01035 + 0.01554	5.6616 ± 0.0638
0.5	0.7828 ± 0.0289	4.3755 ± 0.2226	NA	NA
0.2	0.2374 ± 0.0188	2.5445 ± 0.2291	NA	NA

(b)

C1q concentration, nM	D_f	D_{fd}	D_{fd1}	D_{fd2}
2	2.0336 ± 0.02248	1.1900 ± 0.2922	0.8686 ± 0.4268	2.5412 ± 0.1944
1	1.8524 ± 0.0519	1.0222 ± 0.5778	0 + 1.3524	2.7268 ± 0.02174
0.5	1.7140 ± 0.0307	2.8325 ± 0.0070	NA	NA
0.2	1.5336 ± 0.0646	2.8180 ± 0.0792	NA	NA

It should be borne in mind that the fractal dimension D_f is based on a log scale, and even small changes in the fractal dimension value indicate significant changes in the degree of heterogeneity on the sensor chip surface.

Figures 7.11a–d show the binding and dissociation of (1) 2, (2) 1, (3) 0.5, and (4) 0.2 μM C1q domain in solution in the presence of 10 μM $CuSO_4$ to mouse prion protein immobilized on a sensor chip surface. In cases (1) and (2), single-fractal analysis is adequate to describe the binding and dissociation kinetics. The values of (a) the binding rate coefficient k and the fractal dimension D_f for a single-fractal analysis, and (b) the dissociation rate coefficient k_d and the fractal dimension D_{fd} for a single-fractal analysis are given in Table 7.5.

In cases (3) and (4), dual-fractal analysis is required to describe the binding kinetics. A single-fractal analysis is still adequate to describe the dissociation kinetics. The values of (a) the binding rate coefficient k and the fractal dimension D_f for a single-fractal analysis, (b) the binding rate coefficients, k_1 and k_2, and the fractal dimensions, D_{f1} and D_{f2}, for a dual-fractal analysis, and (c) the dissociation rate coefficient k_d and the fractal dimension D_{fd} for a single-fractal analysis are given in Table 7.5.

It is of interest to note that in the absence of 10 μM $CuSO_4$ in solution at the lower C1q domain concentration (0.2 and 0.5 μM), the dissociation kinetics is described by a single-fractal analysis. At the higher C1q domain concentration in solution (1.0 and 2.0 μM) the dissociation kinetics is described by a dual-fractal analysis. For all of four C1q domain concentrations the binding kinetics is described by a single-fractal analysis.

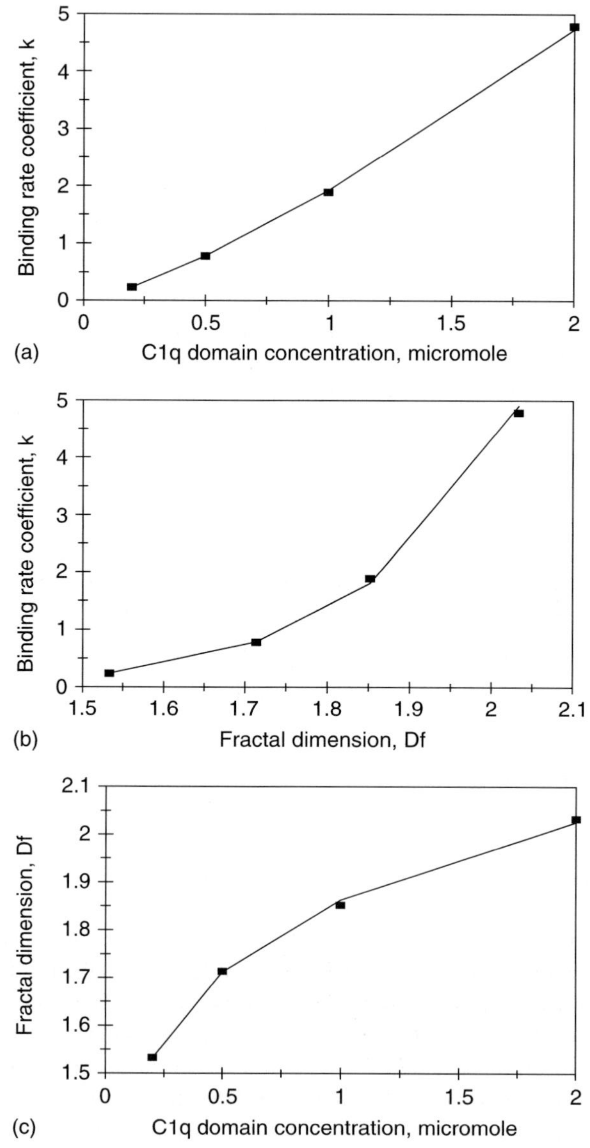

Figure 7.10 (a) Increase in the binding rate coefficient *k* with an increase in the C1q concentration in solution, (b) increase in the binding rate coefficient *k* with an increase in the fractal dimension D_f.

When 10 μM $CuSO_4$ is present in solution the situation is reversed as far as the binding and dissociation kinetics are concerned. Now, for all the C1q domain concentrations in solution (0.2, 0.5, 1.0, and 2.0 μM) the dissociation kinetics is described by a single-fractal analysis. At the higher C1q domain concentrations (1.0 and 2.0 μM) the binding kinetics is described by a single-fractal analysis, and at the lower C1q domain concentrations

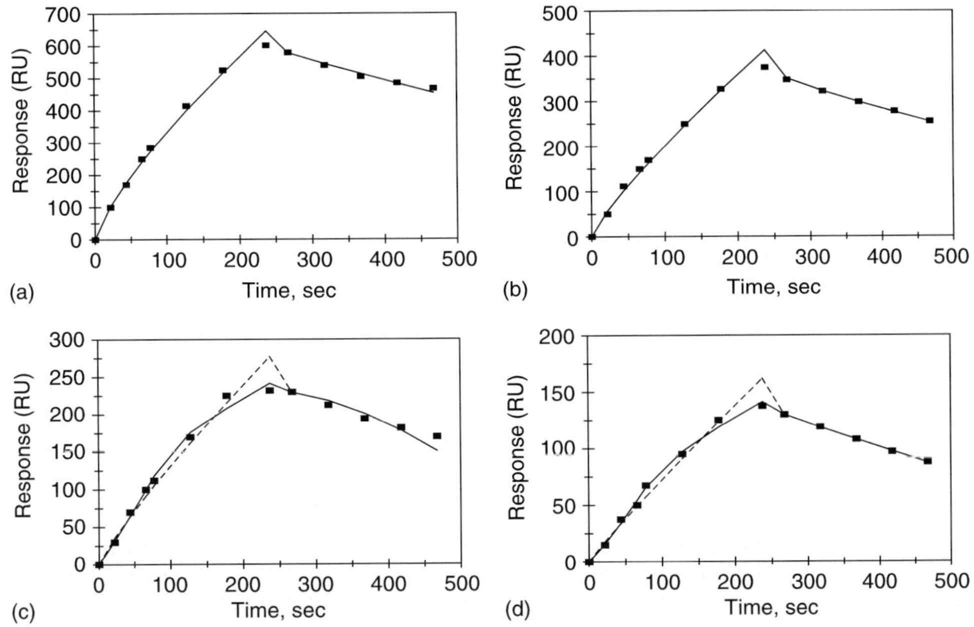

Figure 7.11 Binding and dissociation curves for different concentrations (in μM) of C1q domain in the presence of 10 μM $CuSO_4$ in solution to mouse prion protein PrP immobilized on a sensor chip surface (Blanquet-Grossard *et al.*, 2005): (a) 2, (b) 1, (c) 0.5, (d) 0.2. When only a solid line is used then a single-fractal analysis applies. When both a dashed (-------) and a solid (____) line are used, then the dashed line is for a single-fractal analysis, and the solid line is for a dual-fractal analysis. In this case, the solid line provides the better fit.

(0.5 and 0.2 μM) the binding kinetics is described by a dual-fractal analysis. No explanation is offered at present to describe the behavior.

In the 0.2 to 2.0 μM C1q domain concentration in solution and in the presence of 10 μM $CuSO_4$, Figure 7.12 and Table 7.5 show the increase in the dissociation rate coefficient k_d with an increase in the fractal dimension in the dissociation phase D_{fd}. In this concentration range, the dissociation rate coefficient k_d is given by

$$k_d = (0.6753 + 0.7966) D_{fd}^{0.9739 \pm 0.1863} \tag{7.7}$$

There is quite a bit of scatter in the fit of the data. This is also indicated in the error in the value of the dissociation rate coefficient. The error is large, thus only the positive error is presented, since the dissociation rate coefficient cannot be negative. The availability of more data points would perhaps help strengthen the relationship. The dissociation rate coefficient k_d exhibits less than a first-order dependence (=0.9739) on the fractal dimension D_{fd} or the degree of heterogeneity on the sensor chip surface.

Figures 7.13a–c show the binding of (1) 1.5 μM IgG, (2) 0.1 μM PTX3, and (3) 1 μM recombinant murine PrP in solution to C1q (12,000 RU) immobilized on the sensor chip

Table 7.5

(a) Binding and dissociation rate coefficients and (b) fractal dimensions for the binding and dissociation phase for the interaction of different concentrations of C1q globular domain in solution in the presence of 10 μM $CuSO_4$ to mouse prion protein (PrP) immobilized on a sensor chip surface (Blanquet-Grossard *et al.*, 2005)

(a)

C1q concentration, nM	k	k_1	k_2	k_d
2	9.4747 ± 0.4724	NA	NA	1.0984 ± 0.1157
1	4.3197 ± 0.4095	NA	NA	1.6619 ± 0.0712
0.5	2.4170 ± 0.3283	1.2032 ± 0.0912	14.8993 ± 1.4657	1.00786 ± 0.0030
0.2	1.0090 ± 0.1482	0.4478 ± 0.0434	5.2594 ± 0.3449	0.3090 ± 0.0033

(b)

C1q concentration, nM	D_f	D_{f1}	D_{f2}	D_{fd}
2	1.4568 ± 0.0480	NA	NA	1.1974 ± 0.1244
1	1.3334 ± 0.0894	NA	NA	1.4244 ± 0.0521
0.5	1.2674 ± 0.1258	0.8948 ± 0.1498	1.9818 ± 0.4276	0 + 0.402
0.2	1.1442 ± 0.1354	0.7086 ± 0.1897	1.7972 ± 0.2894	1.1272 ± 0.0134

surface. Bottazzi *et al.* (1997) indicate that both IgG and long pentraxin (PTX3) are known C1q ligands. A single-fractal analysis is adequate to describe the binding kinetics. The values of the binding rate coefficient k and the fractal dimension D_f for a single-fractal analysis are given in Table 7.6.

For the data presented in Table 7.6 for rabbit IgG, human PTX3, and recombinant murine PrP in solution, Figure 7.14 shows the increase in the binding rate coefficient k

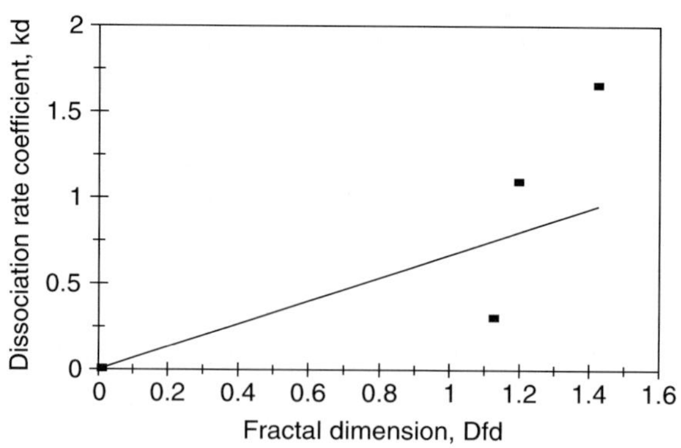

Figure 7.12 Increase in the dissociation rate coefficient k_d with an increase in the fractal dimension in the dissociation phase D_{fd}.

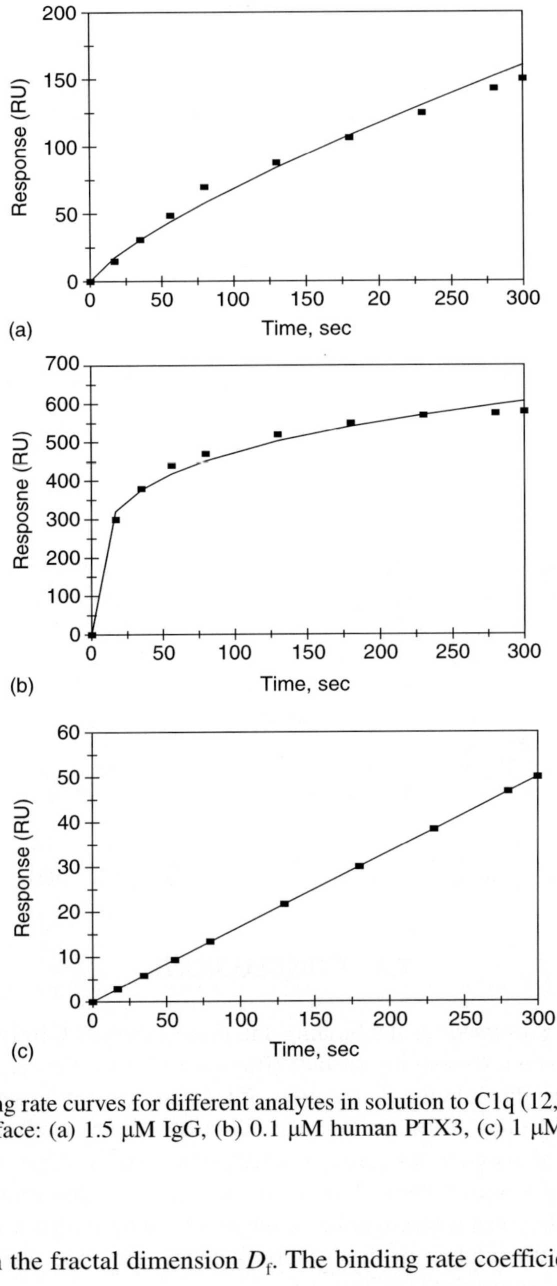

Figure 7.13 Binding rate curves for different analytes in solution to C1q (12,000 RU) immobilized on a sensor chip surface: (a) 1.5 μM IgG, (b) 0.1 μM human PTX3, (c) 1 μM recombinant murine PrP.

with an increase in the fractal dimension D_f. The binding rate coefficient k is given by

$$k = (0.1462 \pm 0.0430)D_f^{7.4360 \pm 0.3871} \quad (7.8)$$

The fit is quite good. Only three data points were available. The availability of more data points would lead to a more reliable fit. For the data presented, the binding rate coefficient k

Table 7.6

Binding rate coefficients and fractal dimensions for the interaction between immobilized C1q (12000 RU) and soluble PrP. Three soluble proteins are used: 1.5 μM rabbit IgG, 0.1 μM human PTX3, and 1.0 μM recombinant murine PrP (Blanquet-Grossard *et al.*, 2005)

Analyte in solution	D_f	k
Rabbit IgG	1.4614 ± 0.7784	1.9914 ± 0.2318
Human PTX3	2.5770 ± 0.3092	171.21 ± 7.651
Recombinant murine PTX3	0.1674 ± 0.0003	1.0014 ± 0.0011

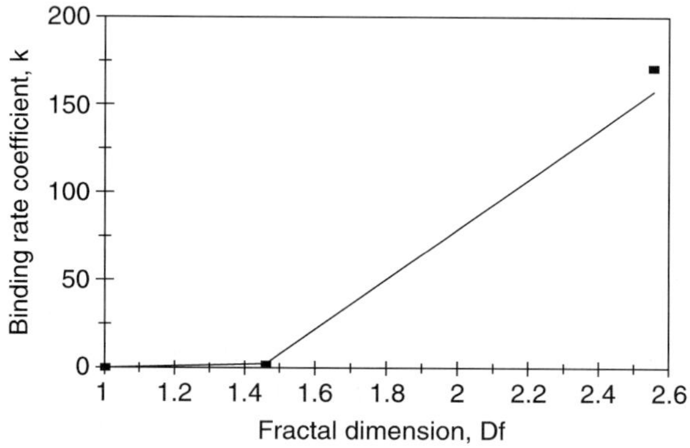

Figure 7.14 Increase in the binding rate coefficient k with an increase in the fractal dimension D_f.

is extremely sensitive to the fractal dimension D_f or the degree of heterogeneity present on the sensor chip surface as noted by the very high order between 7 and 8 (=7.436) exhibited.

7.4 CONCLUSIONS

A fractal analysis is presented for the binding and dissociation of C1q in solution to the PrP protein immobilized on a sensor chip surface (Blanquet-Grossard *et al.*, 2005). The analysis provides physical insights into the interactions of C1q and the PrP protein occurring on the sensor chip surface. Values of the binding and dissociation rate coefficient k and k_d respectively, and the state of disorder or degree of heterogeneity (fractal dimension) on the sensor chip surface are provided for both the binding phase D_f and the dissociation phase D_{fd}. Both types of examples are given wherein either a single- or a dual-fractal analysis is required to adequately describe the binding or dissociation kinetics. This was done by the regression analysis provided by Sigmaplot (1993).

In accordance with the pre-factor analysis for fractal aggregates (Sorenson and Roberts, 1997), quantitative (predictive) expressions are developed for (a) the binding rate coefficient k as a function of the C1q concentration in solution, (b) the fractal dimension D_{fd} as a function of the C1q concentration in solution, (c) the dissociation rate coefficient k_d as a function of the fractal dimension D_{fd}, (d) the affinity $K=k/k_d$ as a function of the ratio of fractal dimensions D_f/D_{fd} and also as a function of the C1q concentration in solution, (e) the binding rate coefficient k_2 as a function of the fractal dimension D_{f2}, and (f) the affinity K_2 ($=k_2/k_d$) as a function of the ratio of fractal dimensions D_{f2}/D_{fd}. In each of the examples analyzed it was assumed that diffusional limitations are present. This is a conservative approach since it is difficult to judge if the binding kinetics have been obtained under diffusion-free conditions for the data available in the literature. There are indications, and the manufacturers (Biacore, Uppsala, Sweden) state this that if the SPR biosensor is used properly, then the binding (and the dissociation) occurs under diffusion-free conditions.

The fractal dimension D_f is not a classical independent variable such as analyte (antigen, antibody or other biological molecule) concentration in solution. Nevertheless, the expressions obtained for the binding rate coefficient for a single- and a dual-fractal analysis as a function of the fractal dimension indicate the high sensitivity of the binding and dissociation rate coefficient on the fractal dimension on the SPR biosensor surface. For example, eq. 7.3c indicates that the dissociation rate coefficient k_d exhibits an 8.133 order of dependence on D_{fd}, and thc binding rate coefficient k_2 exhibits a 3.2725 order of dependence on D_{f2}. Please see eq. 7.5a. The affinity K_2 ($=k_2/k_d$) exhibits an order of dependence close to 2 (=1.9738) on the ratio of the fractal dimensions D_{f2}/D_{fd}. Please see eq. 7.5b.

The fractal analysis provides unique physical insights (namely the degree of heterogeneity present on the surface and relating it to the rate coefficients) into the interactions between C1q and the PrP protein. As indicated earlier C1q is one of the 19 proteins that exhibit an increase in synthesis in scrapie-infected brain tissues (Dandoy-Dron *et al.*, 1998; Riemer *et al.*, 2000). The quantitative expressions developed for the C1q–PrP interactions should assist in the better control of the SPR biosensor performance parameters such as stability, sensitivity, and response time. More importantly, they should provide much needed insights into the types of interactions. Rate coefficient values are provided in the original paper by Blanquet-Grossard *et al.* (2005). These authors have used the Biacore X (Biacore, Uppsala, Sweden), and have provided values for the rate coefficient, k_{on} and k_{off}. However, it should be pointed out, once again, that the software used in these Biacore SPR biosensors to determine the rate coefficients does not take into account either the diffusional limitations or the degree of heterogeneity that exists on the biosensor chip surface. It is appropriate to indicate that one may consider the kinetic analysis presented in this chapter as an alternate analysis.

Our analysis does indicate the importance of heterogeneity on the sensing surface, and its effect on binding and dissociation rate coefficients. If one may take the liberty to extend this to cellular surfaces, where presumably these reactions are taking place, then one has access to a possible avenue by which one may control these reactions in desired directions. If this is true, then surely, this is worth exploring further. Any effort, no matter how small that assists in providing better insights into these debilitating and neurodegenerative disorders such as Alzheimer's is definitely worth the effort.

REFERENCES

Aguzzi, A., and Polymenidou, M. (2004). Mammalian prion biology: one century of evolving concepts, *Cell*, **116**, 313–327.

Arispe, N., Rojas, E., and Pollard, H.B. (1993). Alzheimer disease amyloid beta protein forms calcium channels in bilayer membranes: blockade by tromethamine and aluminum. *Proceedings of the National Academy of Sciences*, **90**, 567–571.

Bhatia, R., Lin, H., and Lal, R. (2000). Fresh and globular amyloid beta protein (1–42) induces rapid cellular degeneration: evidence for a beta P channel-mediated toxicity. *FASEB J*, **14**, 1233–1243.

Blanquet-Grossard, F., Thielens, N.M., Vendrely, C., Jamin, M., and Arlaud, G.J. (2005). Complement Protein C1q recognizes conformationally modified form of prion protein, *Biochemistry*, **44**, 4349–4356.

Bottazzi, B., Vouret-Craviari, V., Bastone, A., De Gioia, L., Matteucci, Peri, G., Spreafico, F., Pausa, M., D'Ettorre, C., Gianazza, E., Tagliabue, A., Salmona, M., Tedesco, F., Introna, M., and Mantovani, A. (1997). Multimer formation and ligand recognition by the long pentraxin PTX3. Similarities and differences with the short pentaxins C-reactive protein and serum amyloid P components. *Journal of Biological Chemistry*, **272**, 32817–32823.

Carrell, R.W. (2004). Prion dormancy and disease. *Science*, **36**(3), 1692–1693.

Corel Quattro Pro 8.0 (1997). Corel Corporation Limited, Ottawa, Canada.

Dandoy-Dron, F., Guillo, F., Benboudjema, L., Deslys, J.P., Lasmezas, C., Dormont, D., Tovey, M.G., and Dron, M. (1998). Gene expression in scrapie-responsive gene and the identification of increased levels of seven other mRNA transcripts. *Journal of Biological Chemistry*, **273**, 7691–7697.

Dobson, C.M. (2003). Protein folding and misfolding. *Nature*, **426**, 884–884.

Dobson, C.M. (2005). Letter response to "Amyloidosis and protein folding." *Science*, **307**, 43–44.

Hirakura, Y., and Kagan, B.L. (2001). *Amyloid*, **8**, 94.

Hirakura, Y., Carreras, I., Sipe, J.D., and Kagan, B.L. (2002). *Amyloid*, **9**, 13.

Ishii, T., Haga, S., Yagishita, S., and Tateishi, J. (1984). The presence of complements in amyloid plaques of Creutzfeldt-Jakob disease and Gerstmann–Straussler–Scheinker disease. *Applied Pathology*, **2**, 370–379.

Kagan, B.L. (2005). Amyloidosis and protein folding. *Science*, 7 January, 42–43.

Kovacs, G.G., Gasque, P., Strobel, T., Lindeck-Pozza, E., Strohschneider, M., Ironside, J.W., Budka, H., and Guentchev, M. (2004). Complement activation in human prion disease. *Neurobiological Disorders*, **15**, 21–28.

Lin, M.C., Mirzabekov, T., and Kagan, B.L. (1997). Channel formation by a neurotoxic prion protein fragment. *Journal of Biological Chemistry*, **272**, 44–47.

Marx, J. (2005). Neurodegeneration: Huntington's research points to possible new therapies. *Science*, 7 October, 43–45.

Mirzabekov, T., Lin, M., and Kagan, B.L. (1996). Pore formation by the cytotoxic islet amyloid peptide amylin. *Journal of Biological Chemistry*, **271**, 1988–1992.

Ramakrishnan, A., and Sadana, A. (2001). A single-fractal analysis of cellular analyte-receptor binding kinetics utilizing biosensors. *Biosystems*, **59**(1), 35–45.

Riemer, C., Queck, I., Simon, D., Kurth, R., and Baier, M. (2000). Identification of upregulated genes in scrapie-infected brain tissue. *Journal of Virology*, **74**, 10245–10248.

Sadana, A. (2001). A fractal analysis approach for the evaluation of hybridization kinetics in biosensors. *Journal of Colloid and Interface Science*, **234**, 9–18.

Sigmaplot, (1993). Scientific Graphing Software, Users Manual, Jandel Scientific, San Rafael, California, USA.

Sorenson, C.M., and Roberts, G.C. (1997). The prefactor of fractal aggregates. *Journal of Colloid and Interface Science*, **186**, 447–452.

Thomson, S.J., and Webb, G, *Heterogeneous Catalysis*, Wiley, p. 23, New York, 1968.

– 8 –

Detection of Environmental Contaminants by Biosensors

8.1 INTRODUCTION

The detection of environmental contaminants in water and in air is an important area of biosensor application. Paitan *et al.* (2004) emphasize that the pollution emitted by agricultural and chemical industries has been steadily increasing, which creates a strong demand for sensors and biosensors. Different types of pollutants require different types of sensors and biosensors. Kohler *et al.* (2000) indicate that organic compounds are one of the major contributors to environmental hazards with some of the more toxic compounds, which include polyaromatic hydrocarbons (PAHs). These organic compounds also include halogenated hydrocarbons, herbicides, pesticides, mutagens, and carcinogens. Paitan *et al.* (2004) also indicate that organic pollutants comprise an important environmental problem.

Skladal *et al.* (2002) indicate that phenolic compounds are a widespread pollution problem. These compounds may be detected by spectrometric and chromatographic methods (Chiriswell *et al.*, 1975; Corcia *et al.*, 1993). The partially oxidized products of phenol such as catechols and hydroquinones are more difficult to detect than phenol itself (Sadana, 1972). Skladal *et al.* (2002) also indicate that tyrosinase is the biorecognition element most frequently used for the detection of phenols. These authors developed an amperometric biosensor for the detection of phenol using chemically modified electrodes. Three out of eight strains of *Pseudomonas* that were adsorbed on the mediator-modified screen-printed electrodes (SPEs) were able to detect phenol using the mediators: ferrocene, duroquinone, and dimethylferrocene.

Swearingen *et al.* (2005) indicate that lead is a persistent environmental contaminant. It is particularly deleterious for children due to their small size and developing bodies. The Centers for Disease Control (CDC) (1991) indicate that high levels of lead in blood ($>18\ \mu M$) can cause coma and death. This report indicates that lead concentration as low as 500 nM in blood decreased intelligence, led to behavioral problems, and impaired growth. Swearingen *et al.* (2005) recently developed a catalytic DNA molecular beacon on Au for Pb(II) detection.

Swearingen *et al.* (2005) indicate that until now the laboratory determination of Pb was done by atomic adsorption spectrometry and by ICP (Ma *et al.*, 1994; Elfering *et al.*, 1998). The recent emphasis on nanotechnology has also led to the development of nanoparticle-based sensors for the detection of Pb (Liu and Lu, 2003, 2004). These are colorimetric methods with detection limits as low as ~100 nM. Another heavy metal in the aqueous environment that is extremely hazardous to human health and to the ecosystem is copper. Forstner and Wittman (1981) indicate that copper is essential for life, but at high doses it is extremely toxic to algae, fungi, bacteria, and viruses. A European Commission Water Quality Directive (1998) indicates that copper concentration in drinking water should not exceed 2 mg/l (31 μM). Mayr *et al.* (2002) developed a dual-lifetime referenced optical sensor membrane for the determination of copper(II) ions.

Wygladacz *et al.* (2005) very recently indicate that silver is a good measure of aquatic environmental pollution. These authors point out that though silver can be used as a disinfectant in water due to its anti-bacterial properties, an excess of silver is toxic to fish and microorganisms. An EPA (Environmental Protection Agency) Drinking Water Criteria Document for Silver (1989) indicates that levels as low as 0.17 μg/l is toxic to fish and microorganisms. Wygladacz *et al.* (2005) indicate that the flame atomic absorption spectrometry (Bermejo-Barrera *et al.*, 1996; Dadfornia *et al.*, 2004) and plasma atomic emission spectrometry (Wang *et al.*, 2003; Yang *et al.*, 2002) have been used to detect trace amounts of silver in different samples. Wygladacz *et al.* (2005) very recently developed a microsensor array for the detection of trace amounts of silver ions based on fluorescent bulk optode microspheres. They were able to detect silver in parts per trillion. Shephard and Toomey (2005) recently used a quartz crystal microbalance using peptide-modified hydrogel layers to detect heavy metal ions.

Hanumegowda and White (2005) recently used a microsphere optical ring resonator sensor to detect mercuric ions in aqueous solution. Lippard and Nolan (2003) developed a fluorescent sensor for the detection of Hg(II) in water. These authors add that their sensor is highly selective for Hg(II) in the presence of contaminants found in aquatic streams. A sulfur-containing moiety is incorporated in the ligand (receptor) framework since Hg(II) has a high affinity for sulfur. There is a fluorescent enhancement as the sulfur binds to the Hg(II) as a 1:1 complex. These authors further add that bacteria convert the Hg(II) to methylmercury, which accumulates in the food chain. This methylmercury particularly affects the large edible fish. Analytical Currents (2004) indicates that the allowable limit for Hg(II) in drinking water is 2 ppb, which is the level set by the United States Environmental Protection Agency (US EPA).

Richardson and Ternes (2005) very recently reviewed the environmental contaminants and the current issues involved in water pollution. These authors emphasize that pollution due to herbicides and pesticides is an area that is most frequently studied and analyzed. They point out the emphasis on degradation products of the original (parent) pesticide. Quite often, the concentration of these degradation products (very often hydrolysis products) is greater than those of the initial parent pesticide itself. Battaglin *et al.* (2003) indicate that there is evidence for this.

Organohalides or halogenated hydrocarbons are frequently used as biocides for agricultural purposes or for domestic purposes (Wiyaratn *et al.*, 2005). These pollutants are a significant health hazard, and have been involved in cancer and deformity even at low levels. Peter *et al.* (1996, 1997) and Hutter *et al.* (1995) indicate that there is a critical need for the development of inexpensive analytical equipment for the rapid and sensitive detection of organohalides. This is true, especially, since they have been included in the list of toxic pollutants of the US EPA. Wiyaratn *et al.* (2005) recently developed a Pt-Zn porphyrin nanocomposite biosensor for the electrochemical detection of organohalides. Their modified glassy carbon electrode was used to detect carbon tetrachloride, chloroform, pentachlorophenol, chlorobenzene, and hexachlorobenzene as test models.

Gonzalez-Martinez *et al.* (2005) recently developed a glyphosate immunosensor for water and soil analysis applications. The sensor developed by these authors has a detection limit (they claim) of 0.021 μg/l, and their biosensor takes 25 min. These authors indicate that the steady increase in the use of glyphosate (*N*-phosphonomethyl) glycine) has made it the most frequently used herbicide in the world. Williams *et al.* (2000) indicated that glyphosate may be considered toxicologically harmless. However, Marc *et al.* (2004) recently indicated that it affects the cell-cycle regulation. Gonzalez-Martinez *et al.* (2005) indicate that the molecule is difficult to detect since it is a very polar molecule, and its solubility in water is high. Its solubility in organic solvents is low. In the United States, the allowable limit of glyphosate is 0.70 mg/l (Cox, 1998), whereas in the European Union this limit is 0.1 μg/l (EEC Council Directive, 1998). Matvienko and Mandelis (2005) recently developed an ultra-high thermal wave sensor to detect water–ethanol mixtures.

Okuyama *et al.* (2004) indicate that polychlorinated dibenzo-*p*-dioxins (PCDDs) and dibenzofurans (PCDFs) are pollutants that exhibit tertogenic, immunotoxic, and carcinogenic effects. These authors emphasize that it is of particular importance to monitor the trace amounts of these chemicals since they may contaminate human breast milk for newborn children. They developed an enzyme-linked immunosorbent assay (ELISA) using a novel monoclonal anti-dioxin antibody to detect toxic dioxin congeners in milk. Okuyama *et al.* (2004) were able to establish a hybridoma clone that secretes an antibody D9-36 group. This D9-36 group, which these authors claim, specifically recognizes the major toxic congeners: 2,3,7, 8-tetrachlorodibenzo-*p*-dioxin (2,3,7,8-TCDD); 1,2,3,7,8-pentachlorodibenzo-*p*-dioxin; and 2,3,4,7,8-pentachlorodibenzofuran.

Capitan-Vallvey *et al.* (2005) recently developed a receptor-based optical sensor for nitrate detection in aqueous solution such as sewer, well, and spring waters. These authors indicate that the nitrate-sensitive element is a bicyclic cyclophane receptor. This is next to a pH-sensitive lipophilic dye immobilized on a plasticized polymeric membrane. These authors further add that their sensor response time is around 5 min, and the sensor dynamic range is from 26 μM to 63 mM. A United Nations Environment Program (UNEP) Millenium (1999) report indicates that the common source of nitrates in water is the runoff from fertilizer use, sewage, and leaching from septic tanks. Bruning-Fann *et al.* (1993) indicate that nitrate causes (a) sanitary problems, and (b) the formation of nitrosamines on the reaction of nitrates with secondary amines found in food products. These nitrosamines have been related presumably to stomach cancer (Capitan-Vallvey *et al.*, 2005).

Environmental pollutants in the gas phase, as is well known, also constitute a very important source of environmental pollution. Seyama *et al.* (2005) very recently developed an array of plasma-polymerized film (PPF)-coated quartz crystal resonators (QCRs). These authors indicate that the PPF sensor array contains PPFs prepared from amino acids and synthetic polymers. These arrays exhibit different response patterns to mono- or mixed volatile sulfur compounds (VSCs) and exhibit good discrimination applications.

Staples and Viswanathan (2005) recently developed a SAW (surface acoustic wave)/GC (gas chromatograph) vapor analyzer to detect contrabands and drugs of abuse. Williams and Bhethanabotla (2005) recently developed a high-frequency thickness shear mode device to detect organic vapors. Their device is able to discriminate between the different compounds of interest.

In this chapter, we analyze the binding and dissociation kinetics using fractals of (a) different concentrations of toluene in solution to MC1061-pXylRS-lacZ in a whole-cell electrochemical biosensor system (Paitan *et al.*, 2004), (b) different aromatic and non-aromatic compounds in solution to MC1016-pXylRS-lacZ whole cells immobilized on an electrochemical biosensor (Paitan *et al.*, 2004), (c) different concentrations of xylene in solution to MC1061-pXylRS-lacZ or MC1061-pXylRS-AP on a whole-cell electrochemical biosensor (Paitan *et al.*, 2004), (d) urea in solution to a urease enzyme on an ion-selective optode membrane (Kovacs *et al.*, 2003), (e) 10 μM Pb(II) in solution to a catalytic DNA beacon on an Au (silver) electrode sensor, (f) 0.5 mM phenol in solution to *Pseudomonas* on a 74-III-modified SPE (Skladal *et al.*, 2002), (g) Cu(II) in solution to Lucifer yellow (LY) immobilized on cellulose particles in a polyurethane hydrogel (M2 membrane) (Mayr *et al.*, 2002), (h) volatile sulfur compounds (hydrogen sulfide and methanethiol) in the gas phase to an array of PPF-coated QCRs (Seyama *et al.*, 2005), (i) methyl parathion to an electrochemical biosensor surface (Liu and Lin, 2005), (j) benzo[a]pyrene (BP) and its derivative benzo[a]pyrene diolepoxide (BPDE) to a PDDA (poly(diallydimethylammonium chloride)-DNA)$_2$ film assembled on an electrode array, (k) deposition of copper and cadmium on an indium tin oxide (ITO) optically transparent electrode (OTE), and (l) toluene and ethanol in the vapor phase to capacitive chemical microsensors coated with the polymer PEUT (Kummer *et al.*, 2004). Binding and dissociation rate coefficients as well as fractal dimensions in the binding and in the dissociation phase are provided. As indicated elsewhere in the different chapters in the book, the fractal analysis is only one method of analyzing the analyte–receptor reactions occurring on biosensor surfaces. The analyte–receptor examples presented in this chapter should only be taken as a very small set of environmental pollutants present either in the air or in aquatic environments that may be detected and analyzed by biosensors.

8.2 THEORY

Havlin (1989) has reviewed and analyzed the diffusion of reactants toward fractal surfaces. The details of the theory and the equations involved for the binding and the dissociation phases for analyte–receptor binding are available (Sadana, 2001, Sadana and Ramakrishnan, 2002). The details are not repeated here, except that the equations are given to permit an easier reading. These equations have been applied to other biosensor systems (Sadana, 2001, 2005; Ramakrishnan and Sadana, 2001). For most of the applications, a single- or a dual-fractal analysis is often adequate to describe the binding and the dissociation kinetics. Peculiarities in the values of the binding and the dissociation rate coefficients as well as in the values of the fractal dimensions with regard to the dilute analyte systems being analyzed will be carefully noted, if applicable.

8.2.1 Single-fractal analysis

Binding rate coefficient

Havlin (1989) indicates that the diffusion of a particle (analyte [Ag]) from a homogeneous solution to a solid surface (e.g., receptor [Ab]-coated surface) on which it reacts to form a product (analyte–receptor complex; (Ab·Ag)) is given by

$$(\text{Analyte}\cdot\text{Receptor}) \sim \begin{cases} t^{(3-D_{\text{f,bind}})/2} = t^{p} & (t < t_{\text{c}}) \\ t^{1/2} & (t > t_{\text{c}}) \end{cases} \tag{8.1}$$

Here $D_{\text{f,bind}}$ or D_{f} (used later on in the chapter) is the fractal dimension of the surface during the binding step, t_{c} is the cross-over value. Havlin (1989) indicates that the cross-over value may be determined by using $r_{\text{c}}^{2} \sim t_{\text{c}}$. Above the characteristic length r_{c}, the self-similarity of the surface is lost and the surface may be considered homogeneous. Above time, t_{c}, the surface may be considered homogeneous, since the self-similarity property disappears and 'regular' diffusion is now present. For a homogeneous surface where $D_{\text{f}}=2$ and when only diffusional limitations are present, $p = ½$ as it should be. Another way of looking at the $p = ½$ case (where $D_{\text{f,bind}}=2$) is that the analyte in solution views the fractal object; in our case, the receptor-coated biosensor surface from a 'large distance.' In essence, in the association process, the diffusion of the analyte from the solution to the receptor surface creates a depletion layer of width $(Ɖt)^{½}$, where $Ɖ$ is the diffusion constant. This gives rise to the fractal power law

$$(\text{Analyte}\cdot\text{Receptor}) \sim t^{(3-D_{\text{f,bind}})/2}$$

For the present analysis, t_{c} is arbitrarily chosen and we assume that the value of the t_{c} is not reached. One may consider the approach as an intermediate 'heuristic' approach that may be used in the future to develop an autonomous (and not time-dependent) model for diffusion-controlled kinetics.

Dissociation rate coefficient

The diffusion of the dissociated particle (receptor [Ab] or analyte [Ag]) from the solid surface (e.g., analyte [Ag]–receptor [Ab]) complex coated surface) into solution may be given, as a first approximation by

$$(\text{Analyte}\cdot\text{Receptor}) \sim -t^{(3-D_{\text{f,diss}})/2} \qquad (t > t_{\text{diss}}) \tag{8.2}$$

Here $D_{\text{f,diss}}$ is the fractal dimension of the surface for the dissociation step. This corresponds to the highest concentration of the analyte–receptor complex on the surface. Henceforth, its concentration only decreases. The dissociation kinetics may be analyzed in a manner 'similar' to the binding kinetics.

8.2.2 Dual-fractal analysis

Binding rate coefficient

Sometimes, the binding curve exhibits complexities and two parameters (k, D_f) are not sufficient to adequately describe the binding kinetics. This is further corroborated by low values of r^2 factor (goodness-of-fit). In that case, one resorts to a dual-fractal analysis (four parameters: k_1, k_2, D_{f1}, and D_{f2}) to adequately describe the binding kinetics. The single-fractal analysis presented above is thus extended to include two fractal dimensions. At present, the time ($t = t_1$) at which the 'first' fractal dimension 'changes' to the 'second' fractal dimension is arbitrary and empirical. For the most part, it is dictated by the data analyzed and experience gained by handling a single-fractal analysis. A smoother curve is obtained in the 'transition' region if care is taken to select the correct number of points for the two regions. In this case, the product (antibody—antigen, or analyte–receptor complex; Ab·Ag, or analyte·receptor) is given by

$$(\mathrm{Ab}\cdot\mathrm{Ag}) \approx \begin{cases} t^{(3-D_{\mathrm{f1,bind}})/2} = t^{p1} & (t < t_1) \\ t^{(3-D_{\mathrm{f2,bind}})/2} = t^{p2} & t_1 < t < t_2 = t_c \\ t^{1/2} & (t > t_c) \end{cases} \tag{8.3}$$

In some cases, as mentioned above, a triple-fractal analysis with six parameters (k_1, k_2, k_3, D_{f1}, D_{f2}, and D_{f3}) may be required to adequately model the binding kinetics. This is when the binding curve exhibits convolutions and complexities in its shape perhaps due to the very dilute nature of the analyte (in some of the cases to be presented) or for some other reasons. Also, in some cases, a dual-fractal analysis may be required to describe the dissociation kinetics.

8.3 RESULTS

Alternate expressions for fitting the data are available that include saturation, first-order reaction, and no diffusion limitations, but these expressions are apparently deficient in describing the heterogeneity that inherently exists on the surface. One might justifiably argue that the appropriate modeling may be achieved by using a Langmuirian or other approach. The Langmuirian approach may be used to model the data presented if one assumes the presence of discrete classes of sites (for example, double exponential analysis as compared with a single-fractal analysis). Lee and Lee (1995) indicate that the fractal approach has been applied to surface science, for example, adsorption and reaction processes. These authors emphasize that the fractal approach provides a convenient means to represent the different structures and morphology at the reaction surface. These authors also emphasize to use the fractal approach to develop optimal structures and as a predictive approach. Another advantage of the fractal technique is that the analyte–receptor association (as well as the dissociation reaction) is a complex reaction, and the fractal analysis via the fractal dimension and the rate coefficient provides a useful lumped parameter(s) analysis of the diffusion-limited reaction occurring on a heterogeneous surface.

In a classical situation, to demonstrate fractality, one should make a log–log plot, and one should definitely have a large amount of data. It may be useful to compare the fit to some other forms, such as exponential or one involving saturation, etc. At present, we do not present any independent proof or physical evidence of fractals in the examples presented. It is a convenient means (since it is a lumped parameter) to make the degree of heterogeneity that exists on the surface more quantitatively. Thus, there is some arbitrariness in the fractal model to be presented. The fractal approach provides additional information about interactions that may not be obtained by conventional analysis of biosensor data.

There is no non-selective adsorption of the analyte. The present system (environmental pollutants in the aqueous or in the gas phase) being analyzed may be typically very dilute. Non-selective adsorption would skew the results obtained very significantly. In these types of systems, it is imperative to minimize this non-selective adsorption. We also do recognize that, in some cases, this non-selective adsorption may not be a significant component of the adsorbed material and that this rate of association, which is of a temporal nature, would depend on the surface availability. If we were to accommodate the non-selective adsorption into the model, there would be an increase in the heterogeneity on the surface, since, by its very nature, non-specific adsorption is more homogeneous than specific adsorption. This would lead to higher fractal dimension values since the fractal dimension is a direct measure of the degree of heterogeneity that exists on the surface.

Paitan *et al.* (2004) have developed a bacterial whole-cell electrochemical biosensor system that they claim can be used to monitor aromatic hydrocarbons. Their system involves the fusion of a promoter that is sensitive to aromatic compounds and to reporter genes that may be monitored electrochemically on a real-time basis and also is online. The authors emphasize that their constructs reacted specifically to aromatic compounds but not to non-aromatic compounds.

Figures 8.1a and b show the binding rate curve for the binding of 0.25 and 0.50 mM toluene in solution to MC1061-pXylRS-lacZ in whole-cell electrochemical biosensor systems. In both the cases, a single-fractal analysis is adequate to describe the binding kinetics. The values of (a) the binding rate coefficient k and the fractal dimension D_f for a single-fractal analysis are given in Table 8.1. The values of the binding rate coefficient(s) k and the fractal dimension(s) for binding presented in Table 8.1 were obtained from a regression analysis using Corel Quattro Pro 8.0 (1997) to model the data using eq. 8.1 wherein $(\mathrm{Ab{\cdot}Ag}) = kt^{(3-D_{f,bind})/2}$. The binding rate coefficient(s) presented in Table 8.1 are within 95% confidence limits. For example, for the binding of 0.25 mM toluene in solution to MC1061-pXylRS-lacZ in a whole-cell electrochemical biosensor system, the binding rate coefficient for a single-fractal analysis is 0.003246±0.000322. The 95% confidence limit indicates that the k value lies between 0.002924 and 0.003568. This indicates that the values are precise and significant.

It is of interest to note that an increase in the toluene concentration by a factor of 2 from 0.25 to 0.50 mM leads to (a) a decrease in the binding rate coefficient k by a factor of 3.2% from a value of 0.003246 to 0.003142, and to (b) a decrease in the fractal dimension

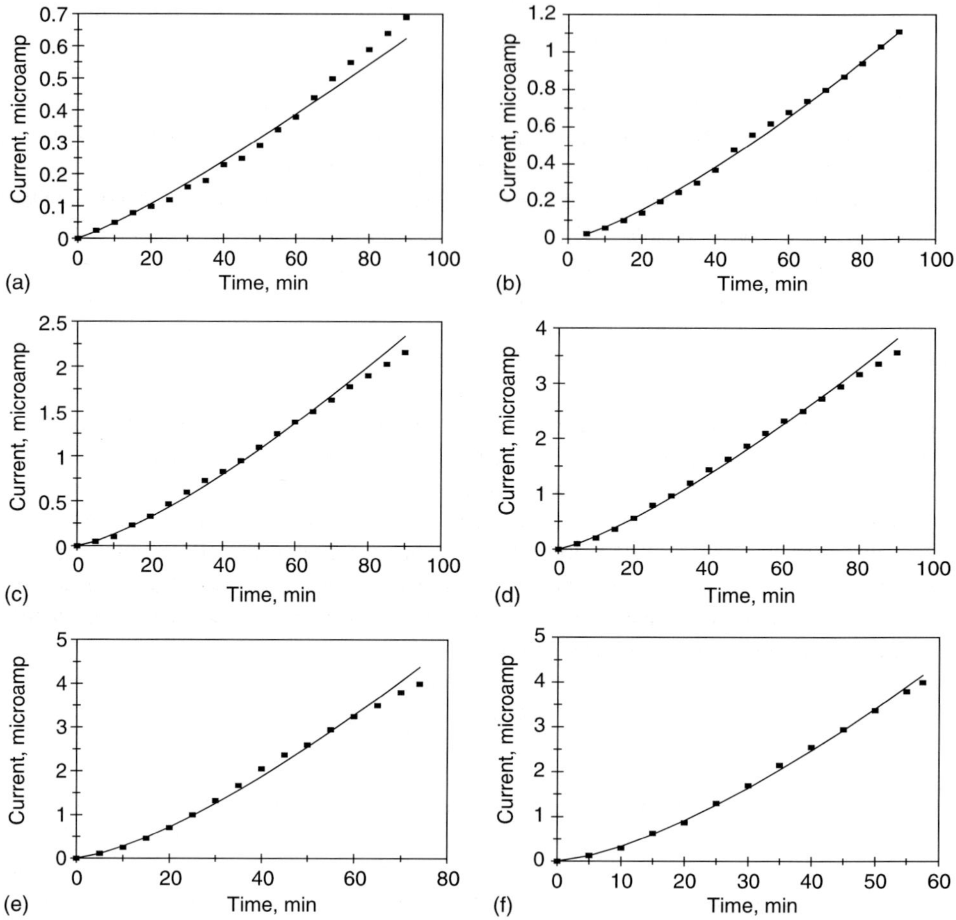

Figure 8.1 Binding of different concentrations (in mM) of toluene in solution to MC1061-pXylRS-lacZ in a whole-cell electrochemical biosensor system (Paitan *et al.*, 2004): (a) 0.25, (b) 0.50, (c) 1.0, (d) 2.5, (e) 5.0, (f) 10.0.

Table 8.1

Fractal dimensions and binding rate coefficients for different concentrations of toluene in solution to MC1061-pXylRS-lacZ in a whole-cell electrochemical biosensor system (Paitan *et al.*, 2004)

Toluene concentration in solution (mM)	k	D_f
0.25	0.003246 ± 0.000322	0.6634 ± 0.0572
0.50	0.003142 ± 0.00021	0.39366 ± 0.0414
1.0	0.005942 ± 0.000504	2.6556 ± 0.04922
2.5	0.012183 ± 0.000734	0.4466 ± 0.03538
5.0	0.011562 ± 0.00074	0.2416 ± 0.0428
10.0	0.012552 ± 0.000716	0.1358 ± 0.0445

D_f value by 40.7% from a value of D_f=0.6634–0.3937. Note that changes in the fractal dimension, and in the binding rate coefficient k are in the same direction, at least for this concentration range.

Figures 8.1c–f show the binding rate curves for the binding of 1.0–10.0 mM toluene in solution to MC1061-pXylRS-lacZ in whole-cell electrochemical biosensor systems. Since the binding rate curve in each case is described by a single-fractal analysis and this is similar to Figures 8.1a and b, each toluene concentration is not individually addressed to avoid repetition.

Figure 8.2a and Table 8.1 show the decrease in the fractal dimension D_f with an increase in the toluene concentration in solution. For the 0.25–10.0 mM toluene concentration in solution, the fractal dimension D_f is given by

$$D_f = (0.3871 \pm 0.1329)[\text{toluene concentration, mM}]^{-0.3411 \pm 0.09393} \quad (8.4)$$

The fit is good. The fractal dimension D_f exhibits less than negative one-half (=−0.3411) order of dependence on the toluene concentration (in mM) in solution. Once again, as indicated previously in the different chapters in the book, the fractal dimension is based on a log scale. Thus, even small changes in the value of the fractal dimension D_f lead to significant changes in the degree of heterogeneity on the electrochemical biosensor surface.

Figure 8.2b and Table 8.1 show the decrease in the binding rate coefficient k with an increase in the fractal dimension D_f. For the 0.25–10 mM toluene concentration in solution, the binding rate coefficient k is given by

$$k = (0.002859 \pm 0.002095) D_f^{-0.7958 \pm 0.4492} \quad (8.5)$$

The fit is good. There is some scatter in the data. The binding rate coefficient k decreases with an increase in the fractal dimension D_f or the degree of heterogeneity on the biosensor surface. The binding rate coefficient k exhibits an order of dependence between negative one-half and negative first order (=−0.7958) on the fractal dimension D_f. This is contrary to the different examples presented in the different chapters in the book. More often, the binding rate coefficient k exhibits an increase with an increase in the fractal dimension D_f value or the degree of heterogeneity on the biosensor surface.

Figure 8.2c and Table 8.1 show the increase in the binding rate coefficient k with an increase in the toluene concentration in solution. For the 0.25–10 mM toluene concentration in solution, the binding rate coefficient k is given by

$$k = (0.005639 \pm 0.001653)[\text{toluene concentration, mM}]^{0.4378 \pm 0.0818} \quad (8.6)$$

The fit is good, though there is some scatter in the data. The availability of more data points would lead to a more reliable fit. The binding rate coefficient k exhibits less than one-half (=0.4378) order of dependence on the toluene concentration in solution. The non-integer order of dependence exhibited by the binding rate coefficient k on the toluene concentration in solution lends support to the fractal nature of the system.

Paitan *et al.* (2004) also monitored the binding of aromatic and non-aromatic compounds using the MC1061-pXylRS-lacZ electrochemical biosensor. Figures 8.3a–f show

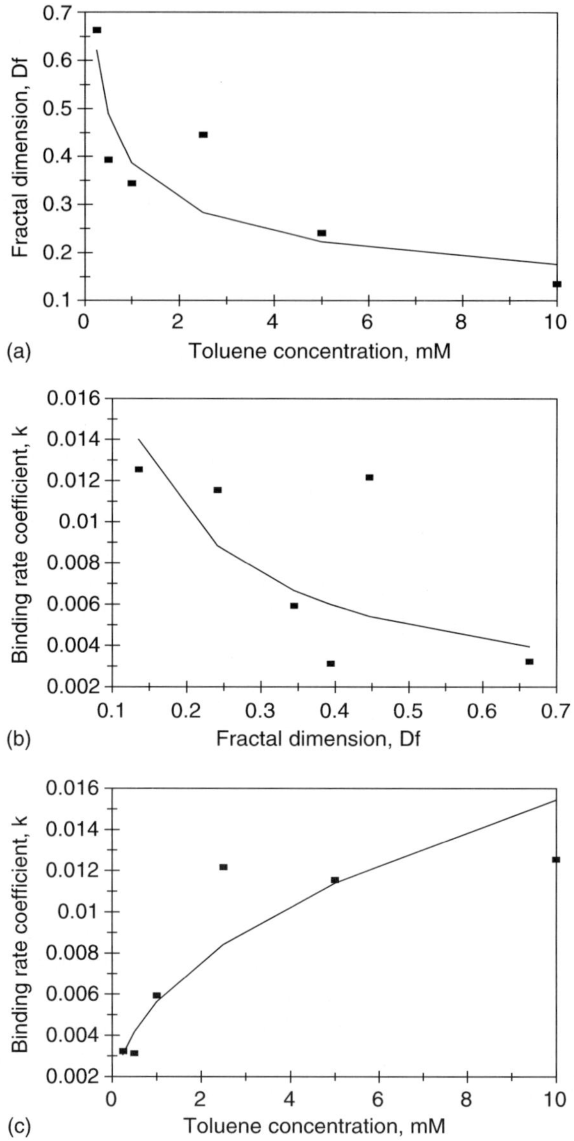

Figure 8.2 (a) Decrease in the fractal dimension D_f with an increase in the toluene concentration (in mM) in solution, (b) decrease in the binding rate coefficient k with an increase in the fractal dimension D_f and (c) increase in the binding rate coefficient k with an increase in the toluene concentration (in mM) in solution.

the binding of 2.5 mM 2-chloronitrotoluene, *m*-xylene, naphthalene, 3-methylbenzylchloride, 3-nitrotoluene, and nonane in solution to the MC1061-pXylRS-lacZ biosensors. In all the cases, a single-fractal analysis is adequate to describe the binding kinetics. The values of the binding rate coefficient k and the fractal dimension D_f are given in Table 8.2.

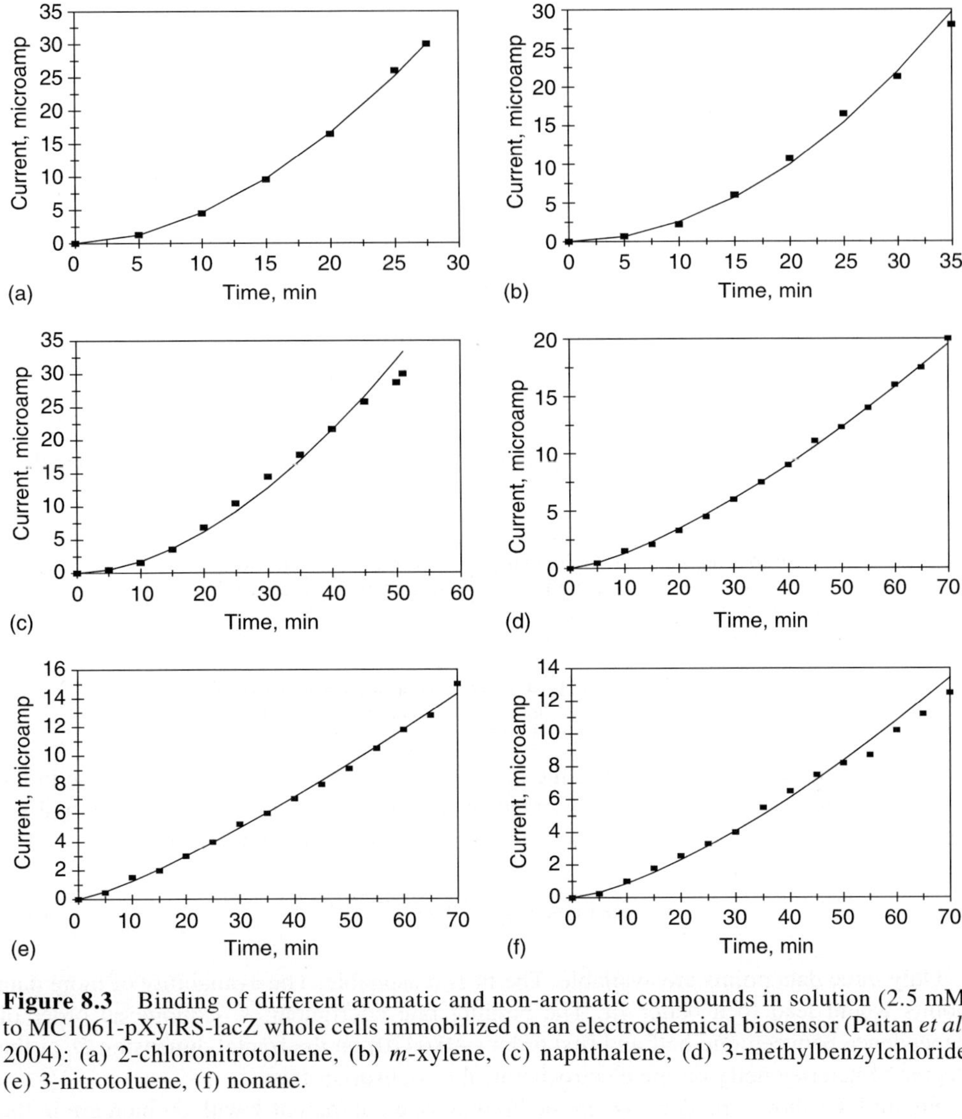

Figure 8.3 Binding of different aromatic and non-aromatic compounds in solution (2.5 mM) to MC1061-pXylRS-lacZ whole cells immobilized on an electrochemical biosensor (Paitan *et al.*, 2004): (a) 2-chloronitrotoluene, (b) *m*-xylene, (c) naphthalene, (d) 3-methylbenzylchloride, (e) 3-nitrotoluene, (f) nonane.

It is of interest to note that whether the compound is aromatic or non-aromatic the binding kinetics is adequately described by a single-fractal analysis when the MC1061-pXylRS-lacZ electrochemical biosensor is used. Attempts made to relate the binding rate coefficient k with the fractal dimension D_f or with the molecular weight for the different compounds presented in Table 8.2 were unsuccessful. Note that the estimated values of the fractal dimension D_f for some of the compounds presented in Table 8.2 are estimated to be equal to zero. Let us ignore these and use the last three items in Table 8.2 with non-zero D_f values. These last three compounds are 3-methylbenzylchloride, 3-nitrotoluene, and nonane. An estimated zero value of the fractal dimension indicates that the biosensor surface in this case acts as a 'Cantor-like'

Table 8.2

Binding rate coefficients and fractal dimensions for aromatic and non-aromatic compounds in solution (2.5 mM) to MC1061-pXylRS-lacZ whole cells immobilized on an electrochemical biosensor (Paitan *et al.*, 2004)

Analyte in solution (2.5 mM)	k	D_f
2-Chloronitrotoluene	0.06447 ± 0.00158	0 + 0.03362
m-Xylene	0.028966 ± 0.00288	0 + 0.1126
Naphthalene	0.02844 ± 0.00287	0 + 0.0823
3-Methylbenzylchloride	0.05505 ± 0.00335	0.2356 ± 0.0422
3-Nitrotoluene	0.07202 ± 0.00516	0.5082 ± 0.04952
Nonane	0.03423 ± 0.00428	0.1888 ± 0.08440

dust. The molecular weight, binding rate coefficient, and the fractal dimension values for these last three compounds are as follows:

Compound	Molecular weight	Fractal dimension	Binding rate coefficient
3-methylbenzyl-chloride	140.61	0.2356	0.05505
3-nitrotoluene	121.14	0.5082	0.07202
nonane	128.26	0.03423	0.1888

Figure 8.4a shows the increase in the binding rate coefficient k with an increase in the fractal dimension D_f. For the data presented above for the three compounds, 3-methylbenzylchloride, 3-nitrotoluene, and nonane, the binding rate coefficient k is given by

$$k = (0.1164 \pm 0.0316) D_f^{\,0.6470 \pm 0.3267} \tag{8.7}$$

Only three data points are available. The fit is reasonable. The availability of more data points would lead to a better fit. The binding rate coefficient, k, exhibits an order of dependence between one-half and first order (=0.6470) on the fractal dimension D_f or the degree of heterogeneity on the electrochemical biosensor surface.

Figure 8.4b shows the decrease in the binding rate coefficient k with an increase in the molecular weight of the compounds presented above. For the data presented above for the three compounds, 3-methylbenzylchloride, 3-nitrotoluene, and nonane, the binding rate coefficient k is given by

$$k = (311.699 \pm 7.876)\{\mathrm{MW}\}^{-1.74837 \pm 0.2304} \tag{8.8}$$

Only three data points are available. The fit is very good. The availability of more data points would lead to a better fit. The binding rate coefficient k exhibits an order of dependence between one and one-half and second (=–1.74837) order on the molecular weight of

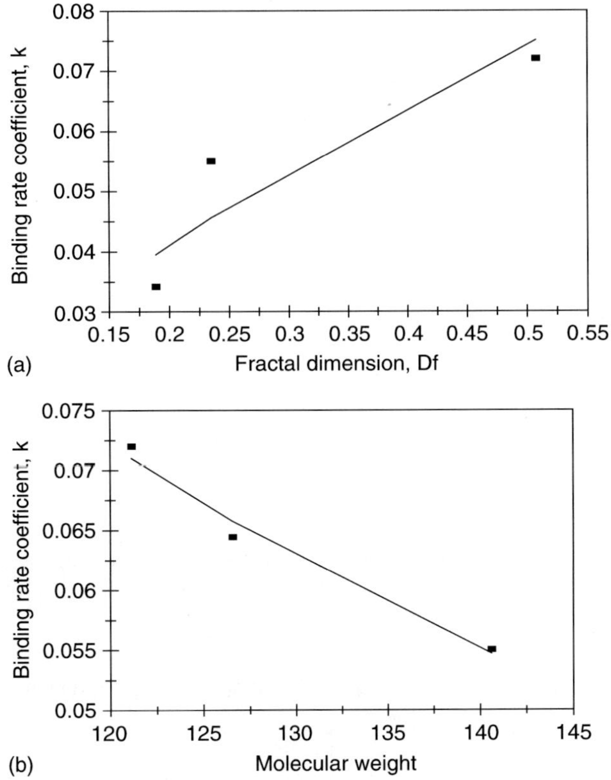

Figure 8.4 (a) Increase in the binding rate coefficient k with an increase in the fractal dimension D_f. (b) Decrease in the binding rate coefficient k with an increase in the molecular weight.

the above-mentioned compounds. The availability of more data points would help to substantially improve the reliability of the above predictive equation.

Figures 8.5a–c show the binding of (1) 0.05 (2) 0.25 and (3) 0.50 mM xylene in solution to MC1061-pXylRS-lacZ on whole-cell electrochemical biosensors (Paitan *et al.*, 2004). In all the three cases, a single-fractal analysis is adequate to describe the binding kinetics, and the values of the binding rate coefficient k and the fractal dimension D_f are given in Table 8.3a.

It is of interest to note that in case (2) the estimated value of the fractal dimension is equal to zero. This indicates that the surface of the electrochemical biosensor acts as a 'Cantor-like' dust. Vicsek (1989) in his book indicates that, and we quote, "One of the simplest and best known fractals is the so called triadic Cantor set which is a finite size fractal consisting of disconnected parts embedded into one dimensional space (d =1)." This author indicates that various Cantor-like sets with the same fractal dimension can be constructed.

As in case (2), the estimated value of the fractal dimension in case (3) is also estimated to be equal to zero indicating that the surface of the electrochemical biosensor acts as a 'Cantor-like' dust.

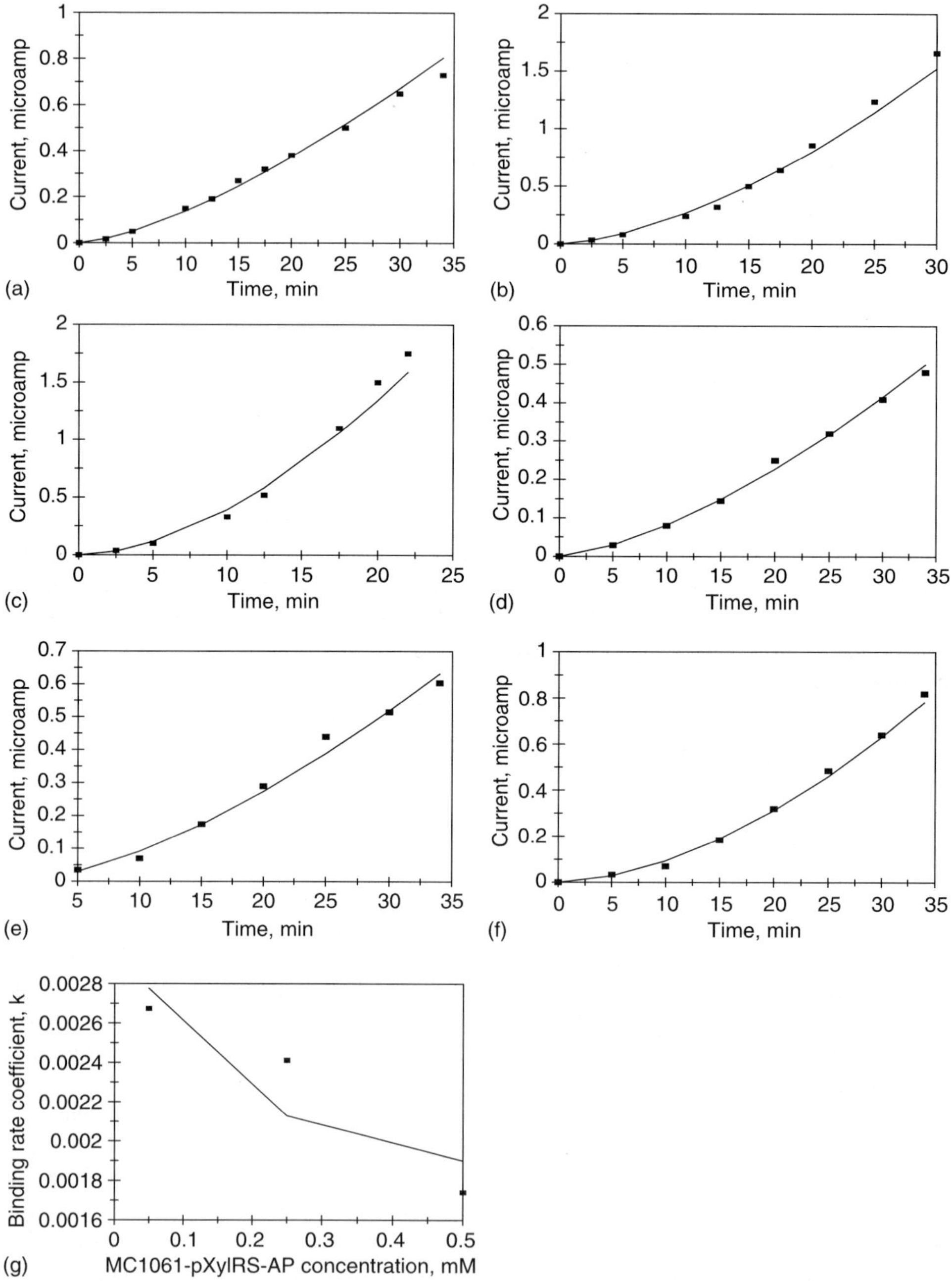

Figure 8.5 Binding of different concentrations (in mM) of xylene in solution to MC1061-pXylRS-lacZ on a whole-cell electrochemical biosensor (Paitan *et al.*, 2004): (a) 0.05, (b) 0.25, (c) 0.5. Binding of different concentrations (in mM) of xylene in solution to MC1061-pXylRS-AP on a whole-cell electrochemical biosensor (Paitan *et al.*, 2004): (d) 0.05, (e) 0.25, (f) 0.5.

Table 8.3

Binding rate coefficients and fractal dimensions for different concentrations of xylene in solution to (a) MC1061-pXylRS-lacZ or (b) MC1061-pXylRS-AP on a whole-cell electrochemical biosensor (Paitan *et al.*, 2004)

Xylene concentration in solution (mM)	MC1061-pXylRS-lacZ or MC1061-pXylRS-AP cultures	k	D_f
(a) 0.05	MC1061-pXylRS-lacZ	0.004833±0.000341	0.1046±0.05508
0.25	MC1061-pXylRS-lacZ	0.006887±0.000872	0
0.5	MC1061-pXylRS-lacZ	0.006598±0.001085	0
(b) 0.05	MC1061-pXylRS-AP	0.002675±0.000133	0.0330±0.05796
0.25	MC1061-pXylRS-AP	0.002412±0.000386	0
0.5	MC1061-pXylRS-AP	0.001742±0.000301	0

Figures 8.5d–f show the binding of (4) 0.05, (5) 0.25, and (6) 0.50 mM xylene in solution to MC1061-pXylRS-AP on whole-cell electrochemical biosensors (Paitan *et al.*, 2004). In all these cases, a single-fractal analysis is adequate to describe the binding kinetics. Note that in this case, AP (alkaline phosphatase) is the reporter gene. The values of (a) the binding rate coefficient k and the fractal dimension D_f are given in Table 8.3a.

It is of interest to note that the estimated value of the fractal dimension in case (5) is estimated to be equal to zero. This indicates that the surface of the electrochemical biosensor acts as a 'Cantor-like' dust. As indicated above, similar behavior was observed at the same xylene concentration (0.25 mM) in solution to MC1061-pXylRS-lacZ.

As in case (5), the estimated value of the fractal dimension in case (6) is also estimated to be equal to zero indicating that the surface of the electrochemical biosensor acts as a 'Cantor-like' dust. Note that for each of the xylene concentrations used, the corresponding binding rate coefficient, k is lower when alkaline phosphatase is used as the reporter gene when compared with β-galactosidase (lacZ).

Figure 8.5g shows the increase of the binding rate coefficient k with an increase in the xylene concentration in solution in the range 0.05–0.5 mM when MC1061-pXylRS-AP on a whole-cell electrochemical biosensor is used. In the 0.05–0.5 mM xylene concentration range, the binding rate coefficient k is given by

$$k = (0.001695 \pm 0.000287)[\text{xylene}]^{-0.16479 \pm 0.09351} \tag{8.9}$$

The fit is reasonable. Only three data points are available. The availability of more data points would lead to a more reliable fit. The binding rate coefficient k exhibits only a mild negative order (=−0.16479) of dependence on the xylene concentration in solution in the range 0.05–0.5 mM. The very low fractional (or non-integer) order of dependence exhibited by the binding rate coefficient k on the xylene concentration in solution lends support to the fractal nature of the system.

Kovacs *et al.* (2003) developed an optical sensor for the detection of urea by coating the surface of a glass waveguide slide with poly(vinylchloride) (PVC) matrix. This PVC matrix is a thin ammonium optode membrane and contains a urease enzyme reaction layer. The authors claim that the response is rapid, and that the detection is sensitive. Figure 8.6a shows the binding and the dissociation of 0.1 mM urea in solution to the urease enzyme on an ion-selective optode membrane. A dual-fractal analysis is required to adequately describe the binding and the dissociation kinetics. The values of (a) the binding rate coefficient k and the fractal dimension D_f for a single-fractal analysis, (b) the binding rate coefficients, k_1 and k_2, and the fractal dimensions, D_{f1} and D_{f2}, for a dual-fractal analysis, (c) the dissociation rate coefficient k_d and the fractal dimension D_{fd} for a single-fractal analysis, and (d) the dissociation rate coefficients, k_{d1} and k_{d2}, and the fractal dimensions, D_{fd1} and D_{fd2}, for a dual-fractal analysis are given in Tables 8.4a and b.

For the binding of 0.1 mM urea in solution to the urease enzyme on an ion-selective optode membrane and for a dual-fractal analysis, it is of interest to note that as the fractal dimension increases by 25.16% from a value of D_{f1}=2.2718 to D_{f2}=2.8436, the binding rate coefficient k increases by a factor of 2.232 from a value of k_1=0.3454 to k_2=0.7710.

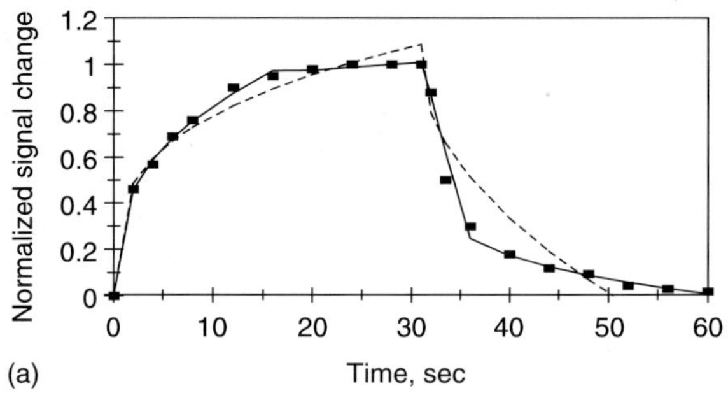

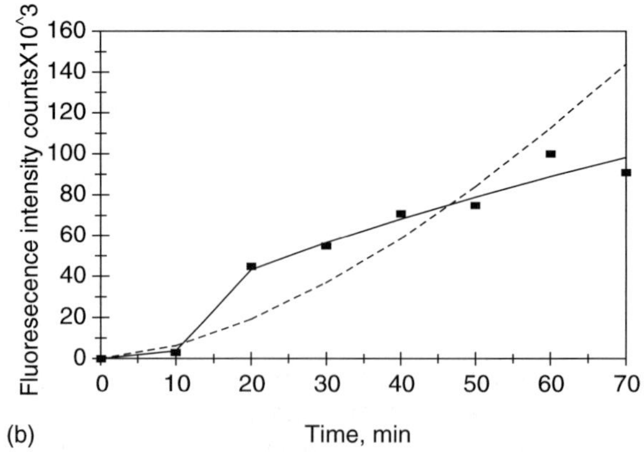

Figure 8.6 (a) Binding of 0.1 mM urea in solution to a urease enzyme on an ion-selective optode membrane (Kovacs *et al.*, 2003), and (b) binding of 10 μM Pb(II) in solution to a catalytic DNA beacon on an Au sensor (Swearingen *et al.*, 2005).

Table 8.4

(a) Binding and dissociation rate coefficients and (b) fractal dimensions for the binding and the dissociation phase for different environmental pollutants in solution to different receptors immobilized on different biosensor surfaces

(a)

Analyte in solution/receptor on surface	k	k_1	k_2	k_d	k_{d1}	k_{d2}	Reference
0.1 mM urea/urease enzyme on ion-selective optode membrane	0.3949±0.0248	0.3454 ±0.0180	0.7710 ± 0.0078	0.2079 ± 0.0791	0.1359 ± 0.05565	0.5850 ± 0.0063	Kovacs *et al.* (2003)
10 μM Pb(II)/catalytic DNA beacon on Au sensor	0.1563±0.1241	0.006096 + 0.006239	6.101 ± 0.5803	na	na	na	Swearingen *et al.* (2005)
0.5 mM phenol/ *Pseudomonas* on 74-III-modified screen-printed electrode (SPE)	0.1828±0.0681	0.08595 ±0.0111	0.8149 ± 0.0255	na	na	na	Skladal *et al.* (2002)
Cu(II)/Lucifer yellow immobilized on cellulose particles in polyurethane hydrogel (M2 membrane)	0.1039±0.0260	0.079791 ±0.01632	0.8447 ± 0.01576	na	na	na	Mayr *et al.* (2002)

(b)

	D_f	D_{f1}	D_{f2}	D_{fd}	D_{fd1}	D_{fd2}	Reference
0.1 mM urea/urease enzyme on ion-selective optode membrane	2.4096 ±0.0446	2.2718 ± 0.02932	2.8436 ±0.03784	1.9428±0.119	0.7602±0.6032	1.6872±0.02172	Kovacs *et al.* (2003)
10 μM Pb(II)/catalytic DNA beacon on Au sensor	0 + 0.6970	0 + 1.7938	1.6910 ±0.2716	na	na	na	Swearingen *et al.* (2005)
0.5 mM phenol/ *Pseudomonas* on 74-III-modified SPE	1.6788 ±0.2274	0.5350 ± 0.1920	2.7804 ±0.05370	na	na	na	Skladal *et al.* (2002)
Cu(II)/Lucifer yellow immobilized on cellulose particles in polyurethane hydrogel (M2 membrane)	0.4198 ±0.1459	0 + 0.2220	2.1450 ±0.05512	na	na	na	Mayr *et al.* (2002)

Similarly, as the fractal dimension, for dissociation increases by a factor of 2.219 from a value of D_{fd1}=0.7602 to D_{fd2}=1.6872, the dissociation rate coefficient increases by a factor of 4.30 from a value of k_{d1}=0.1359 to k_{d2}=0.5850. Note that an increase in the degree of heterogeneity on the biosensor surface (increase in the fractal dimension value) leads to an increase in the binding and in the dissociation rate coefficients.

Swearingen *et al.* (2005) recently developed a catalytic molecular beacon on an Au sensor for the detection of Pb(II). A thiol moiety was used to modify the DNAzyme fluorescent sensor to immobilize it on an Au surface. These authors used a mercaptohexanol (MCH)-DNA mixed monolayer to immobilize the DNAzyme on the Au surface. Figure 8.6b shows the binding of 10 μM Pb(II) in solution to the catalytic beacon on an Au sensor. A dual-fractal analysis is required to adequately describe the binding kinetics. The values of (a) the binding rate coefficient k and the fractal dimension D_f for a single-fractal analysis, and (b) the binding rate coefficients, k_1 and k_2, and the fractal dimensions, D_{f1} and D_{f2}, for a dual-fractal analysis are given in Tables 8.4a and b.

Skladal *et al.* (2002) developed novel bacterial electrodes for the detection of phenol using three *Pseudomonas* strains and mediator-modified SPEs. Figure 8.7a shows the binding of 0.5 mM phenol in solution to *Pseudomonas* on an SPE. A dual-fractal analysis is required to adequately describe the binding kinetics. The values of (a) the binding rate coefficient k and the fractal dimension D_f for a single-fractal analysis, and (b) the binding rate coefficients, k_1 and k_2, and the fractal dimensions, D_{f1} and D_{f2}, for a dual-fractal analysis are given in Tables 8.4a and b. It is of interest to note that as the fractal dimension increases by a factor of 5.19 from D_{f1}=0.5350 to D_{f2}=2.7804, the binding rate coefficient increases by a factor of 9.48 from a value of k_1=0.08595 to k_2=0.8149. Once again, an increase in the degree of heterogeneity or the fractal dimension on the biosensor surface leads to an increase in the binding rate coefficient.

Mayr *et al.* (2002) developed a sensor membrane for the detection of Cu(II) ions. These authors use Lucifer yellow (LY) as an indicator and an inert reference luminophore, which is a ruthenium complex entrapped in polyacrylonitrile beads. This indicator and the reference luminophore were enclosed in a hydrogel matrix. Figure 8.7b shows the binding of Cu(II) in solution to the dual lifetime referenced optical sensor membrane. A dual-fractal analysis is required to adequately describe the binding kinetics. The dissociation kinetics is adequately described by a single-fractal analysis. The values of (a) the binding rate coefficient k and the fractal dimension D_f for a single-fractal analysis, (b) the binding rate coefficients, k_1 and k_2, and the fractal dimensions, D_{f1} and D_{f2}, for a dual-fractal analysis, and (c) the dissociation rate coefficient k_d and the fractal dimension for dissociation D_{fd} are given in Tables 8.4a and b.

Seyama *et al.* (2005) developed a PPF-based QCR sensor array for the detection of volatile sulfur compounds. These authors installed this sensor array in a desktop-size relative humidity controller. The volatile sulfur compounds (VSCs) detected were hydrogen sulfide and methanethiol.

Figures 8.8a–c show the binding of 84 ppb hydrogen sulfide to (1) a PhE (phenylalanine), (2) PE (polyethylene), and (3) a His (D-histidine) sensor. PhE, PE, and D-histidine, respectively, for the three cases are the sputtering targets. In all the cases, a single-fractal analysis is adequate to describe the binding kinetics. The values of the binding rate coefficient k and the fractal dimension D_f for a single-fractal analysis are given in Tables 8.5a and b.

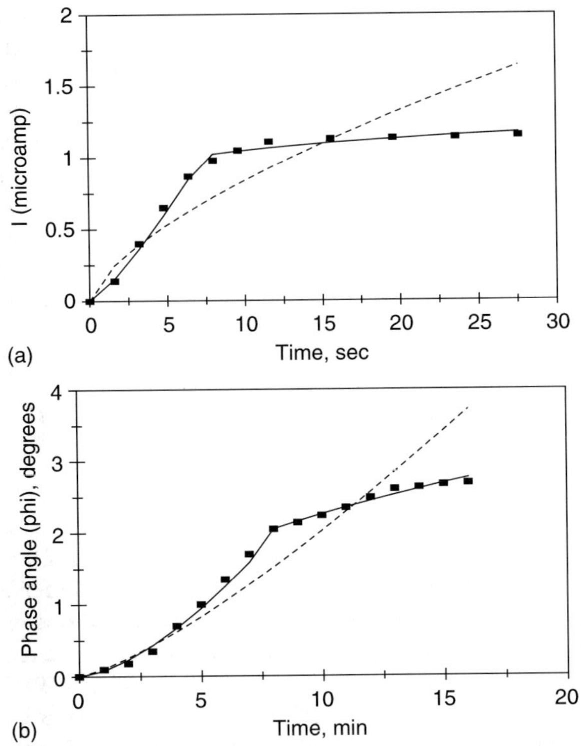

Figure 8.7 (a) Binding of 0.5 mM phenol in solution to *Pseudomonas* on a 74-III-modified SPE (Skladal *et al.*, 2002), and (b) binding of Cu(II) in solution to Lucifer yellow (LY) immobilized on cellulose particles in a polyurethane hydrogel (M2 membrane) (Mayr *et al.*, 2002). When only a solid (____) line is used, then a single-fractal analysis applies. When a dashed (------) line and a solid (____) line are used, then the dashed line represents a single-fractal analysis, and the solid line represents a dual-fractal analysis. In this case, the solid line provides the better fit.

Figures 8.9a–c show the binding of 84 ppb methanethiol to (1) a PE (polyethylene), (2) PhE (phenylalanine), and (3) a His (D-histidine) sensor. In cases (1) and (3), a single-fractal analysis is adequate to describe the binding kinetics and in case (2) a dual-fractal analysis is required to describe the binding kinetics. The values of (a) the binding rate coefficient k and the fractal dimension D_f for a single-fractal analysis, and (b) the binding rate coefficients, k_1 and k_2, and the fractal dimensions, D_{f1} and D_{f2}, for a dual-fractal analysis are given in Tables 8.5a and b.

Figures 8.10a and b show the binding and dissociation of humid air to a PhE and a PE sensor. In both the cases, a single-fractal analysis is required to adequately describe the binding and the dissociation kinetics. The values of (a) the binding rate coefficient k and the fractal dimension for binding D_f for a single-fractal analysis, and (b) the dissociation rate coefficient k_d and the fractal dimension for dissociation D_{fd} for a single-fractal analysis are given in Tables 8.5a and b.

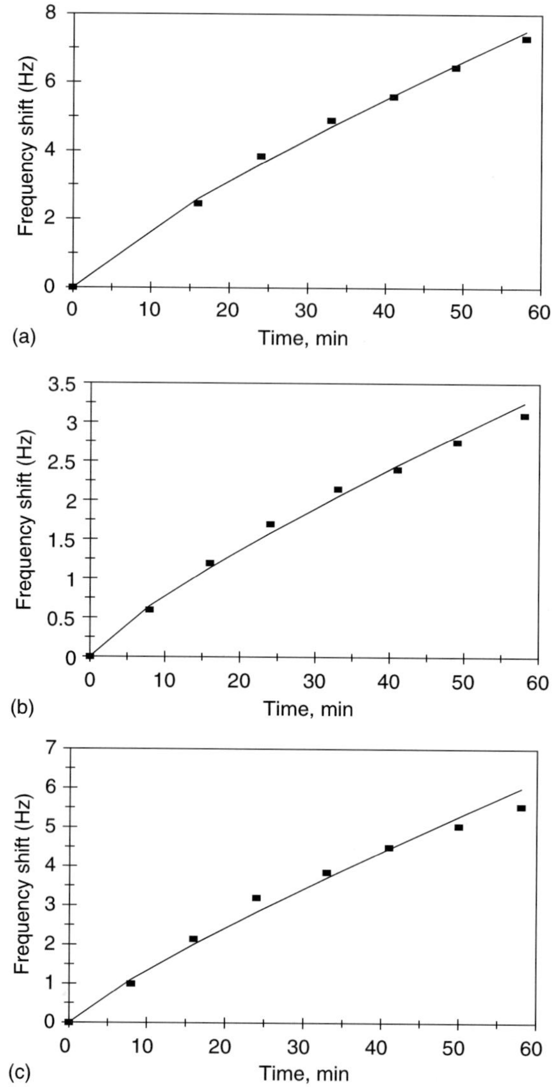

Figure 8.8 Binding and dissociation of volatile sulfur compounds (VSCs) (hydrogen sulfide and methanethiol) in the gas phase to an array of plasma-polymerized film (PPF)-coated quartz crystal resonators (QCRs). Sensors used were PE (polyethylene), PhE (phenylalanine), and His (D-histidine) (Seyama *et al.*, 2005): (a) 84 ppb hydrogen sulfide/PhE sensor, (b) 84 ppb hydrogen sulfide/PE sensor, (c) 84 ppb hydrogen sulfide/His sensor.

Note that the binding rate coefficients for both the PE sensor and the PhE sensor are almost the same. The PhE sensor exhibits a slightly higher binding rate coefficient than the PE sensor. As expected, its fractal dimension value is also slightly higher than that of the PE sensor. The dissociation rate coefficient for the PE sensor ($k_d = 4.0938$) is about four times higher than that of the PhE sensor ($k_d = 1.0226$).

Table 8.5a

Binding and dissociation rate coefficients for volatile sulfur compounds (VSCs) (hydrogen sulfide and methanethiol) in the gas phase to an array of plasma-polymerized film (PPF)-coated quartz crystal resonator (QCRs). Sensors used were PE (polyethylene), PhE (phenylalanine), and His (D-histidine). PE, PhE, and His are the sputtering targets (Seyama *et al.*, 2005).

Analyte in solution/ sensor type	k	k_1	k_2	k_d
84 ppb hydrogen sulfide/ PhE sensor	0.2630±0.0130	na	na	na
84 ppb hydrogen sulfide/ PE sensor	0.1185±0.0075	na	na	na
84 ppb hydrogen sulfide/ His sensor	0.1895±0.0179	na	na	na
84 ppb methanethiol/PE sensor	0.1536±0.0058	na	na	na
84 ppb methanethiol/PhE sensor	0.0437±0.00846	0.0696±0.00258	0.2146±0.0067	na
84 ppb methanethiol/His sensor	0.13591±0.00832	na	na	na
Humid air/PhE sensor	48.0659±3.1916	na	na	1.0226±0.0907
Humid air/PE sensor	47.0726±2.1139	na	na	4.0938±0.4248

Table 8.5b

Fractal dimensions for the binding and the dissociation phase for volatile sulfur compounds (VSCs) (hydrogen sulfide and methanethiol) in the gas phase to any array of plasma-polymerized film (PPF)-coated quartz crystal resonator (QCRs). Sensors used were a PE (polyethylene), PhE (phenylalanine) and His (D-histidine). PE, PhE, and His are the sputtering targets (Seyama *et al.*, 2005).

Analyte in solution/ sensor type	D_f	D_{f1}	D_{f2}	D_{fd}
84 ppb hydrogen sulfide/ PhE sensor	1.3498±0.09056	na	na	na
84 ppb hydrogen sulfide/ PE sensor	1.3686±0.07252	na	na	na
84 ppb hydrogen sulfide/ His sensor	1.2978±0.1050	na	na	na
84 ppb methanethiol/PE sensor	1.3942±0.03350	na	na	na
84 ppb methanethiol/PhE sensor	1.00308±0.04858	0.1680±0.3466	1.7656±0.1542	na
84 ppb methanethiol/His sensor	1.6378±0.04858	na	na	na
Humid air/PhE sensor	2.3310±0.1470	na	na	0.9548±0.1814
Humid air/PE sensor	2.1118±0.10044	na	na	1.0984±0.1189

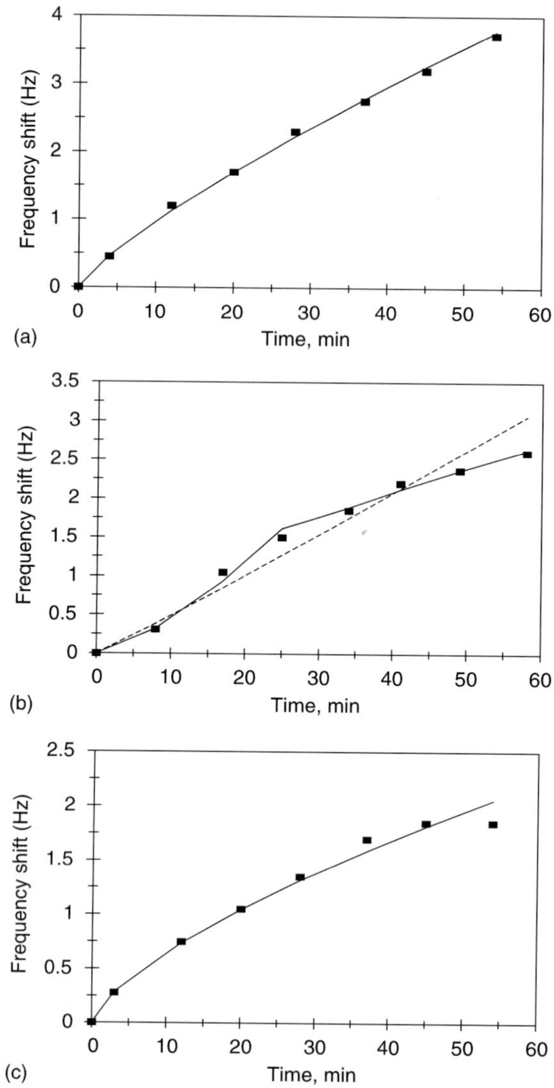

Figure 8.9 Binding and dissociation of volatile sulfur compounds (VSCs) (hydrogen sulfide and methanethiol) in the gas phase to an array of plasma-polymerized film (PPF)-coated quartz crystal resonators (QCRs). Sensors used were PE (polyethylene), PhE (phenylalanine), and His (D-histidine) (Seyama *et al.*, 2005): (a) 84 ppb methanethiol/PE sensor, (b) 84 ppb methanethiol/PhE sensor, (c) 84 ppb methanethiol/His sensor.

Liu and Lin (2005) very recently developed an electrochemical sensor for the detection of organophosphate pesticides and nerve agents. They used zirconia nanoparticles as selective sorbents. These authors deposited zirconia particles electrodynamically on a polycrystalline gold electrode surface by cyclic voltametry. Zirconia exhibits a strong affinity for the phosphate group. Thus, these authors state that nitroaromatic organophosphates (OPs) bind

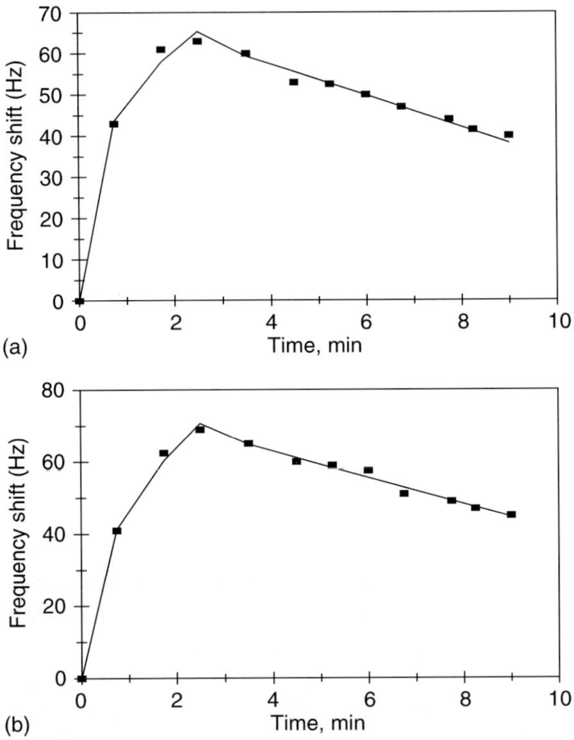

Figure 8.10 Binding and dissociation of volatile sulfur compounds (VSCs) (hydrogen sulfide and methanethiol) in the gas phase to an array of plasma-polymerized film (PPF)-coated quartz crystal resonators (QCRs). Sensors used were PE (polyethylene), and PhE (phenylalanine) (Seyama *et al.*, 2005): (a) humid air/PhE sensor, (b) humid air/PE sensor.

strongly to the zirconia surface. Liu and Lu (2005) also state that organophosphates are highly neurotoxic. They disrupt the cholinesterase enzyme that regulates acetylcholine (Rosenberry, 1975; Zhang *et al.*, 2001; Fennouh *et al.*, 1997; Creminisi *et al.*, 1995; Guerrieri *et al.*, 2002). Acetylcholine is a neurotransmitter and is required for the appropriate function of the nervous system. Liu and Lin (2005) further state that these organophosphates are widely used as pesticides and as biological warfare agents since they are neurotoxic. Thus, there is a need to develop rapid and reliable detection devices for the quantitative determination of these organophosphates.

Figure 8.11 shows the binding (adsorption) of methyl parathion to the electrochemical biosensor surface. A single-fractal analysis is adequate to describe the binding kinetics. The values of the binding rate coefficient k and the fractal dimension D_f are 0.0017 ± 0.00411 and 2.5754 ± 0.08120, respectively.

Wang *et al.* (2005) very recently indicated the bioactivation of lipophilic pollutants and drugs to toxic metabolites in the liver by cytochrome P450 (cyt P450) enzymes. These authors indicate that this is a major mechanism of genotoxicity (Travis and Hester, 1991; Jakoby, 1980; Ortiz de Montellano, 1995; Singer and Grunberger, 1983; Schenkman

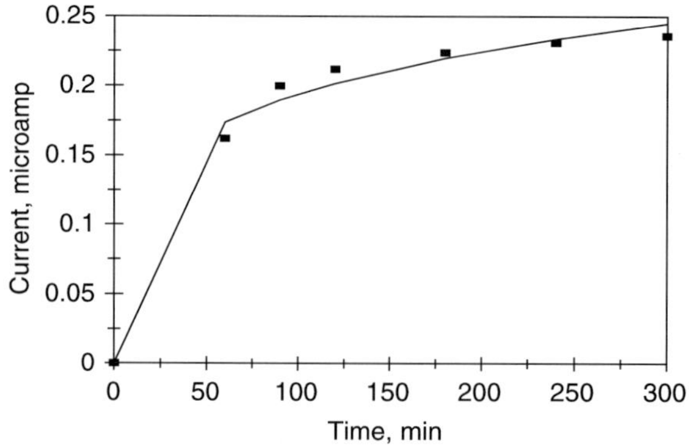

Figure 8.11 Binding (adsorption) of methyl parathion to the electrochemical biosensor surface (Liu and Lin, 2005).

and Greim, 1993). Gonzalez (1991, 1992) indicates that there are over 20 different human cyt P450s that have been expressed. Each of these has different activities. Wang *et al.* (2005) emphasize that there is a need to identify which of the human cyt P450s actively generate DNA-reactive metabolites, which are toxic pollutants and drugs.

Wang *et al.* (2005) very recently developed a sensor array for the detection of DNA damage after incubation with benzo[a]pyrene diolepoxide (BPDE). Initially, ds (double-stranded) DNA films were assembled on electrode arrays using layer-by-layer electrostatic adsorption. The enzymes cyt P450 cam, cyt P450 1A2, and myoglobin were introduced into the DNA films on the array electrodes to produce BPDE from BP (benzo[a]pyrene).

(1) Figure 8.12a shows the binding of 35 μM BPDE in pH 5.5 buffer containing 50 μM $Ru(bpy)_3{}^{2+}$ (ruthenium tris(2,2′-bipyridyl) to PDDA (poly(diallydimethylammonium chloride)-DNA)$_2$ film assembled on an electrode array (Wang *et al.*, 2005).
(2) Figure 8.12b shows the binding of 35 μM BP+1 mM H_2O_2 in pH 5.5 containing 50 μM $Ru(bpy)_3{}^{2+}$ in solution to PDDA-DNA/(horse-heart myoglobin (MB)/DNA)$_2$ film assembled on an electrode array (Wang *et al.*, 2005). In both the cases, a single-fractal analysis is adequate to describe the binding kinetics. The values of the binding rate coefficient k and (b) the fractal dimension D_f for a single-fractal analysis are given in Table 8.6. These authors indicate that the enzymes, cyt P450 cam, cyt P40 1A2, and myoglobin on activation by H_2O_2 metabolize BP to genotoxic metabolites. $Ru(bpy)_3{}^{2+}$ is used as a catalyst to detect DNA damage by the metabolites.

Figures 8.13a–c show the binding of 35 μM BP+1 mM H_2O_2 in solution to (1) PDDA-DNA/(MB·cyt P450 1A2/DNA)$_2$, (2) PDDA-DNA/(cyt P450 cam/DNA)$_2$, and (3) PDDA-DNA/(cyt P450 1A2/DNA)$_2$ film assembled on an electrode array (Wang *et al.*, 2005). In all the cases, a single-fractal analysis is adequate to describe the binding kinetics. The values of (a) the binding rate coefficient k and the fractal dimension D_f for a single-fractal analysis are given in Table 8.6.

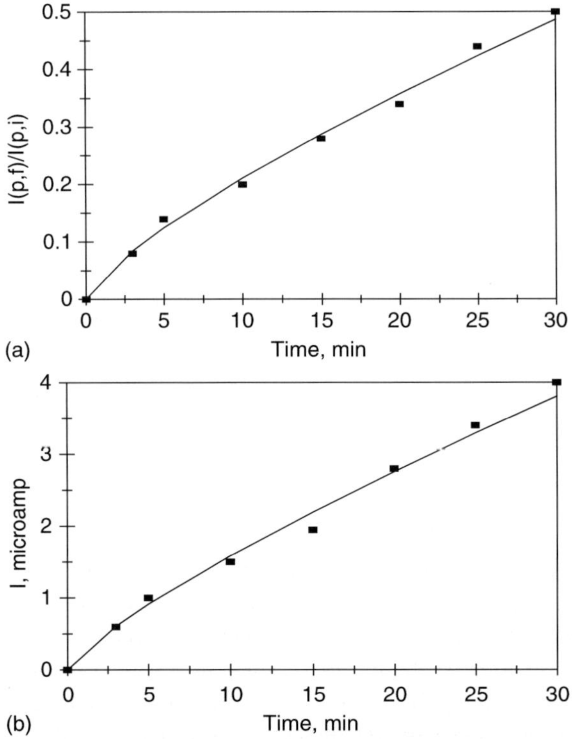

Figure 8.12 (a) Binding of 35 μM benzo[a]pyrene diolepoxide (BPDE) in pH 5.5 buffer containing 50 μM $Ru(bpy)_3^{2+}$ (ruthenium tris(2,2′-bipyridyl) to a PDDA (poly(diallydimethylammonium chloride)-DNA$)_2$ film assembled on an electrode array (Wang *et al.*, 2005), and (b) binding of 35 μM benzo[a]pyrene (BP)+1 mM H_2O_2 in pH 5.5 containing 50 μM $Ru(bpy)_3^{2+}$ in solution to a PDDA-DNA/(horse-heart myoglobin (MB)/DNA$)_2$ film assembled on an electrode array (Wang *et al.*, 2005).

Shtoyko *et al.* (2004) recently used stripping voltametry (SV)-attenuated total internal reflectance (ATR) spectroscopy to detect the deposition of copper (as thin films) on an indium tin oxide (ITO) optically transparent electrode (OTE). These authors emphasize that stripping voltametry is an economical method to determine trace amounts of metal (Wang, 1985, 1996, 2000; Heinemann *et al.*, 1984). Shtoyko *et al.* (2004) indicate that SV involves the following two steps: (a) a pre-concentration step wherein the analyte accumulates on the working electrode. During this step the metal is reduced at a certain potential, and (b) in the stripping step the deposited metal is oxidized (Wang, 1985, 1996, 2000).

Figures 8.14a and b show the binding of 0.1 mM $CuSO_4$ and 0.1 mM $CdSO_4$ in 0.1 mM sodium acetate at pH 7.3 to ITE-OTE (mercury-platinum) transparent electrodes (Shtoyko *et al.*, 2004). A dual-fractal analysis is required to adequately describe the binding kinetics. The values of (a) the binding rate coefficient k and the fractal dimension D_f for a single-fractal analysis, and (b) the binding rate coefficients, k_1 and k_2, and the fractal dimensions, D_{f1} and D_{f2}, for a dual-fractal analysis are given in Table 8.7a.

Table 8.6

Binding rate coefficients and fractal dimensions for the binding of benzo[a]pyrene (BP) and BP metabolites to individually addressable, demountable electrodes coated with ultra-thin DNA/enzyme films (Wang *et al.*, 2005)

Analyte in solution/receptor on surface	k	D_f
35 μM benzo[a]pyrene diolepoxide (BPDE) in pH 5.5 buffer containing 50 μM $Ru(bpy)_3^{2+}$ (ruthenium tris(2,2′-bipyridyl)/PDDA (poly(diallydimethylammonium chloride)-DNA)$_2$	0.03653 ± 0.00267	1.4770 ± 0.06710
35 μM BP+1mM H_2O_2 in pH 5.5 buffer containing 50 μM $Ru(bpy)_3^{2+}$ /PDDA film	0.2534 ± 0.0200	1.4066 ± 0.07242
35 μM BP+1mM H_2O_2/PDDA-DNA/ (Mb·cyt P450 1A2/DNA)$_2$ film	0.02535 ± 0.00135	1.4092 ± 0.04966
35 μM BP+1 mM H_2O_2/PDDA-DNA/ (cyt P450 cam /DNA)$_2$ film	0.02245 ± 0.0148	1.6792 ± 0.0606
35 μM BP+1mM H_2O_2/PDDA-DNA/ (cyt P450 1A2/DNA)$_2$ film	0.03205 ± 0.00509	2.0812 ± 0.1456

Figures 8.15a and b show the binding of 1×10^{-4} M $CuSO_4$ +0.1 M KNO_3 and 5×10^{-4} M $CuSO_4$ +0.1 M KNO_3 in solution to bare ITO-OTE (mercury-platinum) transparent electrodes (Shtoyko *et al.*, 2004). In both the cases, a dual-fractal analysis is required to adequately describe the binding kinetics. The values of (a) the binding rate coefficient k and the fractal dimension D_f for a single-fractal analysis, and (b) the binding rate coefficients, k_1 and k_2, and the fractal dimensions, D_{f1} and D_{f2}, for a dual-fractal analysis are given in Table 8.7b.

Figures 8.15c and d show the binding of 1×10^{-5} M $CuSO_4$ +0.1 M KNO_3 and 5×10^{-6} M $CuSO_4$ +0.1 M KNO_3 in solution to bare ITO-OTE (mercury-platinum) transparent electrodes (Shtoyko *et al.*, 2004). In both the cases, a single-fractal analysis is adequate to describe the binding kinetics. The values of the binding rate coefficient k and the fractal dimension D_f for a single-fractal analysis are given in Table 8.7b.

Kummer *et al.* (2004) recently indicate that complementary metal oxide semiconductor (CMOS) microsensors have been used for different applications (Kakerov *et al.*, 1994; Cane *et al.*, 1996; Qiu *et al.*, 2001; Hierlemann and Baltes, 2003). Kummer *et al.* (2004) also indicate that there is a change in the dielectric properties of the sensing material on exposure to analytes for chemocapacitors (dielectrometers). These authors emphasize that CMOS capacitive microsensors with sensitive polymeric coatings lead to effective chemical sensing devices. They developed a CMOS-based chemical microsensing system for the detection of volatile organics in ambient air.

Figures 8.16a and b show the binding and dissociation of 1000–5000 ppm ethanol (100–500 Pa) in the vapor phase to a capacitor coated with 0.3- and 2.3-μM-thick film of the polymer, PEUT. In the first case, a single-fractal analysis is adequate to describe the binding kinetics, and a dual-fractal analysis is required to adequately describe the dissociation kinetics. In the second case, a single-fractal analysis is adequate to describe the binding as well as the dissociation kinetics. The values of (a) the binding rate coefficient k and the fractal dimension D_f for a single-fractal analysis, (b) the dissociation rate coefficient k_d and the fractal

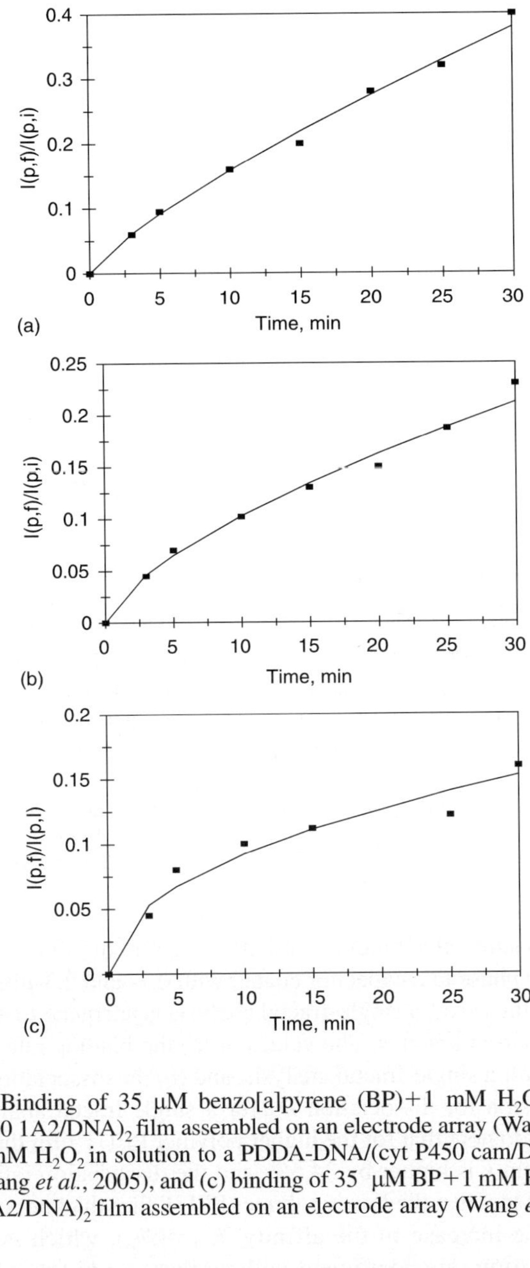

Figure 8.13 (a) Binding of 35 μM benzo[a]pyrene (BP)+1 mM H_2O_2 in solution to a PDDA-DNA/(MB·cyt P450 1A2/DNA)$_2$ film assembled on an electrode array (Wang *et al.*, 2005), (b) binding of 35 μM BP+1 mM H_2O_2 in solution to a PDDA-DNA/(cyt P450 cam/DNA)$_2$ film assembled on an electrode array (Wang *et al.*, 2005), and (c) binding of 35 μM BP+1 mM H_2O_2 in solution to a PDDA-DNA/(cyt P450 1A2/DNA)$_2$ film assembled on an electrode array (Wang *et al.*, 2005).

dimension D_{fd} for a single-fractal analysis, and (c) the dissociation rate coefficients, k_{d1} and k_{d2}, and the fractal dimensions for dissociation, D_{fd1} and D_{fd2}, for a dual-fractal analysis are given in Table 8.8. Note that when a higher (2.3 μm) film thickness is used, the dissociation is described by a single-fractal analysis, whereas when the lower (0.3 μM) film thickness is used a dual-fractal analysis is required to describe the dissociation kinetics.

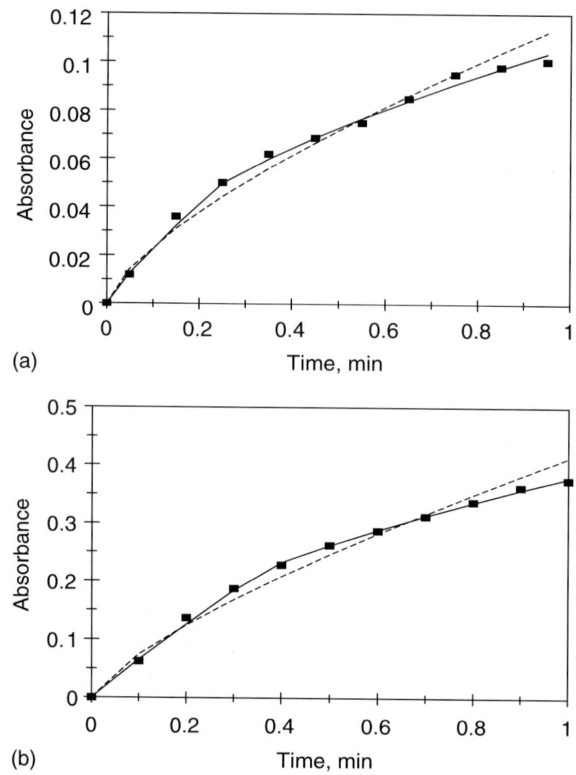

Figure 8.14 (a) Binding of 0.1 mM $CuSO_4$ in 0.1 mM sodium acetate at pH 7.3 to an ITE-OTE (mercury-platinum) transparent electrode (Shtoyko *et al.*, 2004), and (b) binding of 0.1 mM $CdSO_4$ in 0.1 mM sodium acetate at pH 7.3 to an ITE-OTE (mercury-platinum) transparent electrode (Shtoyko *et al.*, 2004).

Figures 8.16c and d show the binding and dissociation of 600–3000 ppm toluene (60–300 Pa) in the vapor phase to a capacitor coated with 0.3- and 2.3-μm-thick film of the polymer, PEUT. In both the cases, a single-fractal analysis is adequate to describe the binding as well as the dissociation kinetics. The values of (a) the binding rate coefficient k and the fractal dimension D_f for a single-fractal analysis, and (b) the dissociation rate coefficient k_d and the fractal dimension for dissociation D_{fd} for a single-fractal analysis are given in Table 8.8. It is of interest to note that for the higher polymer PEUT film thickness (2.3 μm), the binding rate coefficient k is higher by 24.5%, and the dissociation rate coefficient k_d is lower by 4.1% compared to when the lower polymer PEUT film thickness (0.3 μm) is used.

Figure 8.17 shows the increase in the affinity, K $(=k/k_d)$, which is the ratio of the binding and the dissociation rate coefficient with an increase in the ratio of the fractal dimensions in the binding and in the dissociation phase D_f/D_{fd}. For the data presented in Table 8.8, the affinity K $(=k/k_d)$ is given by

$$k/k_d = (1.3077 \pm 0.1275)(D_f/D_{fd})^{0.6757 \pm 0.3067} \tag{8.10}$$

Table 8.7

Binding of 0.1 mM $CuSO_4$ and 0.1 mM $CdSO_4$ as thin films on an indium tin oxide (ITO) surface monitored by stripping voltametry-attenuated internal reflectance (Shtoyko *et al.*, 2004): Binding rate coefficients and fractal dimensions

Analyte in solution/ receptor on surface	k	k_1	k_2	D_f	D_{f1}	D_{f2}
(a) 0.1 mM $CuSO_4$ in 0.1 mM sodium acetate at pH 7.3/ ITO-OTE (mercury-platinum) transparent electrode	0.1167 ± 0.01369	0.1623 ± 0.0169	0.1066 ± 0.0032	1.6024 ± 0.0806	1.2942 ± 0.1342	1.9024 ± 0.09416
0.1 mM $CdSO_4$ in 0.1 mM sodium acetate at pH 7.3/ ITO-OTE (mercury-platinum) transparent electrode	0.4132 ± 0.0427	0.5739 ± 0.0490	1.5110 ± 0.08946	1.1150 ± 0.1573	1.1150 ± 0.1573	1.9372 ± 0.02918
(b) 1×10^{-4} M $CuSO_4$+0.1 M KNO_3 /bare ITO/OTE	0.03925 ± 0.02186	0.02831 ± 0.00122	0.1076 ± 0.0048	1.0918 ± 0.01933	0.5868 ± 0.04144	1.7092 ± 0.06088
5×10^{-5} M $CuSO_4$+0.1 M KNO_3 /bare ITO/OTE	0.006975 ± 0.000475	0.006632 ± 0.00050	0.01453 ± 0.000388	0.4010 ± 0.02872	0.3410 ± 0.05214	0.8778 ± 0.06242
1×10^{-5} M $CuSO_4$+0.1 M KNO_3 /bare ITO/OTE	0.004241 ± 0.000284	na	na	0.7490 ± 0.02832	na	na
5×10^{-6} M $CuSO_4$ 0.1 M KNO_3/bare ITO/OTE	0.055 ± 0	na	na	2.0 ± 7.4 E -17	na	na

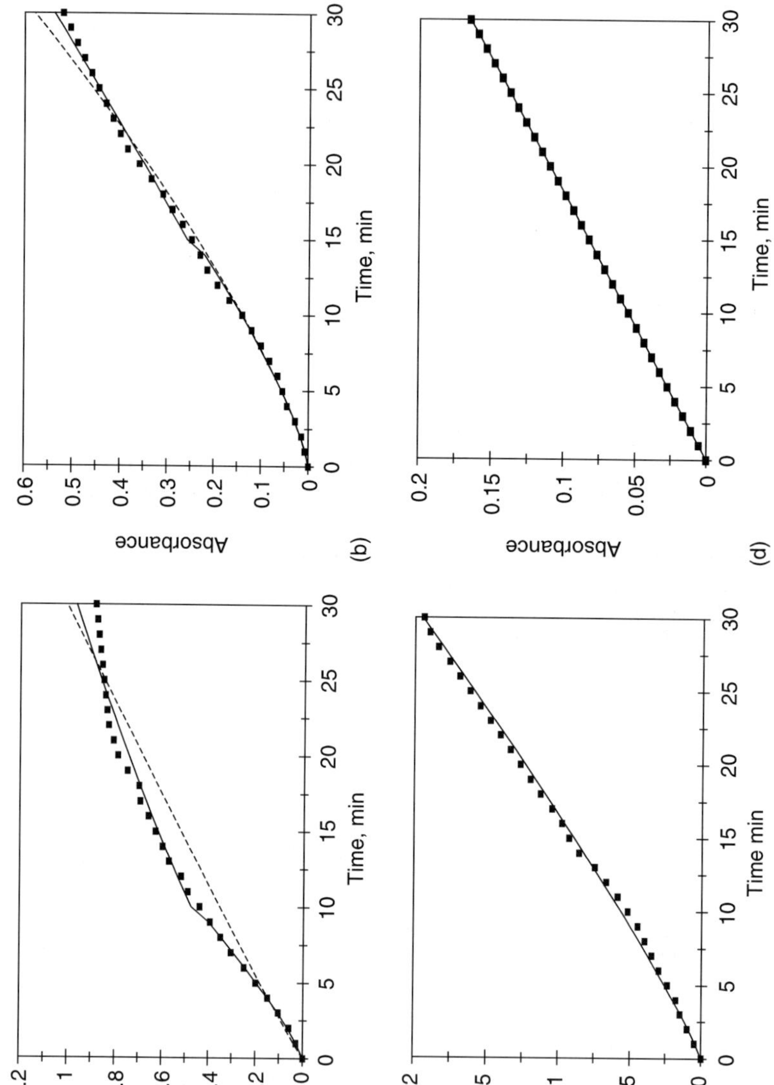

Figure 8.15 Binding of different concentrations (in M) of $CuSO_4$+0.1 M KNO_3 to a bare ITE-OTE (mercury-platinum) transparent electrode (Shtoyko *et al.*, 2004): (a) 1×10^{-4}, (b) 5×10^{-5}, (c) 1×10^{-5}, (d) 5×10^{-6}.

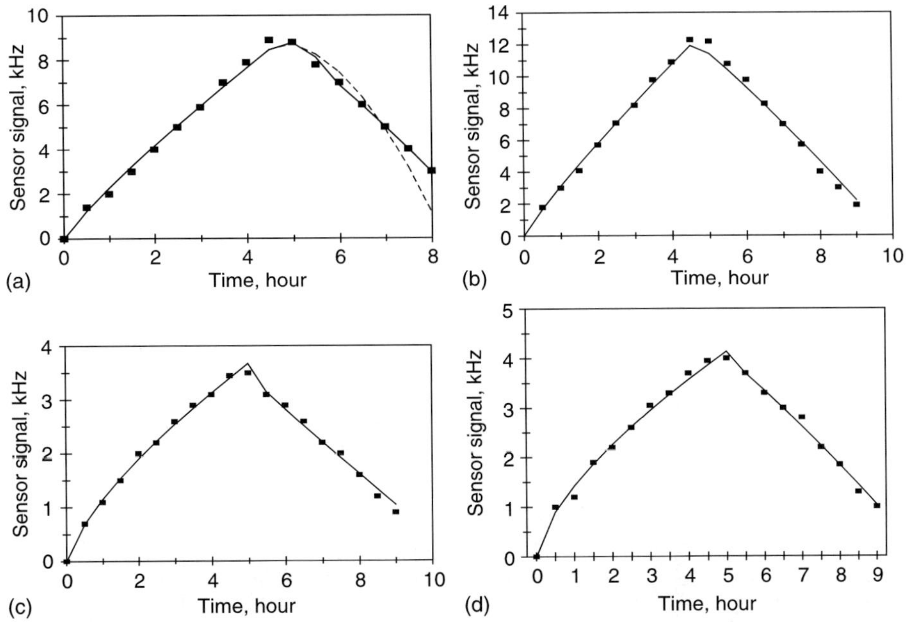

Figure 8.16 Binding of ethanol (100–500 Pa; 1000–5000 ppm) to two different thicknesses of polymer PEUT coated on capacitors: (a) PEUT thickness, 0.3 μM, (b) PEUT thickness, 2.3 μM. Binding of toluene (60–300 Pa; 600–3000 ppm) to two different thicknesses of polymer PEUT coated on capacitors (c) PEUT thickness, 0.3 μM (d) PEUT thickness, 2.3 μM.

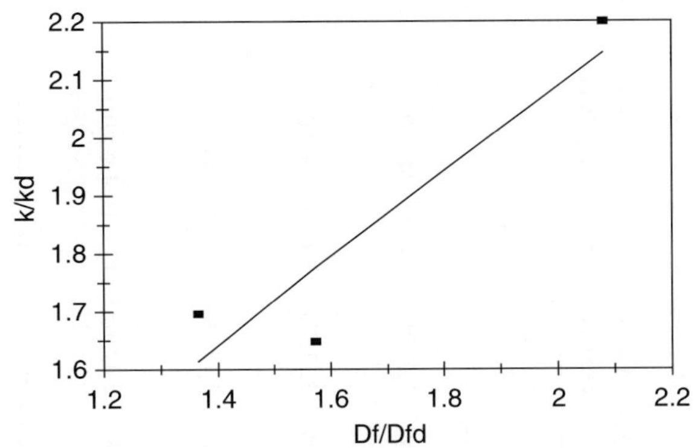

Figure 8.17 Increase in the affinity, K (= k/k_d), with an increase in the ratio of fractal dimensions D_f/D_{fd}.

Table 8.8

Binding rate coefficients and fractal dimensions for the binding of 1000–5000 ppm (100–500 Pa) ethanol and 600–3000 ppm (60–300 Pa) toluene in the vapor phase to capacitors coated with a polymer, PEUT thin film of different thicknesses (0.3 and 2.3 μM) (Kummer *et al.*, 2004)

Analyte in the vapor phase/thickness of the PEUT polymer film	k	k_d	k_{d1}	k_{d2}	D_f	D_{fd}	D_{fd2}	D_{fd2}
1000–5000 ppm ethanol/ 0.3 μM-thick PEUT polymer film	2.2579±0.1846	0.6442±0.2958	0.7735±0.4229	1.2093±0.0101	1.7574±0.07726	0.0240+0.4504	0+1.1100	0.4606 ± 0.03962
1000–5000 ppm ethanol/ 2.3 μM-thick PEUT polymer film	3.1137±0.1540	1.8890±0.2951	na	na	1.2142±0.04746	0.7720±0.1427	na	na
600–3000 ppm toluene/ 0.3 μM-thick PEUT polymer film	1.1439±0.0393	0.6744±0.0624	na	na	1.5504±0.03074	1.1354±0.03074	na	na
600–3000 ppm toluene/ 2.3 μM-thick PEUT polymer film	1.4236±0.1119	0.6468±0.0476	na	na	1.6754±0.06878	0.8052±0.07622	na	na

The fit is quite good. Only three data points are available. The availability of more data points would lead to a more reliable fit. The affinity, K $(=k/k_d)$, exhibits an order of dependence between one-half and first order on the ratio of fractal dimensions in the binding and in the dissociation phase.

8.4 CONCLUSIONS

A fractal analysis is presented for the binding and dissociation of different environmental pollutants either in solution or in the gas phase to receptors immobilized on different biosensor surfaces. The kinetic analysis provides a better picture of the analyte–receptor reactions occurring on the biosensor surface. The analysis provides a quantitative indication of the state of disorder or the degree of heterogeneity on the biosensor surface. The fractal analysis is offered as an alternate way to improve the understanding of the kinetics in the heterogeneous case with diffusion-limited reactions occurring on structured surfaces.

Binding and dissociation (where ever applicable) rate coefficients as well as fractal dimensions in the binding and in the dissociation phase are provided. Both types of examples are presented wherein either a single- or a dual-fractal analysis is required to describe the binding and/or the dissociation kinetics. The dual-fractal analysis is used only when the single-fractal analysis did not provide an adequate fit (sum of least squares less than 0.98). This was done by the regression provide by Quattro Pro 8.0 (Corel Quattro Pro, 1997).

In accord with the prefactor analysis of aggregates (Sorenson and Roberts, 1997), quantitative and predictive equations are developed for (a) the binding rate coefficient k as a function of the fractal dimension D_f or the toluene concentration in solution for the binding of toluene in solution to MC1061-pXylRS-lacZ (Paitan *et al.*, 2004), (b) the binding rate coefficient k as a function of the fractal dimension D_f for the binding of 3-methylbenzylchloride, 3-nitrotoluene, and nonane to the MC1061-pXylRS-lacZ biosensor (Paitan *et al.*, 2004), (c) the binding rate coefficient k as a function of the molecular weight of the aromatic and non-aromatic compounds in solution (Paitan *et al.*, 2004), and (d) the binding rate coefficient k as a function of xylene concentration in solution (Paitan *et al.*, 2004). In the cases presented above, the binding rate coefficient k is only mildly sensitive to the degree of heterogeneity or the fractal dimension present on the biosensor surface. The order of dependence on the fractal dimension D_f is -0.7958 for the binding of different concentrations of toluene in solution to the MC1061-pXylRS-lacZ electrochemical biosensor, and it is equal to 0.6470 for the binding of aromatics and non-aromatics (3-methylbenzylchloride, 3-nitrotoluene, and nonane) in solution to the MC1061-pXylRS-lacZ electrochemical biosensor.

The binding and dissociation (if applicable) kinetics of other environmental pollutants present either in the gas phase or in solution need to be analyzed, since one needs to pay more and more attention to the nature of the biosensor surface, and how it influences the biosensor performance parameters such as selectivity, sensitivity, response time, regenerability, and stability. The biosensor surface is an additional parameter that may be manipulated to advantage biosensor applications. As in medical and biomedical applications this aspect is important here too for environmental applications.

REFERENCES

Analytical Currents. (2004). Water soluble Hg(II) sensor. *Analytical Chemistry*, 8A, January 1.

Battaglin, W.A., Thurman, E.M., Kalkhoff, S.J., and Porter, S.D. (2003). *Journal of the American Water Works Association*, **39**(4), 743–756.

Bermejo-Barrera, P., Moreda-Pineiro, J., Moreda-Pinero, A., and Bermejo-Barrerra, A. (1996). Study of chemical modifiers for direct determination of silver in sea-water by ETA-AAS with deuterium background correction. *Talanta*, **43**, 35–44.

Bruning-Fann, C.S., and Kaneene, J.B. (1993). The effects of nitrate, nitrite, and N-nitroso compounds on human health. *Vetenarian and Human Toxicology*, **35**, 521.

Cane, C., Gotz, A., Merlos, A., Garcia, I., Errachid, A., Losantos, P., and LoraJamayo, E. (1996). Multilayer ISFET membranes for microsystems applications. *Sensors & Actuators B*, **35**, 136–140.

Capitan-Vallvey, L.F., Arroyo-Guerrero, E., Fernandez-Ramos, M.D., Santoyo-Castle, L.A., Siehl, D.L., Gorton, R., Patten, P.A., Chen, Y.H., Bertain, S., Cho, H., Duck, N., Wong, J., Liu, D., and Lassner, M.W. (2005). *Science*, 1151–1154.

Centers for Disease Control (U.S.). (1991). *Preventing Lead Poisoning in Young Children: A Statement* (4th revision), The Centers, Atlanta, Georgia.

Chiriswell, C.D., Chang, R.C., and Fritz, J.S. (1975). Chromatographic determination of phenols in water. *Analytical Chemistry*, **47**, 1325–1329.

Corcia, A.D., Marchese, S., and Sampri, R. (1993). Selective determination of phenols in water by a two-trap tandem extraction system followed by liquid chromatography. *Journal of Chromatography*, **642**, 125–184.

Corel Corporation Limited. (1997). *Corel Quattro Pro 8.0*, Ottawa, Canada.

Cox, C.J. (1998). Pesticide fact handbook. *Journal of Pesticide Reform*, **18**(3), 40.

Creminisi, C., Disario, S., Mela, J., Pilloton, R., and Palleschi, G. (1995). Evaluation of the use of free and immobilized acetylcholinesterase for paraoxon detection with amperometric choline sensor. *Analytica Chimica Acta*, **311**, 273–280.

Dadfornia, S., Haji Shabani, A.M., and Gohari, M. (2004). Trace enrichment and determination of silver by immobilized DDTC microcolumn and flow injection atomic absorption spectrometry. *Talanta*, **64**(3), 682–687.

EEC Council Directive. (1998). 98/83/EEC, OJ L 330 12.05.1998, pp. 32–54.

Elfering, H., Andersson, J.T., and Poll, K.G. (1998). Determination of organic lead in soils and waters by hydride generation inductively coupled plasma atomic emission spectrometry. *Analyst*, **123**, 669–674.

EPA Drinking Water Criteria Document for Silver. (1989). Environmental Protection Agency, Washington, DC, p^pp EPA CASRN 7440-7422-7444.

European Commission, Water Quality Directive. (1998). 98/83/EU.

Fennouh, S., Casimiri, V., and Burstein, C. (1997). Increased paraxon detection with solvents using acetylcholinesterase in activation measured with a choline oxidase biosensor. *Biosensors & Bioelecronics*, **12**(2), 97–104.

Forstner, U., and Wittman, G.T. (1981). *Metal Pollution in the Aquatic Environment*. Springer, Berlin.

Gonzalez, F. (2005). Disposable receptor-based optical sensor for nitrate. *Analytical Chemistry*, **77**, 4459–4466.

Gonzalez, F.J. (1991). *New Horizons in Biological Diversity* (eds. B.I. Gladhill, and I. Mauro), Wiley-Liss, New York, pp. 11–20.

Gonzalez, F.J. (1992). *Trends in Pharmaceutical Sciences*, **13**, 346–352.

Gonzalez-Martinez, M.A., Brun, E.M., Puchades, R., Maquiera, A., Ramsey, K., and Fernando, R. (2005). Glyphosphate immunosensor. Application for water and soil analysis. *Analytical Chemistry*, **77**, 4219–4227.

Guerrieri, A., Monaci, L., Wqunito, M., and Palmisano, T. (2002). A disposable amperometric biosensor for rapid screening of anticholinesterase activity in soil extracts. *Analyst*, **127**, 5–7.

Hanumegowda, N.M., and White, I. (2005). Aqueous mercuric ion detection with microsphere optical ring resonators. Paper 563a, Annual American Institute of Chemical Engineers Meeting, Cincinnati, Ohio, October 31–November 4.

Havlin, S. (1989). Molecular diffusion and reactions. In: *The Fractal Approach to Heterogeneous Chemistry: Surfaces, Colloids, Polymers* (ed. D. Avnir), Wiley, New York, pp. 251–269.

Heinemann, W.R., Mark, H.B., Jr., and Wise, J.A. (1984). In: *Laboratory Techniques in Electroanalytical Chemistry* (eds. P.T. Kissinger, and W.R. Heinemann), Marcel Dekker, New York, pp. 499–538.

Hierlemann, A., and Baltes, H. (2003). CMOS-based chemical microsensors. *Analyst*, **128**, 15–28.

Hutter, W., Peter, J., Swoboda, H., Hampel, W., Rosenberg, E., Kramer, D., and Kellner, R. (1995). Development of a microbial bioassay for chlorinated and brominated hydrocarbons. *Analytica Chimica Acta*, **306**, 237–241.

Jakoby, W.B., ed., (1980). *Enzymatic Basis of Detoxification*, Academic, New York, Volumes I and II.

Kakerov, R., Manoli, Y., Mokwa, W., Rospert, M., Meyer, H., Drewer, H., Krause, J., and Cammann, J. (1994). *Sensors and Actuators*, **43**, 296–301.

Kohler, S., Belkin, S., and Schmid, R.D. (2000). Reporter gene biassays in environmental analysis. *Journal of Fresenius Analytical Chemistry*, **366**, 769–779.

Kovacs, B., Nagy, G., Dombi, R., and Toth, K. (2003). Optical biosensor for urea with improved response time. *Biosensors & Bioelectronics*, **18**, 111–118.

Kummer, A.M., Hierlemann, A., and Baltes, H. (2004). Tuning sensitivity and selectivity of complementary metal oxide semiconductor-based capacitive chemical microsensors. *Analytical Chemistry*, **76**, 2470–2477.

Lee, C.K., and Lee, S.L. (1995). Multi-fractal scaling analysis of reactions over fractal surfaces. *Surface Science*, **325**, 294–306.

Lippard, S.J., and Nolan, E.M. (2003). Water soluble Hg(II) sensor. *Journal of the American Chemical Society*, **125**, 12749–12758, also in Analytical Currents, *Analytical Chemistry*, January 1, 2004, 8A.

Liu, G., and Lin, Y. (2005). Electrochemical sensor for organophosphate pesticides and nerve agents using zirconia nanoparticles as selective sorbents. *Analytical Chemistry*, **77**, 5894–5901.

Liu, J., and Lu, Y. (2003). A colorimetric lead biosensor using DNAzyme-directed assembly of gold nanoparticles. *Journal of the American Chemical Society*, **125**, 6642–6643.

Liu, J.W., and Lu, Y. (2004). Colorimetric biosensors based on DNAzyme-assembled gold nanoparticles. *Journal of Fluorescence*, **14**, 343–354.

Ma, R., Mol, W.V., and Adams, F. (1994). Determination of cadmium, copper, and lead in environmental samples: an evaluation of flow injection on-line sorbent. *Analytica Chimica Acta*, **285**, 33–43.

Marc, J., Mulner-Lorillon, O., and Belle, R. (2004). *Biological Cell*, **96**, 245–249.

Martin, J.S., Frye, G.C., Ricco, A.J., and Senturia, A.D. (1993). Effect of surface roughness on the response of thickness-shear mode resonators in liquids. *Analytical Chemistry*, **65**(20), 2910–2922.

Matvienko, A., and Mandelis, A. (2005). Ultra-high thermal wave sensor: application to water–ethanol mixtures. Paper 563f, Annual American Institute of Chemical Engineers Meeting, Cincinnati, Ohio, October 31–November 4.

Mayr, T., Klimant, I., Wolfbeis, O.S., and Werner, T. (2002). Dual lifetime referenced optical sensor membrane for the determination of copper(II) ions. *Analytica Chimica Acta*, **462**, 1–10.

Okuyama, M., Kobayashi, N., Takeda, W., Anjo, T., Matsuki, Y., Goto, J., Kimbegawa, A., and Hori, S. (2004). Enzyme-linked immunosorbent assay for monitoring toxic dioxin congeners in milk based on a newly generated monoclonal anti-dioxin antibody. *Analytical Chemistry*, **76**, 1948–1956.

Ortiz de Montellano, P.R., ed., (1995). *Cytochrome P450*, Plenum, New York.

Paitan, Y., Biran, I., Schecter, N., Biran, D., Rishpon, J., and Ron, E.Z. (2004). Monitoring aromatic hydrocarbons by whole cell electrochemical biosensors. *Analytical Biochemistry*, **335**, 175–183.

Peter, J., Hutter, W., and Hampel, W. (1996). Detection of chlorinated and brominated hydrocarbons by an ion sensitive whole cell biosensor. *Biosensors & Bioelectronics*, **11**, 1215–1219.

Peter, J., Hutter, W., Stollenberger, W., Karner, F., and Hampel, W. (1997). Semicontinuous detection of 1,2-dichloroethane in water samples using *Xanthobacter autotrophicus* GJ 10 encapsulated in chitosan beads. *Analytical Chemistry*, **69**, 2077–2079.

Qiu, Y.Y., Azerdo-Leme, C., Alcacer, L.R., and Franca, J.E. (2001). *Sensors & Actuators*, **92**, 80–87.

Richardson, S.D., and Ternes, T.A. (2005). Water analysis: contaminants and current issues. *Analytical Chemistry*, **77**, 3807–3838.

Rosenberry, T.L. (1975). *Advances in Enzymology and Related Areas of Molecular Biology*, Wiley, New York.

Sadana, A. (1972). The catalytic oxidation of phenol in aqueous solution by CuO, M.Chem. E. thesis, University of Delaware.

Sadana, A. (2001). A fractal analysis for the evaluation of hybridization kinetics in biosensors. *Journal of Colloid and Interface Science*, **151**(1), 166–177.

Sadana, A. (2005). *Fractal Binding and Dissociation Kinetics for Different Biosensor Applications*, Elsevier, Amsterdam.

Sadana, A., and Ramakrishnan, A. (2002). A kinetic study of analyte–receptor binding and dissociation in biosensor applications: a fractal analysis for cholera toxin and peptide –protein interactions. *Sensors & Actuators B*, **85**, 61–72.

Schenkman, J.B., and Greim, H., ed., (1995). *Cytochrome P450*, Plenum, New York.

Seyama, M., Iwasaki, Y., Ogawa, S., Sugimoto, I., Tate, A., and Niwa, O. (2005). Discriminative detection of volatile sulfur compound mixtures with a plasma-polymerized film-based sensor array installed in a humidity-controlled system. *Analytical Chemistry*, **77**, 4228–4234.

Shephard, J., and Toomey, O.B. (2005). Heavy metal ion detection using peptide-modified hydrogel layers on a quartz crystal microbalance. Paper 563e, Annual American Institute of Chemical Engineers Meeting, Cincinnati, Ohio, October 31–November 4.

Shtoyko, T., Conklin, A.T., Maghari, A.T., Richardson, T.N., Piruska, A., Seliskar, C.J., and Heinemann, W.R. (2004). *Analytical Chemistry*, **76**, 1466–1473.

Singer, B., and Grunberger, D. (1983). *Molecular Biology of Mutagens and Carcinogens*, Plenum, New York.

Skladal, P., Morozova, N.O., and Reshetilov, A.N. (2002). Amperometric biosensors for detection of phenol using chemically modified electrodes containing immobilized bacteria. *Biosensors and Bioelectronics*, **17**, 867–873.

Sorenson, C.M., and Roberts, G.C. (1997). The prefactor of fractal aggregates. *Journal of Colloids and Interface Science*, **186**, 447–453.

Staples, E.J., and Viswanathan, S. (2005). Detection of contrabands and drugs of abuse using an ultrafast SAW/GC vapor analyzer. Paper 563a, Annual American Institute of Chemical Engineers Meeting, Cincinnati, Ohio, October 31–November 4.

Swearingen, C.B., Wernette, D.P., Cropek, D.M., Sweedler, J.V., and Bohn, P.W. (2005). Immobilization of a catalytic DNA molecular beacon on Au for Pb(II) detection. *Analytical Chemistry*, **77**, 442–448.

Travis, C.C., and Hester, S.T. (1991). Global chemical pollution. *Environmental Science and Technology*, **25**, 814–819.

United Nations Environment Programme (UNEP) Millenium (1999). *Report on the Environment, Global Environmental Outlook*. Earthscan Publications, London, England.

Vicsek, T. (1989). *Fractal Growth Phenomena*, World Scientific Publishing Company, Singapore.

Wang, B., Jansson, I., Schenkman, J.B., and Rushing, J.F. (2005). Evaluating enzymes that generate genotoxic benzo[a]pyrene metabolites using sensor arrays. *Analytical Chemistry*, **77**, 1361–1367.

Wang, J. (1985). *Stripping Analysis: Principles, Instrumentation, and Applications*, VCH Publishers, Deerfield Beach, Florida.

Wang, J. (1996). Electrochemical preconcentration. In: *Laboratory Techniques in Electroanalytical Chemistry* (eds P.T. Kissinger, and W.R. Heinemann), 2nd edn., Marcel Dekker, New York, pp. 719–738.

Wang, J. (2000). In: *Electrochemistry*, 2nd edn., Wiley, New York, pp. 75–84.

Wang, T., Jia, X., and Wu, J. (2003). Direct determination of metals in organics by inductively coupled plasma atomic emission spectrometry in aqueous matrices. *Pharmaceutical and Biomedical Analysis*, **33**, 639–646.

Williams, G.M., Kroes, R., and Munro, I.C. (2000). *Regulatory Toxicology and Pharmacology*, **31**, 117–165.

Williams, R.D., and Bhethanabotla, V.R. (2005). Organic vapor sensing and discrimination using high frequency shear mode devices. Paper 563d, Annual American Institute of Chemical Engineers Meeting, Cincinnati, Ohio, October 31–November 4.

Wiyaratn, W., Hrapovic, S., Liu, Y., Surareungchai, W., and Luong, J.H.T. (2005). Light-assisted synthesis of Pt-Zn porphyrin nanocomposites and their use for electrochemical detection of organohalides. *Analytical Chemistry*, **77**, 5742–5749.

Wygladacz, K., Radu, A., Xu, C., Qin, Y., and Bakker, E. (2005). Fiber-optic microsensor array based on fluorescent bulk optode microspheres for the trace analysis of silver ions. *Analytical Chemistry*, **77**, 4706–4712.

Yang, X.J., Foley, R., and Low, G.K.C. (2002). A modified digestion procedure for analyzing silver in environmental samples. *Analyst*, **127**, 315–318.

Zhang, S., Zhao, H., and John, R. (2001). Development of a quantitative relationship between inhibition percentage and both incubation time and inhibitor concentration for inhibition biosensors-theoretical and practical considerations. *Biosensors & Bioelectronics*, **16**, 1119–1126.

– 9 –

Fractal Analysis of Proteins Involved in Drug Design on Biosensor Surfaces

9.1 INTRODUCTION

Drug–lipid interactions are an important and crucial step in drug discovery. Drug candidates should be soluble in cell membranes. Abdiche and Myszka (2004) indicate that if one were able to predict this, then this would permit a rational method of drug design. These authors further add that the surface plasmon resonance (SPR) biosensor permits the evaluation of these interactions in an automated mode. Furthermore, Markgren *et al.* (2000) indicate that an important step in drug discovery is the identification, design, synthesis, and characterization of enzyme inhibitors. They provide the example of drugs against HIV. This approach has resulted in quite a few potent inhibitors of HIV-1 reverse transcriptase and protease. They further indicate the iterative optimization process involved that includes the identification of (a) a lead compound with proven inhibitory effect, and (b) the subsequent systematic structural variations of the lead compound to provide the required properties.

Nordin *et al.* (2005) recently analyzed the kinetics of small molecule interactions with kinases. These authors indicate that kinases are a large family of proteins. These proteins catalyze the transfer of γ-phosphate group of ATP or GTP to the hydroxyl group of tyrosine, threonine, or serine. Davies *et al.* (2000) and Cohen (1999) indicated that as many as a third of all human proteins may be substrates for protein kinases.

Kamimori *et al.* (2005) indicate that cyclotides exhibit a broad range of biological activities that include antimicrobial (Tam *et al.*, 1999), cytotoxic (Lindholm *et al.*, 2002), and anti-HIV activities (Bokesch *et al.*, 2001). Craik *et al.* (1999) indicate that these cyclotides are circular proteins ($\sim$28–37 amino acids), and have been isolated from various Rubiaceae and Violacea plants. Mozsolits *et al.* (2002) recently analyzed the membrane-binding proteins of antimicrobial peptides using the SPR biosensor to gain insights into their mechanisms of action. Kamimori *et al.* (2005) recently used the SPR biosensor to analyze the binding and dissociation interactions and kinetics of the cyclotides kalata B1 and its naturally occurring analogue kalata B6 with five model lipid membranes. These authors indicate that their results show that these two kalata peptides indicate a high affinity for phosphatidylethanolamine-containing membranes. Thus, these peptides would bind to bacterial membranes.

In this chapter, we use fractals to re-analyze the binding and dissociation kinetics of kalata B1 and kalata B6 in solution to DMPE (dimyristoyl-L-α-phosphatidylethanolamine) liposome immobilized on the L1 sensor chip (Kamimori *et al.*, 2005). Values of the binding and dissociation rate coefficients and affinity values are provided. Fractal dimension values in the binding and in the dissociation phase are also provided. The fractal analysis may be considered as an alternate analysis to the SPR Biacore software analysis. One is unable to compare the analysis presented in this chapter with that provided by the SPR analysis, since the SPR analysis ignores the presence of heterogeneity of receptors on the sensing surface, and presumes that diffusional limitations are not present if the SPR biosensor is run properly.

9.2 THEORY

Havlin (1989) has reviewed and analyzed the diffusion of reactants toward fractal surfaces. The details of the theory and the equations involved for the binding and the dissociation phases for analyte–receptor binding are available (Sadana, 2001). The details are not repeated here, except that just the equations are given to permit an easier reading. These equations have been applied to other biosensor systems (Ramakrishnan and Sadana, 2001; Sadana, 2001). For most applications, a single- or a dual-fractal analysis is often adequate to describe the binding and the dissociation kinetics. Peculiarities in the values of the binding and the dissociation rate coefficients as well as in the values of the fractal dimensions with regard to the dilute analyte systems being analyzed will be carefully noted, if applicable.

9.2.1 Single-fractal analysis

Binding rate coefficient

Havlin (1989) indicates that the diffusion of a particle (analyte [Ag]) from a homogeneous solution to a solid surface (e.g., receptor [Ab]-coated surface) on which it reacts to form a product (analyte–receptor complex; (Ab·Ag)) is given by

$$\text{(Analyte·Receptor)} \sim \begin{cases} t^{(3-D_{\mathrm{f,bind}})/2} = t^{p} & (t < t_{\mathrm{c}}) \\ t^{1/2} & (t > t_{\mathrm{c}}) \end{cases} \tag{9.1a}$$

Here $D_{\mathrm{f,bind}}$ or D_{f} (used later on in the chapter) is the fractal dimension of the surface during the binding step; t_{c} is the cross-over value. Havlin (1989) indicates that the cross-over value may be determined by using $r_c^2 \sim t_{\mathrm{c}}$. Above the characteristic length r_{c}, the self-similarity of the surface is lost and the surface may be considered homogeneous. Above time t_{c}, the surface may be considered homogeneous, since the self-similarity property disappears and 'regular' diffusion is now present. For a homogeneous surface where $D_{\mathrm{f}} = 2$ and when only diffusional limitations are present, $p = ½$, as it should be. Another way of looking at the $p = ½$ case (where $D_{\mathrm{f,bind}} = 2$) is that the analyte in solution views the fractal object: in our case, the receptor-coated biosensor surface from a 'large distance.' In essence, in the association process, the diffusion of the analyte from the solution to the receptor surface creates a depletion layer of width $(Đt)^{1/2}$, where $Đ$ is the diffusion constant. This gives rise to the fractal power law

$$(\text{Analyte}\cdot\text{Receptor}) \sim t^{(3-D_{\text{f,bind}})/2}$$

For the present analysis, t_c is arbitrarily chosen and we assume that the value of the t_c is not reached. One may consider the approach as an intermediate 'heuristic' approach that may be used in the future to develop an autonomous (and not time-dependent) model for diffusion-controlled kinetics.

Dissociation rate coefficient

The diffusion of the dissociated particle (receptor [Ab] or analyte [Ag]) from the solid surface (e.g., analyte [Ag]–receptor [Ab]) complex coated surface) into solution may be given, as a first approximation by

$$(\text{Analyte}\cdot\text{Receptor}) \sim -k' t^{(3-D_{\text{f,diss}})/2} \qquad (t > t_{\text{diss}}) \tag{9.1b}$$

Here $D_{\text{f,diss}}$ is the fractal dimension of the surface for the dissociation step. This corresponds to the highest concentration of the analyte–receptor complex on the surface. Henceforth, its concentration only decreases. The dissociation kinetics may be analyzed in a manner 'similar' to the binding kinetics.

9.2.2 Dual-fractal analysis

Binding rate coefficient

Sometimes, the binding curve exhibits complexities and two parameters (k, D_f) are not sufficient to adequately describe the binding kinetics. This is further corroborated by low values of r^2 factor (goodness-of-fit). In that case, one resorts to a dual-fractal analysis (four parameters: k_1, k_2, D_{f1}, and D_{f2}) to adequately describe the binding kinetics. The single-fractal analysis presented above is thus extended to include two fractal dimensions. At present, the time ($t = t_1$) at which the 'first' fractal dimension 'changes' to the 'second' fractal dimension is arbitrary and empirical. For the most part, it is dictated by the data analyzed and experience gained by handling a single-fractal analysis. A smoother curve is obtained in the 'transition' region if care is taken to select the correct number of points for the two regions. In this case, the product (antibody–antigen, or analyte–receptor complex; Ab·Ag, or analyte·receptor) is given by

$$(\text{Ab}\cdot\text{Ag}) \approx \begin{cases} t^{(3-D_{\text{f1,bind}})/2} = t^{p1} & (t < t_1) \\ t^{(3-D_{\text{f2,bind}})/2} = t^{p2} & t_1 < t < t_2 = t_c \\ t^{1/2} & (t > t_c) \end{cases} \tag{9.1c}$$

In some cases, as mentioned above, a triple-fractal analysis with six parameters (k_1, k_2, k_3, D_{f1}, D_{f2}, and D_{f3}) may be required to adequately model the binding kinetics. This is when the binding curve exhibits convolutions and complexities in its shape perhaps due to the very dilute nature of the analyte (in some of the cases to be presented) or for some other reasons. Also, in some cases, a dual-fractal analysis may be required to describe the dissociation kinetics.

9.3 RESULTS

At the outset, it is appropriate to indicate that a fractal analysis will be applied to the data obtained from the literature for SPR studies of biomolecular interaction studies of cyclotides kalata B1 and its analogue kalata B6 in solution to DMPE liposome immobilized on an L1 sensor chip (Kamimori *et al.*, 2005). This is one possible explanation to analyze diffusion-limited binding kinetics assumed to be present in the system to be analyzed. The parameters thus obtained would provide a useful comparison of different situations. Alternate explanations involving saturation, first-order reaction, and no diffusion limitations are possible, but they are apparently deficient in describing the heterogeneity that inherently exists on the surface. The binding (and dissociation) on the biosensor surface (SPR or other biosensor) is a complex reaction, and the fractal analysis via the fractal dimension and the binding and the dissociation rate coefficient provides a useful lumped parameter(s) analysis of the diffusion-limited situation.

Note that the SPR, as indicated elsewhere in this book (for example, the BIAcore instrumentation manufactured by Biacore, Uppsala, Sweden) utilizes a carboxymethylated dextran surface, which, according to them, under appropriate and careful usage leads to diffusion-free binding and dissociation kinetics. There are references to this effect available in the literature whereby first-order kinetics without heterogeneity on the surface describes the diffusion-free binding kinetics (Karlsson *et al.*, 1997; Lundstrom, 1994). Furthermore, good performance is also demonstrated for small molecules (Karlsson *et al.*, 1995). This is a widely used and expensive biosensor, and as indicated elsewhere in this book, what we are offering or presenting here is an alternate explanation to describe the binding (and dissociation) kinetics that includes both diffusional limitations and heterogeneity on the surface. This would be especially true if the SPR is not carefully utilized. Finally, the analysis to be presented here would be of more value if we could offer an analysis of two different sets of experiments on the same sensing surface, one clearly diffusion limited and one kinetically limited to see if a fractal analysis is really required for the second case. However, since we are analyzing the data available in the literature, we are unable to judge or estimate if the data have been obtained under diffusion-free conditions. Thus, to be conservative, we have assumed that diffusion limitations are present and heterogeneity exists in all of the cases analyzed.

Kamimori *et al.* (2005) analyzed the binding and dissociation kinetics of the cyclotides kalata B1 and its naturally occurring analogue kalata B6 with five model lipid membranes using an SPR biosensor. Figures 9.1a–e show the binding and dissociation kinetics of 10, 20, 30, 40, and 50 μM kalata B1 cyclotide in solution to DMPE liposomes immobilized on L1 sensor chips. In all the cases, a dual-fractal analysis is required to describe the binding

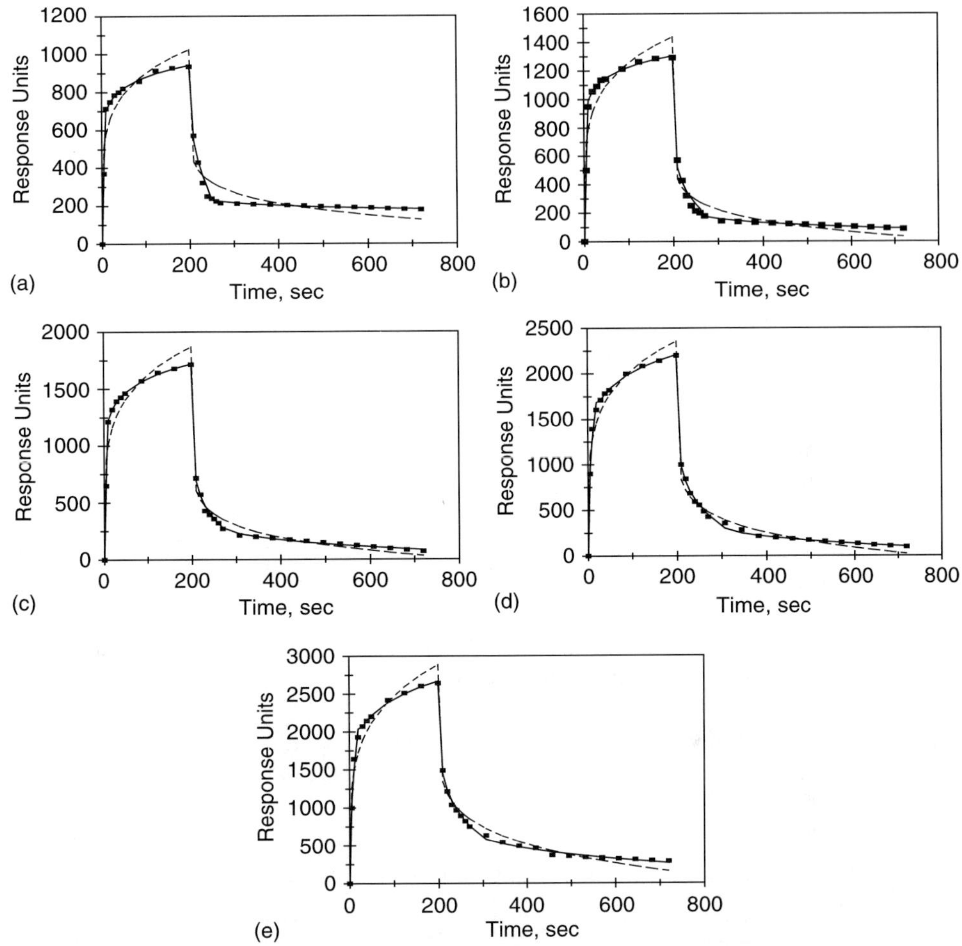

Figure 9.1 Binding and dissociation of different kalata B1 (cyclotide) concentrations (in μM) in solution to DMPE liposomes immobilized on an L1 sensor chip (Kamimori *et al.*, 2005): (a) 10, (b) 20, (c) 30, (d) 40, (e) 50. When only a solid line is used then a single-fractal analysis applies. When both a dashed (-------) and a solid (______) line are used, then the dashed line is for a single-fractal analysis, and the solid line is for a dual-fractal analysis. In this case the solid line provides the better fit.

and the dissociation kinetics. Table 9.1a shows the values of (a) the binding rate coefficient k for a single-fractal analysis, (b) the binding rate coefficients, k_1 and k_2, for a dual-fractal analysis, (c) the dissociation rate coefficient k_d for a single-fractal analysis, and (d) the dissociation rate coefficients, k_{d1} and k_{d2}, for a dual-fractal analysis. Table 9.1b shows the values of (a) the fractal dimension D_f in the binding phase for a single-fractal analysis, (b) the fractal dimensions, D_{f1} and D_{f2}, in the binding phase for a dual-fractal analysis, (c) the fractal dimension in the dissociation phase for a single-fractal analysis D_{fd}, and (d) the fractal dimensions, D_{fd1} and D_{fd2}, in the dissociation phase for a dual-fractal analysis. Corel

Table 9.1

(a) Binding and dissociation rate coefficients and (b) fractal dimensions for the binding and the dissociation phase for different kalata B1 concentrations in solution to DMPE liposomes immobilized on L1 sensor chip surfaces (Kamimori *et al.*, 2005)

(a)

Kalata B1 concentration (μM)	k	k_1	k_2	k_d	k_{d1}	k_{d2}
10	375.94 ± 62.11	180.29 ± 50.81	557.54 ± 5.97	381.89 ± 45.10	170.12 ± 11.46	628.92 ± 2.00
20	505.34 ± 85.82	229.23 ± 55.97	804.45 ± 6.74	681.02 ± 43.64	511.74 ± 26.75	956.54 ± 1.82
30	633.05 ± 98.15	312.14 ± 77.04	943.92 ± 4.73	870.19 ± 35.42	685.31 ± 16.18	1129.39 ± 7.03
40	788.59 ± 80.07	482.19 ± 61.47	1088.15 ± 7.57	1041.49 ± 83.63	782.45 ± 14.98	1476.73 ± 3.39
50	884.65 ± 7.38	493.25 ± 72.53	1308.48 ± 11.30	897.87 ± 42.49	699.53 ± 20.24	1352.42 ± 13.70

(b)

	D_f	D_{f1}	D_{f2}	D_{fd}	D_{fd1}	D_{fd2}
10	2.6216 ± 0.0838	1.9808 ± 0.5064	2.8018 ± 0.0118	2.7602 ± 0.0440	2.2848 ± 0.0776	2.9427 ± 0.0038
20	2.6052 ± 0.4456	1.9202 ± 0.4456	2.8168 ± 0.0092	2.8036 ± 0.0246	2.6428 ± 0.0432	2.9270 ± 0.0037
30	2.5912 ± 0.0790	1.9766 ± 0.4500	2.7734 ± 0.0054	2.7896 ± 0.0156	2.6524 ± 0.0232	2.9416 ± 0.0074
40	2.5866 ± 0.0530	2.1638 ± 0.2448	2.7326 ± 0.0774	2.7640 ± 0.0300	2.6210 ± 0.0190	2.8878 ± 0.0044
50	2.5536 ± 0.0628	2.0522 ± 0.2798	2.7306 ± 0.0096	2.6746 ± 0.0182	2.5332 ± 0.0284	2.8200 ± 0.0154

Quattro Pro (1997) was used to fit the data. The Corel Quattro Pro program provided the parameters presented in Table 9.1a. The values of the parameters presented in Table 9.1a are within 95% confidence limits. For example, the value of the binding rate coefficient k provided for a single-fractal analysis is 375.94 ± 62.11. The 95% confidence limits indicate that 95% of the k values will lie between 313.83 and 438.05. This indicates that the values are precise and significant.

In all fairness, this is one possible way by which to analyze the analyte–receptor binding and the dissociation data. One might justifiably argue that appropriate modeling might be achieved by using a Langmuirian or other approach. The Langmuir approach was originally developed for gases (Thomson and Webb, 1968). Consider a gas at constant pressure, p, in equilibrium with a surface. The rate of adsorption is proportional to the gas pressure and to the fraction of the uncovered surface. Adsorption will only occur when a gas molecule strikes a bare site. Researchers in the past have successfully modeled the adsorption behavior of analytes in solution to solid surfaces using the Langmuir model even though it does not conform to theory. Rudzinski *et al.* (1983) indicate that other appropriate 'liquid' counterparts of the empirical isotherms have been developed. These include counterparts of the Freundlich (Dabrowski and Jaroniec, 1979), Dubinin-Radushkevich (Oscik *et al.*, 1976), and Toth (Jaroniec and Derylo, 1981) empirical equations. These studies with their known constraints have provided some 'restricted' physical insights into the adsorption of adsorbates on different surfaces. The Langmuirian approach may be utilized to model the data presented if one assumes the presence of discrete classes of sites. One might also attempt to model the data to be presented using a fractal (statistical) growth process such as DLA (diffusion-limited aggregation), kinetic gelation, or invasion percolation. However, at present, no such attempt is made.

Kamimori *et al.* (2005) indicate that in all cases there is a proportional increase in the response in the binding phase as the kalata B1 concentration increases in solution. This to them indicated that the system has not reached saturation. This is consistent with the results presented in Table 9.1a wherein there is an increase in the binding rate coefficients, k_1 and k_2, with an increase in the kalata B1 concentration in solution. For the data presented in Table 9.1a, and in the 10–50 μM kalata B1 concentration range in solution, the binding rate coefficients, (1) k_1 and (2) k_2, are given by

$$k_1 = (34.703 \pm 5.114)[\text{kalata B1}]^{0.6759 \pm 0.1081} \tag{9.2a}$$

$$k_2 = (172.64 \pm 6.147)[\text{kalata B1}]^{0.5077 \pm 0.0275} \tag{9.2b}$$

In case (1), the fit is good. The availability of more data points would lead to a more reliable fit. The binding rate coefficient k_1 is only mildly sensitive to the kalata B1 concentration in solution, as noted by the low order (= 0.6759) of dependence exhibited. In case (2), the fit is very good. Once again, the binding rate coefficient k_2 is only mildly sensitive to the kalata B1 concentration in solution, as noted by the low order (= 0.5077) of dependence exhibited.The non-integer order of dependence exhibited by the binding rate coefficients, k_1 and k_2, on the kalata B1 concentration in solution lends support to the fractal nature of the system.

For the data presented in Table 9.1a, and in the 10–50 μM kalata B1 concentration range in solution, Figures 9.2c and d show the decrease in the fractal dimensions, (1) D_{f2} and (2) D_{fd2}, with an increase in the kalata B1 concentration in solution. In the 10–50 μM kalata B1 concentration range in solution, the fractal dimensions, D_{f2} and D_{fd2}, are given by

$$D_{f2} = (2.9597 \pm 0.0244)[\text{kalata B1}]^{-0.0201 \pm 0.00646} \tag{9.2c}$$

$$D_{fd2} = (3.1126 \pm 0.0428)[\text{kalata B1}]^{-0.0213 \pm 0.0107} \tag{9.2d}$$

In case (1) the fit is good and in case (2) the fit is reasonable. The availability of more data points would lead to a more reliable fit. The fractal dimensions, D_{f2} and D_{fd2}, are only slightly sensitive to the kalata B1 concentration in solution, as noted by the low order

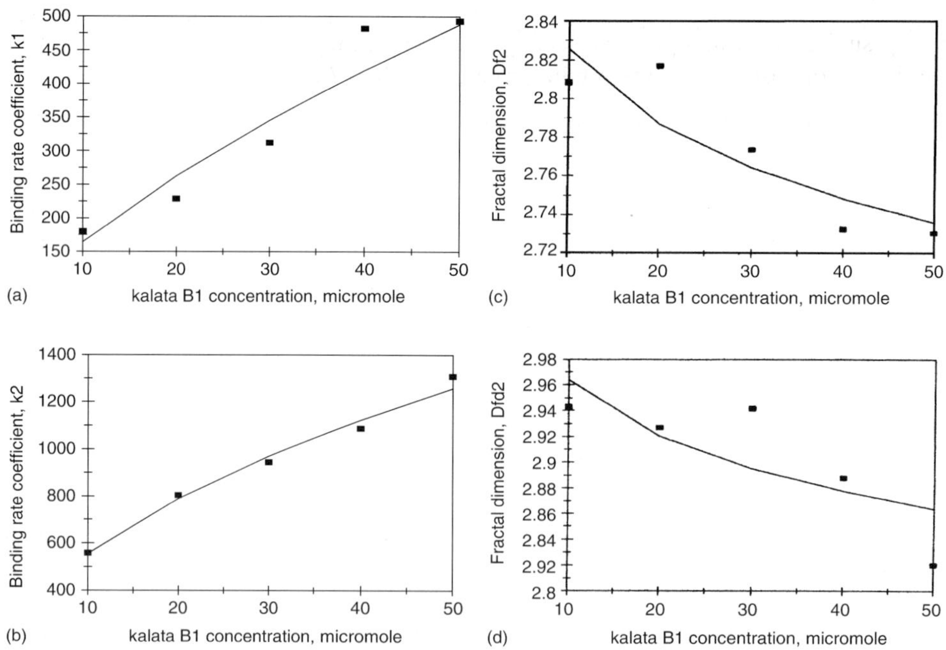

Figure 9.2 Increase in the (a) binding rate coefficient, k_1, with an increase in the kalata B1 concentration in solution, (b) binding rate coefficient, k_2, with an increase in the kalata B1 concentration in solution, (c) decrease in the fractal dimension, D_{f2}, in the binding phase with an increase in the kalata B1 concentration in solution, and (d) decrease in the fractal dimension, D_{fd2}, in the dissociation phase with an increase in the kalata B1 concentration in solution.

(=−0.0201 and −0.0213) of dependence exhibited. Note, however, that the fractal dimension is based on a log scale, and even small changes in the fractal dimension exhibit significant changes in the degree of heterogeneity on the sensor chip surface.

For the data presented in Table 9.1a, and in the 10–50 μM kalata B1 concentration range in solution, Figures 9.3a and b show the increase in the binding rate coefficient, k_1 and decrease in the binding rate coefficient k_2, with an increase in the fractal dimensions, D_{f1} and D_{f2}. In the 10–50 μM kalata B1 concentration range in solution, the binding rate coefficients, k_1 and k_2, are given by

$$k_1 = (1.5421 \pm 0.587) D_{f1}^{\;7.5785 \pm 3.5254} \tag{9.3a}$$

$$k_2 = (2.3\text{E}+11 \pm 0.5\,\text{E}+11) D_{f2}^{\;-18.99 \pm 7.371} \tag{9.3b}$$

In both the cases, the fit is good. The availability of more data points would lead to a more reliable fit. The binding rate coefficients, k_1 and k_2, are very sensitive to the degree of heterogeneity on the surface or the fractal dimensions, D_{f1} and D_{f2}, as noted by the very high order (= 7.5785 and −18.99) of dependence exhibited. This indicates that even very small changes in the value of the fractal dimensions, D_{f1} and D_{f2}, on the sensor chip surface lead to significant changes in the binding rate coefficients, k_1 and k_2. Note that as opposed to k_1, the binding rate coefficient k_2 exhibits a decrease with an increase in the degree of heterogeneity on the surface. No explanation is offered at present for this effect.

Affinity values are of interest to practicing biosensorists. Affinity K_1 may be defined as k_1/k_{d1}, which is the ratio of the binding rate coefficient to the dissociation rate coefficient. For the data presented in Table 9.1a, and in the 10–50 μM kalata B1 concentration range in solution, Figure 9.4a shows the increase in the affinity K_1 with an increase in the ratio D_{f1}/D_{fd1}. In the 10–50 μM kalata B1 concentration range in solution, the affinity K_1 is given by

$$K_1 = (k_1/k_{d1}) = (1.7817 \pm 0.2699)(D_{f1}/D_{fd1})^{4.5330 \pm 0.9571} \tag{9.4a}$$

The fit is good. The availability of more data points would lead to a more reliable fit. The affinity K_1 is very sensitive to the ratio of the fractal dimensions in the binding phase and in the dissociation phase, as noted by the very high order (= 4.5330) of dependence exhibited. This indicates that even very small changes in the ratio of the fractal dimensions on the sensor chip surface leads to very significant changes in the affinity value K_1. This is one way of manipulating the affinity, K_1, value. This can be tricky in the sense that one has to manipulate the ratio of the degree of heterogeneities present in the binding phase and in the dissociation phase. At present, at best one can just do with trial-and-error. It would be very useful to note, especially in this case, if one can, for example, increase the degree of heterogeneity in the binding phase, and simultaneously not change the degree of heterogeneity for the

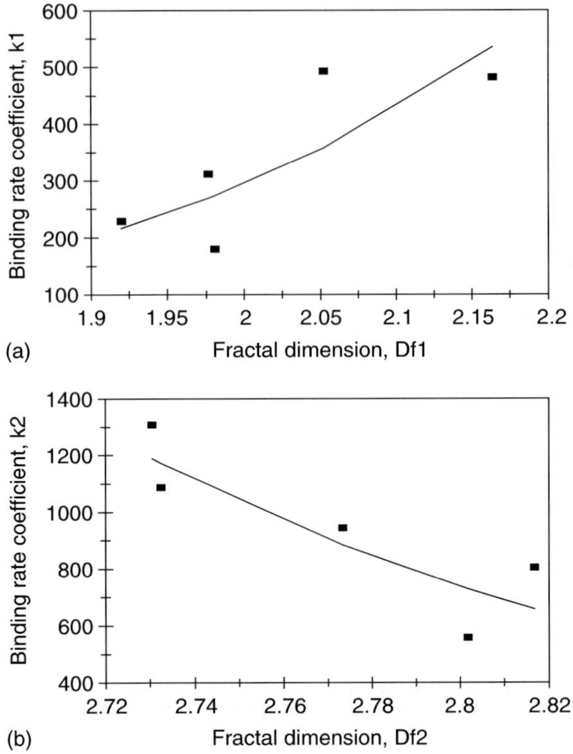

Figure 9.3 (a) Increase in the binding rate coefficient, k_1, with an increase in fractal dimension, D_{f1}, and (b) decrease in the binding rate coefficient, k_2, with an increase in fractal dimension, D_{f2}.

dissociation phase, or perhaps change it very insignificantly. In other words, selectively increase the degree of heterogeneity in the binding phase only. This would help to enhance the affinity, K_1, value. This is presuming that one is looking for higher K_1 values. Surely, if the reverse is required then one should be able to manipulate the degree of heterogeneity on the sensor chip accordingly.

Affinity K_2 may be defined as k_2/k_{d2}, which is the ratio of the binding rate coefficient to the dissociation rate coefficient. For the data presented in Table 9.1a, and in the 10–50 μM kalata B1 concentration range in solution, Figure 9.4b shows the increase in the affinity K_2 with an increase in the ratio, D_{f2}/D_{fd2}. In the 10–50 μM kalata B1 concentration range in solution, the affinity K_2 is given by

$$K_2 = (k_2/k_{d2}) = (1.129 \pm 0.098)(D_{f2}/D_{fd2})^{6.0351 \pm 3.659} \quad (9.4b)$$

The fit is reasonable. The availability of more data points would lead to a more reliable fit. There is scatter in the data. The affinity K_2 is very sensitive to the ratio of the fractal dimensions in the binding phase and the dissociation phase as noted by the very high order

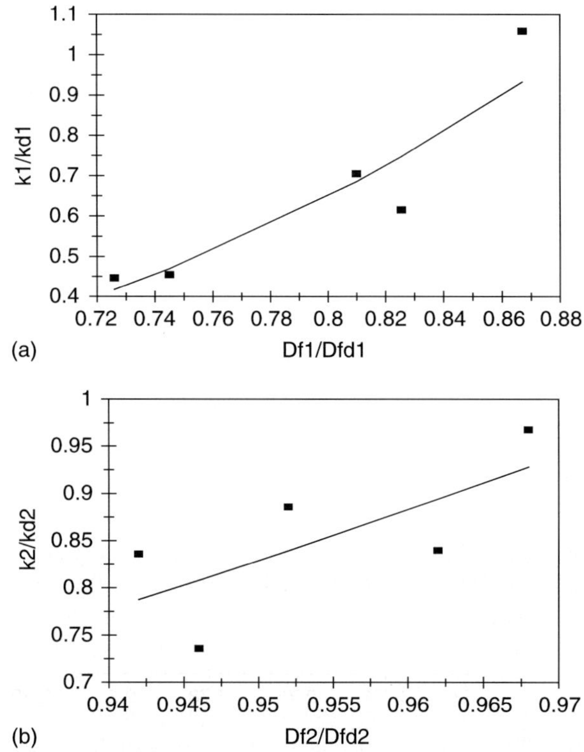

Figure 9.4 (a) Increase in the affinity, K_1 ($= k_1/k_{d1}$), with an increase in fractal dimension ratio, D_{f1}/D_{fd1}, and (b) increase in the affinity, K_2 ($= k_2/k_{d2}$), with an increase in fractal dimension ratio, D_{f2}/D_{fd2}.

(= 6.0351) of dependence exhibited. This indicates that even very small changes in the ratio of the fractal dimensions on the sensor chip surface lead to very significant changes in the affinity value K_2. This is, once again, one way of manipulating the affinity, K_2, value. This can be tricky, as indicated above, in the sense that one has to manipulate the ratio of the degree of heterogeneities present in the binding phase and in the dissociation phase. At present, at best one can just do that, and as indicated above with trial-and-error.

It is of interest to note for a dual-fractal analysis how the binding rate coefficient ratio, k_2/k_1, changes with the change in the ratio of the degrees of heterogeneity in the binding phase, D_{f2}/D_{f1}, on the sensor chip surface. For the data presented in Table 9.1a and 9.1b, and in the 10–50 μM kalata B1 concentration range in solution, Figure 9.5 shows the increase in the ratio of the binding rate coefficients, k_2/k_1, with an increase in the ratio of the fractal dimensions, D_{f2}/D_{f1}. The ratio of the binding rate coefficients, k_2/k_1, is given by

$$k_2/k_1 = (1.1592 \pm 0.0142)(D_{f2}/D_{f1})^{2.8614 \pm 0.1047} \quad (9.5)$$

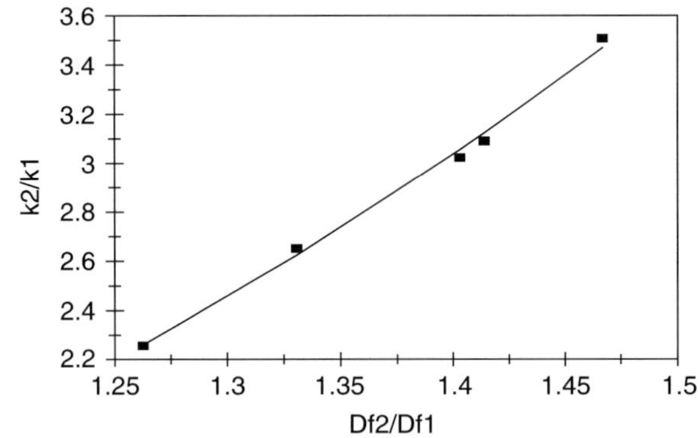

Figure 9.5 Increase in the ratio of the binding rate coefficients, k_2/k_1, with an increase in the fractal dimension ratio, D_{f2}/D_{f1}, for kalata B1 (cyclotide).

The fit is very good. The ratio of the binding rate coefficients, k_2/k_1, exhibits an order of dependence between 2 and 3 (= 2.8614) on the ratio of the fractal dimensions, D_{f2}/D_{f1}. These represent the fractal dimension in the first (D_{f1}) and in the second (D_{f2}) phase of binding. In this case, the binding rate coefficient ratio increases from about 2.2 to 3.6 as the ratio of fractal dimensions increases from 1.26 to 1.46.

Figures 9.6a and b show the binding and dissociation kinetics of 10 and 20 μM kalata B6 cyclotide in solution to DMPE liposomes immobilized on L1 sensor chips. In both the cases, a dual-fractal analysis is required to describe the binding and the dissociation kinetics. Figures 9.6c–e show the binding and dissociation kinetics of 30–50 μM kalata B6 in solution to DMPE liposomes immobilized on L1 sensor chips. A dual-fractal analysis is required to describe the binding kinetics. A single-fractal analysis is adequate to describe the dissociation kinetics. This indicates that there is a change in the dissociation mechanism at the higher (30–50 μM) B6 concentration in solution, since a single-fractal analysis is adequate to describe the dissociation kinetics when compared to the dual-fractal analysis required at the lower (10–20 μM) B6 concentration in solution. Table 9.2a shows the values of (a) the binding rate coefficient k for a single-fractal analysis, (b) the binding rate coefficients, k_1 and k_2, for a dual-fractal analysis, (c) the dissociation rate coefficient k_d for a single-fractal analysis, and (d) the dissociation rate coefficients, k_{d1} and k_{d2}, for a dual-fractal analysis. Table 9.2b shows the values of (a) the fractal dimension D_f in the binding phase for a single-fractal analysis, (b) the fractal dimensions, D_{f1} and D_{f2}, in the binding phase for a dual-fractal analysis, (c) the fractal dimension in the dissociation phase for a single-fractal analysis D_{fd}, and (d) the fractal dimensions, D_{fd1} and D_{fd2}, in the dissociation phase for a dual-fractal analysis.

It is of interest to compare the binding rate coefficient k_1 value when a dual-fractal analysis applies for the cyclotide kalata B1 and its analogue kalata B6, when they are present in solution in the 10–50 μM concentration range. Note that at the respective concentrations, the binding coefficient k_1 value is higher for kalata B1 than for kalata B6 for each of the10–50 μM concentrations used. The corresponding values of the fractal dimension

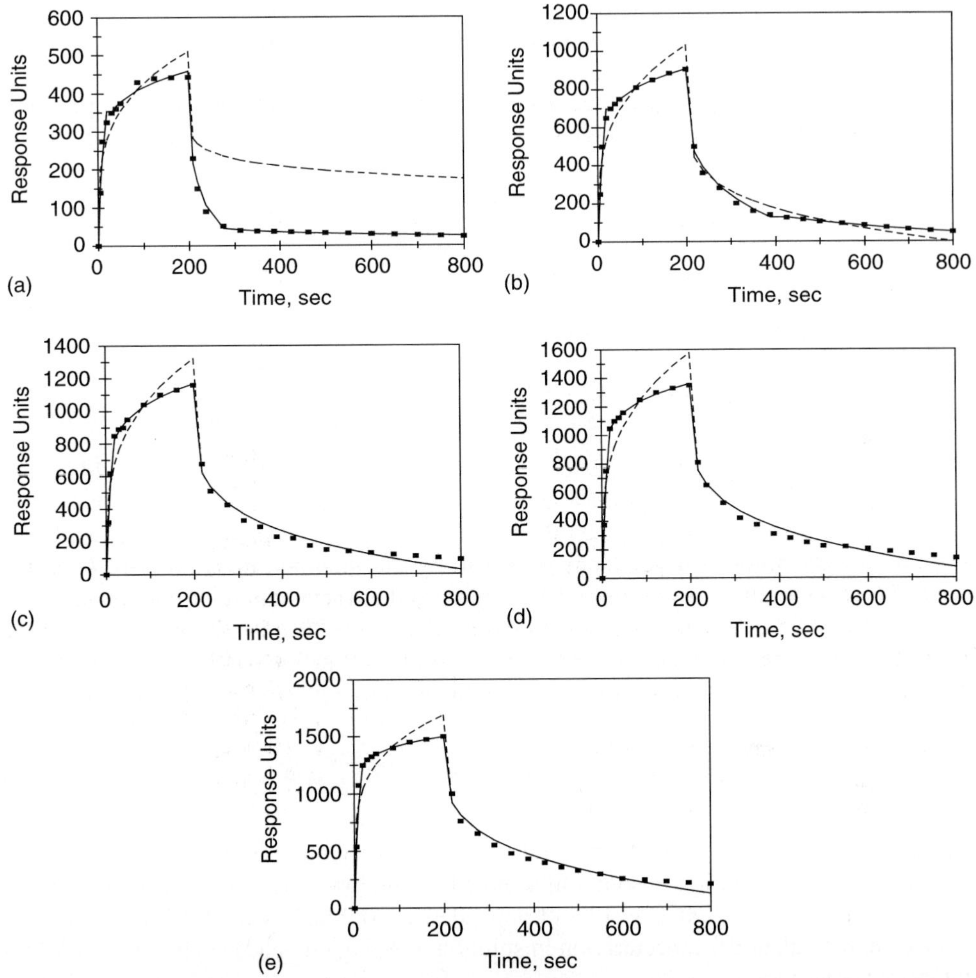

Figure 9.6 Binding and dissociation of different kalata B6 (cyclotide analogue) concentrations (in μM) in solution to DMPE liposomes immobilized on an L1 sensor chip (Kamimori *et al.*, 2005): (a) 10, (b) 20, (c) 30, (d) 40, (e) 50. When only a solid line is used then a single-fractal analysis applies. When both a dashed (------) and a solid (_____) line are used, then the dashed line is for a single-fractal analysis, and the solid line is for a dual-fractal analysis. In this case the solid line provides the better fit.

D_{f1} are also higher for the natural kalata B1 when compared with those obtained for the cyclotide analogue, kalata B6. At the lower end of the concentration (10 μM), the ratio of the corresponding k_1 values is 3.14, and at the higher end of he concentration (50 μM), the ratio of the corresponding k_1 values is 2.2.

Similarly, note that the binding rate coefficient k_2 values are higher for the natural cyclotide kalata B1 when compared with its analogue kalata B6 for each of the corresponding concentrations used in the range 10–50 μM. This would indicate that, at least, in this 10–50 μM concentration range the natural cyclotide kalata B1 binds better that its cyclotide analogue kalata B6.

Kamimori *et al.* (2005) once again indicate that in all cases there is a proportional increase in the response in the binding phase as the kalata B6 concentration increases in solution. This to them indicated, once again, that the system has not reached saturation. This is consistent with the results presented in Table 9.2a wherein there is an increase in the binding rate coefficients, k_1 and k_2, with an increase in the kalata B6 concentration in solution. For the data presented in Table 9.2a, and in the 10–50 μM kalata B6 concentration range in solution, the binding rate coefficient k_1 is given by

$$k_1 = (10.016 \pm 2.181)[\text{kalata B6}]^{0.7301 \pm 0.1551} \tag{9.6a}$$

Please see Figure 9.7a. The fit is good. The availability of more data points would lead to a more reliable fit. The binding rate coefficient k_1 is only mildly sensitive to the kalata B6 concentration in solution, as noted by the low order (= 0.7301) of dependence exhibited. The non-integer order of dependence exhibited by the binding rate coefficient k_1 on the kalata B6 concentration in solution lends support to the fractal nature of the system.

Figure 9.7b shows the increase in the binding rate coefficient k_2 with an increase in the kalata B6 concentration in solution. For the data presented in Table 9.2a, and in the 10–50 μM kalata B6 concentration range in solution, the binding rate coefficient k_2 is given by

$$k_2 = (27.413 \pm 2.424)[\text{kalata B6}]^{0.9027 \pm 0.0667} \tag{9.6b}$$

The fit is very good. Once again, the binding rate coefficient k_2 is only mildly sensitive to the kalata B6 concentration in solution, as noted by the low order (= 0.9027) of dependence exhibited. The non-integer order of dependence exhibited by the binding rate coefficient k_2 on the kalata B6 concentration in solution, once again, lends support to the fractal nature of the system.

For the data presented in Table 9.2a, and in the 10–50 μM kalata B6 concentration range in solution, Figure 9.8 shows the increase in the binding rate coefficient k_1 with an increase in the fractal dimension D_{f2}. In this 10–50 μM kalata B6 concentration range, the binding rate coefficient k_2 is given by

$$k_2 = (7.2\text{E}-07 \pm 4.9\text{E}-07)D_{f2}^{\;20.13 \pm 10.56} \tag{9.7}$$

The fit is reasonable. There is some scatter in the data. The availability of more data points would lead to a more reliable fit. The binding rate coefficient k_2 is extremely sensitive to the fractal dimension D_{f2}, or the degree of heterogeneity present on the sensor chip surface as noted by the very high order (= 20.13) exhibited. It is of interest to note that the binding rate coefficient k_2 exhibits (a) an increase with an increase in the degree of heterogeneity on the sensor chip surface for the cyclotide analogue kalata B6, whereas (b) it

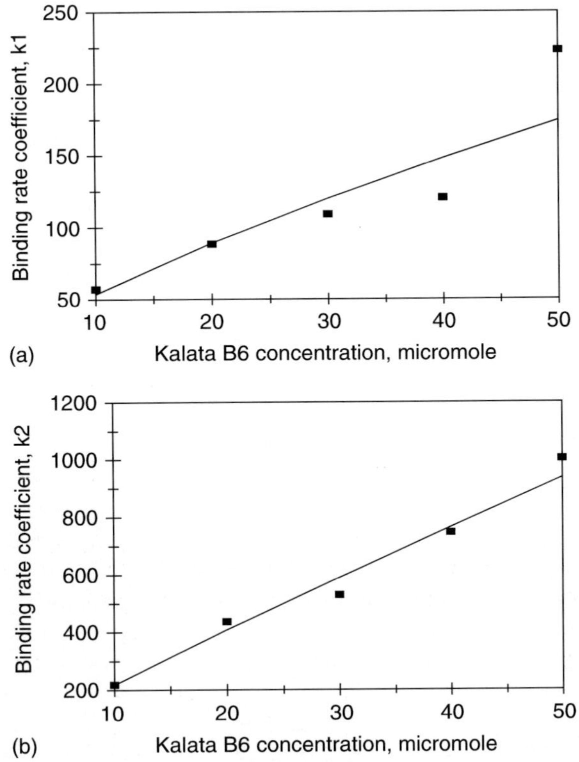

Figure 9.7 Increase in the (a) binding rate coefficient, k_1, with an increase in the kalata B6 concentration in solution, and (b) binding rate coefficient, k_2, with an increase in the kalata B6 concentration in solution.

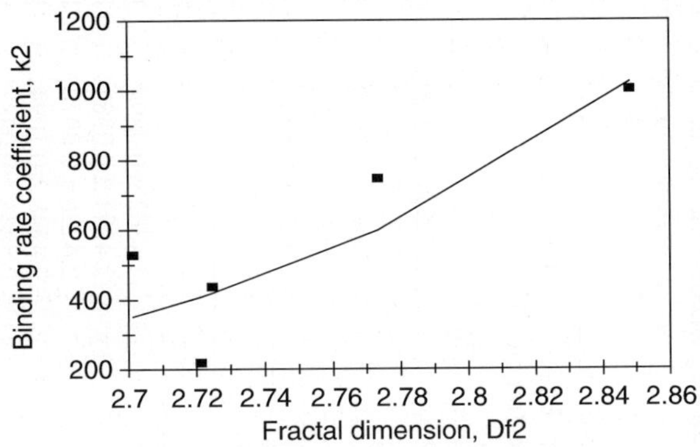

Figure 9.8 Increase in the binding rate coefficient, k_2, with an increase in the fractal dimension, D_{f2}.

Table 9.2

(a) Binding and dissociation rate coefficients and (b) fractal dimensions for the binding and the dissociation phase for different kalata B6 concentrations in solution to DMPE liposomes immobilized on L1 sensor chip surfaces (Kamimori *et al.*, 2005)

(a)

Kalata B6 concentration (μM)	k	k_1	k_2	k_d	k_{d1}	k_{d2}
10	128.688 ± 22.693	57.315 ± 13.212	219.23 ± 6.927	199.22 ± 18.75	114.29 ± 9.82	628.92 ± 2.00
20	227.65 ± 44.84	88.583 ± 17.035	437.51 ± 1.092	262.76 ± 16.00	201.05 ± 9.77	956.54 ± 1.82
30	286.35 ± 55.32	109.049 ± 16.539	529.41 ± 6.54	286.07 ± 13.09	na	na
40	352.99 ± 77.81	120.42 ± 18.88	746.78 ± 5.06	312.05 ± 14.97	na	na
50	560.63 ± 9.87	222.91 ± 54.71	1002.36 ± 2.88	276.34 ± 16.60	na	na

(b)

	D_f	D_{f1}	D_{f2}	D_{fd}	D_{fd1}	D_{fd2}
10	2.4802 ± 0.0894	1.7850 ± 0.4232	2.7214 ± 0.0347	2.7462 ± 0.0376	2.4054 ± 0.1096	2.9531 ± 0.0033
20	2.4294 ± 0.0987	1.6216 ± 0.3588	2.7248 ± 0.0028	2.6134 ± 0.0314	2.4827 ± 0.0429	2.8076 ± 0.0036
30	2.4226 ± 0.0970	1.5906 ± 0.2880	2.7014 ± 0.0138	2.5696 ± 0.0252	na	na
40	2.4348 ± 0.1094	1.5146 ± 0.2970	2.7734 ± 0.0062	2.5602 ± 0.0246	na	na
50	2.5844 ± 0.0982	1.7892 ± 0.4478	2.8482 ± 0.0032	2.4966 ± 0.0306	na	na

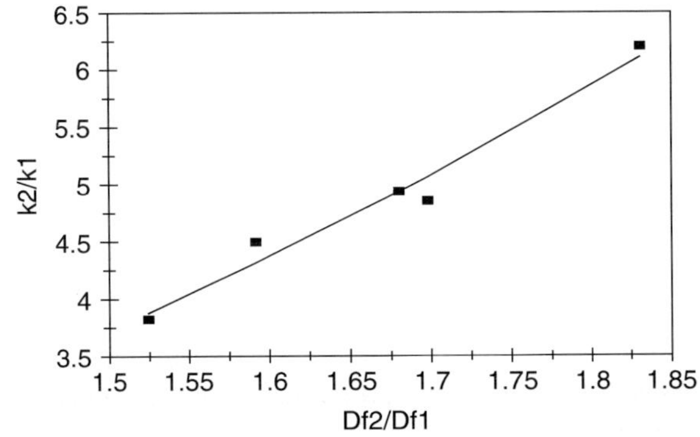

Figure 9.9 Increase in the ratio of the binding rate coefficients, k_2/k_1, with an increase in the fractal dimension ratio, D_{f2}/D_{f1}, for kalata B6 (cyclotide analogue).

exhibits a decrease with an increase in the degree of heterogeneity on the sensor chip surface for the natural cyclotide, kalata B1. No explanation is offered at present for this effect.

It is of interest to note for a dual-fractal analysis how the binding rate coefficient ratio, k_2/k_1, changes with the change in the ratio of the degrees of heterogeneity in the binding phase, D_{f2}/D_{f1}, on the sensor chip surface. For the data presented in Tables 9.1a and 9.1b, and in the 10–50 μM kalata B6 concentration range in solution, Figure 9.9 shows the increase in the ratio of the binding rate coefficients, k_2/k_1, with an increase in the ratio of the fractal dimensions, D_{f2}/D_{f1}. The ratio of the binding rate coefficients, k_2/k_1, is given by

$$k_2/k_1 = (1.364 \pm 0.051)(D_{f2}/D_{f1})^{2.4777 \pm 0.2612} \tag{9.8}$$

The fit is very good. The ratio of the binding rate coefficients, k_2/k_1, exhibits an order of dependence between 2 and 3 (=2.4777) on the ratio of the fractal dimensions, D_{f2}/D_{f1}. These represent the fractal dimension in the first (D_{f1}) and in the second (D_{f2}) phase of binding. In this case, the binding rate coefficient ratio increases from 3.825 to 6.201 as the ratio of fractal dimensions increases from 1.5245 to 1.831.

9.4 CONCLUSIONS

A fractal analysis of the binding (and dissociation) of the natural cyclotide, kalata B1 and its analogue, kalata B6 in solution to DMPE liposome immobilized on an L1 sensor chip (Kamimori *et al.*, 2005) is presented. These authors have indicated that cyclotides exhibit a broad range of biological activities that include antimicrobial (Tam *et al.*, 1999), cytotoxic (Lindholm *et al.*, 2002), and anti-HIV activities (Bokesch *et al.*, 2001). Mozsolits and Aguilar (2001) suggest that a better understanding of the binding (and dissociation) reactions on membrane surfaces would lead to better physical insights into these types of reactions. The fractal analysis is used with this purpose in mind to relate the binding (and dissociation) rate coefficients with the degree of heterogeneity or fractal dimension present on the DMPE liposome immobilized on the L1 sensor chip surface.

The fractal analysis provides a quantitative indication of the state of disorder (fractal dimension) and the binding (and dissociation) rate coefficients on the sensor chip surface. Both types of examples are given wherein either a single- or a dual-fractal analysis was used. The dual-fractal analysis was used only when the single-fractal analysis did not provide an adequate fit. This was done by the regression analysis provided by Corel Quattro Pro 8.0 (1997). In accordance with the prefactor analysis for fractal aggregates (Sorenson and Roberts, 1997), quantitative (predictive) expressions are developed for (a) the binding rate coefficients, $k_1$and k_2, as a function of the natural cyclotide, kalata B1 concentration in solution, (b) the fractal dimension D_f in the binding phase and the fractal dimension in the dissociation phase D_{fd} as a function of the cyclotide, kalata B1 concentration, (c) the binding rate coefficients, k_1 and k_2, as a function of the fractal dimensions, $D_{f1}$and D_{f2}, for the 10–50 μM kalata B1 concentration in solution, (d) the affinity, K_1 $(= k_1/k_{d1})$, as a function of the ratio of fractal dimensions, D_f/D_{fd1}, and (e) the ratio of the binding rate coefficients, k_2/k_1, as a function of the fractal dimensions, D_{f2}/D_{f1}. Similar predictive relations are also developed and presented for the cyclotide analogue, kalata B6.

The fractal dimension is not a classical independent variable such as analyte (antigen, antibody, or other biological molecule) concentration in solution. Nevertheless, the expressions obtained for the binding (and the dissociation) rate coefficients for a single- and a dual-fractal analysis as a function of the fractal dimension indicate a high sensitivity of these rate coefficients on the their respective fractal dimensions on the SPR sensor chip surface. For example, for the binding of 10–50 μM natural cyclotide kalata B1 in solution, the binding rate coefficient k_1 exhibits a 7.585 order of dependence on the fractal dimension D_{f1} or the degree of heterogeneity present on the L1 sensor chip surface. This emphasizes the importance of the extent of heterogeneity on the SPR biosensor chip surface and its impact on the binding rate coefficient.

Note that the data analysis in itself does not provide any evidence for surface roughness or heterogeneity, and the existence of surface roughness or heterogeneity assumed may not be correct. Considering the complexity involved on the SPR chip surface, this is not an unreasonable assumption. Furthermore, there is deviation in the data that may be minimized by providing a correction for the depletion of the analyte (in our case natural cyclotide kalata B1, or analogue B6) in the vicinity of the surface (imperfect mixing).

An increase in the binding rate coefficient value should lead to enhanced sensitivity and to a decrease in the response time of the SPR biosensor. Both of these aspects would be beneficial to the SPR biosensor construction. For a selective (or multiple) reaction system, if an increase in the fractal dimension leads to an increase in the binding rate coefficient (of interest), then this would enhance selectivity. Stability is a more complex issue, and one might intuitively anticipate that a distribution or heterogeneity of the receptor on the biosensor surface would lead to a more stable biosensor. In this case, one type of distribution may be better than the other depending on the biosensor performance of interest. Considering the complexity that is involved on the SPR biosensor chip surface, and if we are looking at stability, then this aspect is particularly valuable here. Similar behavior has been observed for the deactivation of enzymes wherein a distribution of activation energies for deactivation (as compared to a single activation energy for deactivation) leads to a more stable enzyme (Malhotra and Sadana, 1989).

For the most part, and as indicated throughout in different chapters in this book, the analysis is extendable to analyte–receptor reactions occurring on (a) other types of biosensors, and on (b) non-biosensor (for example, cell surface reactions on membranes as emphasized in this chapter) surfaces. Here it is hoped that the biosensor-based analysis would lead to a better understanding of the binding and dissociation interactions of the natural cylotide B1 and its analogue cyclotide B6 on membrane surfaces. This would very significantly facilitate the screening and selection of possible drug candidates with, for example, antimicrobial or other useful medicinal properties.

REFERENCES

Abdiche, Y.N., and Myszka, D.G. (2004). Probing the mechanism of drug/lipid membrane interactions using Biacore. *Analytical Biochemistry*, **328**, 233–243.

Bokesch, H.R., Pannell, L.K., Cochran, P.K., Sowder II, R.C., McKee, T.C., and Boyd, M.R. (2001). A novel anti-HIV macrocyclic peptide from *Palicourea condensata. Journal of Natural Products*, **64**, 249–250.

Cohen, P. (1999). The development and therapeutic potential of protein kinase inhibitors. *Current Opinion in Chemical Biology*, **3**, 459–465.

Corel Corporation. (1997). *Corel Quattro Pro 8.0*, Ottawa, Canada.

Craik, D.J., Daly, N.L., Bond, T.C., and Waine, C. (1999). Plant cyclotides: a unique family of cyclic and knotted proteins that defines cystine knot structural motif. *Journal of Molecular Biology*, **294**, 1327–1336.

Dabrowski, A., and Jaroniec, M. (1979). Effects of surface heterogeneities in adsorption from binary liquid mixtures. III. Analysis of experimental data using Langmuir–Freundlich type equations. *Journal of Colloid and Interface Science*, **73**, 475–482.

Davies, S.P., Reddy, H., Caivano, M., and Cohen, P. (2000). Specificity mechanism of action of some commonly used protein kinase inhibitors. *Biochemical Journal*, **351**, 95–105.

Havlin, S. (1989). Molecular diffusion and reaction. In: *The Fractal Approach to Heterogeneous Chemistry: Surfaces, Colloids, Polymers*, (ed. D. Avnir), Wiley, New York, pp. 251–269.

Jaroniec, M., and Derylo, Y. (1981). Simple relationships for predicting multisolute adsorption from dilute aqueous solutions. *Chemical Engineering Science*, **36**, 1017–1019.

Kamimori, H., Hall, K., Craik, D.J., and Aguilar, M.-I. (2005). Studies on the membrane interactions of the cyclotides kalata B1 and kalata B6 on model membrane systems by surface plasmon resonance. *Analytical Biochemistry*, **337**, 149–153.

Karlsson, R., and Falt, R. (1997). Experimental design for kinetic analysis of protein-protein interactions with suface plasmon resonance biosensors. *Journal of Immunological Methods*, **200**, 121–133.

Karlsson, R., and Stahlberg, R. (1995). Surface plasmon resonance detection and multispot sensing for direct monitoring of interactions involving low molecular weight analytes and for determination of low affinities. *Analytical Biochemistry*, **228**, 274–280.

Lindholm, P., Goransson, U., Johansson, S., Claeson, P., Gulbo, J., Larsson, R., Bohlin, L., and Backlund, A. (2002). Cyclotides: a novel type of cytotoxic agents. *Molecular Cancer Therapy*, **1**, 365–369.

Lundstrom, I. (1994). Real-time biospecific interaction analysis. *Biosensors and Bioelectronics*, **9**(9–10), 725–736.

Malhotra, A., and Sadana, A. (1989). Influence of diffusion on first-order deactivation of microheterogeneous enzyme samples. *Biotechnology and Bioengineering*, **34**(5), 725–730.

Markgren, P.-O., Hamalainenen, M., and Danielson, U.H. (2000). Kinetic analysis of the interaction between HIV-1 protease and inhibitors using optical biosensor technology. *Analytical Biochemistry*, **279**, 71–78.

Mozsolits, H., and Aguilar, M.I. (2002). Surface plasmon resonance spectroscopy: an emerging tool for the study of peptide-membrane interactions. *Biopolymers*, **66**, 3–18.

Mozsolits, H., Unabia, S., Ahmad, A., Morton, C.J., Thomas, W.G., and Aguilar, M.I. (2002). Electrostatic and hydrophobic forces tether the proximal region of the angiotensin II receptor (AT_{1A}) carboxyl terminus to anionic lipids. *Biochemistry*, **41**, 7830–7840.

Nordin, H., Jungnelius, M., Karlsson, R., and Karlsson, O.P. (2005). Kinetic studies of small molecule interactions with protein kinases using biosensor technology. *Analytical Biochemistry*, **340**, 359–368.

Oscik, J., Dabrowski, A., Jaroniec, M., and Rudzinski, W. (1976). Effects of surface heterogeneity in adsorption from binary liquid mixtures. I. Adsorption from ideal solutions. *Journal of Colloid and Interface Science*, **56**, 403–412.

Ramakrishna, A., and Sadana, A., (2001). A single-fractal analysis of cellular analyte-recptor binding kinetics utilizing biosensors. *Biosystems*, **59**(1), 35–45.

Rudzinski, W., Lattar, L., Zajac, J., Wofram, E., and Paszli, J. (1983). Ideal adsorption from binary liquid mixtures on a heterogeneous solid surface: equations for excess isotherms and heats of immersion. *Journal of Colloid and Interface Science*, **96**, 339–359.

Sadana, A. (2001). A fractal analysis approach for the evaluation of hybridization kinetics in biosensors. *Journal of Colloid and Interface Science*, **234**, 9–18.

Sorenson, C.M., and Roberts, G.C. (1997). The prefactor of fractal aggregates. *Journal of Colloid and Interface Science*, **186**, 447.

Tam, J.P., Lu, Y.A., Yang, J.L., and Chiu, K.W. (1999). An unusual structural motif of antimicrobial peptides containing end-to-end macrocycle and cystine-knot disulfides. *Proceedings of the National Academy of Sciences, USA*, **96**, 8913–8918.

Thomson, S.J., and Webb, G. (1968). *Heterogeneous Catalysis*, Wiley, New York, p. 23.

– 10 –

Sol–Gel Biosensor Applications

10.1 INTRODUCTION

In recent years, there has been a surge of biosensor applications in different areas. The more recent techniques have attempted to modify or correct the disadvantages of previously used techniques for biosensor applications besides attempting to use biosensors in newer areas of interest. Jin and Brennan (2002) have indicated some of the requirements of receptors (or bioimmobilisates as they call them), which include a high density of receptors, high activity, easy accessibility, stability, rapid response times, and resistance to leaching/or desorption of the receptor from the (biosensor) surface. As expected, not all receptors for biosensor applications meet satisfactorily the requirements for the above-mentioned parameters. Novel methods are continuously being explored to help satisfy these parameters to a reasonable degree.

The sol–gel method of the entrapment of biological molecules that can be used to enclose receptors for the detection of different analytes by a biosensor is an area of useful investigation (Bronshtein *et al.*, 1997; Nishida *et al.*, 1995; Yamanaka *et al.*, 1995; Akbarian *et al.*, 1997; Narang *et al.*, 1994; Lundgren and Bright, 1996; Jordan *et al.*, 1996; Aylott *et al.*, 1997a,b). Jin and Brennan (2002) indicate that Dickey (1955) had initially demonstrated the entrapment of biologically active materials in silica-derived glasses. Avnir *et al.* (1994) demonstrated the entrapment of proteins into alkoxysilane-derived silica glasses via the sol–gel method.

Jin and Brennan (2002) elucidate the main advantages and disadvantages of the sol–gel method for biosensor applications. The advantages include optical transparency which facilitates the development of sensors that rely on the changes in the absorbance or fluorescence signal, use of different chemical modifiers (such as polymer additives), and tunable pore size and pore distribution. The last quality facilitates size-dependent analysis.

Shin *et al.* (2005) have recently developed a sol–gel derived amperometric nitric oxide microsensor. These authors emphasize that nitric oxide (NO) is involved in quite a few physiological processes such as neurotransmission, vasodilation, blood pressure regulation, angiogenesis, and wound healing (Ignarro *et al.*, 1987; Ignarro, 2000; Lincoln *et al.*, 1997). Shin *et al.* (2005) emphasize that sol–gel derived materials are suitable for biosensor applications since they may be synthesized under mild conditions, are chemically flexible, and adhere to a variety of substrates (Hench and West, 1990; Walcarius, 2001). Furthermore, Walcarius (1998, 2001) indicates that, and as indicated previously, the physical and chemical properties of sol–gel derived biosensors are readily tailored to required needs by attaching biomolecules and electrochemical and optical sensing elements to the polymer backbone.

The disadvantages according to Jin and Brennan (2002) include brittleness, cracking due to hydration stresses, and limited accessibility of analytes to receptors on the surface, or in this case the sol–gels. Furthermore, diffusion may be a problem, unless very thin layers are used to minimize the diffusion length of the analyte. Denaturation of the enclosed receptor may also be a problem, which may perhaps be alleviated with the development of improved organic–inorganic nanocomposite materials. With the recent growing emphasis on nanoengineering and nanobiotechnology, this aspect of compatible and composite material development suitable for sol–gels is bound to gain increasing attention.

In this chapter, we analyze the binding (and dissociation) kinetics if applicable of different analytes in solution to receptors enclosed in sol–gel biosensors. We analyze the binding kinetics of (a) different toxic chemicals to recombinant *E. coli* present in thick silicate films (Premkumar *et al.*, 2002), (b) superoxide radical to superoxide dismutase (SOD) plus horseradish peroxidase (HRP) plus probe (Amplex red) enclosed in sol–gels (Pastor *et al.*, 2004), and (c) cholesterol in solution to cholesterol oxidase (COX) enclosed sol–gel based on chiotsan/silica and a multiwalled carbon nanotube (MWCNTs) organic–inorganic hybrid composite material (Tan *et al.*, 2005).

10.2 THEORY

Havlin (1989) has reviewed and analyzed the diffusion of reactants toward fractal surfaces. The details of the theory and the equations involved for the binding and the dissociation phases for analyte–receptor binding are available. The details are not repeated here, except that just the equations are given to permit an easier reading. These equations have been applied to other biosensor systems (Ramakrishnan and Sadana, 2001). For most applications, a single- or a dual-fractal analysis is often adequate to describe the binding and the dissociation kinetics. Peculiarities in the values of the binding and in the dissociation rate coefficients as well as in the values of the fractal dimensions with regard to the dilute analyte systems being analyzed will be carefully noted, if applicable.

10.2.1 Single-fractal analysis

Binding rate coefficient

Havlin (1989) indicates that the diffusion of a particle (analyte [Ag]) from a homogeneous solution to a solid surface (e.g., receptor [Ab]-coated surface) on which it reacts to form a product (analyte–receptor complex; (Ab · Ag)) is given by

$$(\mathrm{Ab}\cdot\mathrm{Ag}) \approx \begin{cases} t^{(3-D_{\mathrm{f,bind}})/2} = t^{\mathrm{p}} & t < t_{\mathrm{c}} \\ t^{1/2} & t > t_{\mathrm{c}} \end{cases} \tag{10.1a}$$

Here $D_{f,bind}$ or D_f (used later on in the chapter) is the fractal dimension of the surface during the binding step; t_c is the cross-over value. Havlin (1989) indicates that the cross-over value may be determined by using $r_c^2 \sim t_c$. Above the characteristic length r_c, the self-similarity of the surface is lost and the surface may be considered homogeneous. Above time t_c, the surface may be considered homogeneous, since the self-similarity property disappears and 'regular' diffusion is now present. For a homogeneous surface where $D_f = 2$ and when only diffusional limitations are present, $p = ½$ as it should be. Another way of looking at the $p = ½$ case (where $D_{f,bind} = 2$) is that the analyte in solution views the fractal object; in our case, the receptor-coated biosensor surface from a 'large distance.' In essence, in the association process, the diffusion of the analyte from the solution to the receptor surface creates a depletion layer of width $(Đt)^{½}$, where $Đ$ is the diffusion constant. This gives rise to the fractal power law

$$(\text{Analyte} \cdot \text{Receptor}) \sim t^{(3 - D_{f,bind})/2}$$

For the present analysis, t_c is arbitrarily chosen and we assume that the value of the t_c is not reached. One may consider the approach as an intermediate 'heuristic' approach that may be used in the future to develop an autonomous (and not time-dependent) model for diffusion-controlled kinetics.

Dissociation rate coefficient

The diffusion of the dissociated particle (receptor [Ab] or analyte [Ag]) from the solid surface (e.g., analyte [Ag]–receptor [Ab]) complex coated surface) into solution may be given, as a first approximation by

$$(\text{Ab} \cdot \text{Ag}) \approx -t^{(3 - D_{f,diss})/2} = t_p \qquad (t > t_{diss}) \tag{10.1b}$$

Here $D_{f,diss}$ is the fractal dimension of the surface for the dissociation step. This corresponds to the highest concentration of the analyte–receptor complex on the surface. Henceforth, its concentration only decreases. The dissociation kinetics may be analyzed in a manner 'similar' to the binding kinetics.

10.2.2 Dual-fractal analysis

Binding rate coefficient

Sometimes, the binding curve exhibits complexities and two parameters (k, D_f) are not sufficient to adequately describe the binding kinetics. This is further corroborated by low values of the r^2 factor (goodness-of-fit). In that case, one resorts to a dual-fractal analysis (four parameters: k_1, k_2, D_{f1}, and D_{f2}) to adequately describe the binding kinetics. The single-fractal analysis presented above is thus extended to include two fractal dimensions. At present, the time ($t = t_1$) at which the 'first' fractal dimension 'changes' to the 'second' fractal dimension is arbitrary and empirical. For the most part, it is dictated by the data analyzed and experience gained by handling a single-fractal analysis. A smoother curve is obtained in the 'transition' region if care is taken to select the correct number of points for the two regions. In this case, the product (antibody–antigen, or analyte–receptor complex; Ab·Ag or analyte·receptor) is given by

$$(\mathrm{Ab}\cdot\mathrm{Ag}) \approx \begin{cases} t^{(3-D_{\mathrm{f1,bind}})/2} = t^{\mathrm{P}_1} & (t < t_1) \\ t^{(3-D_{\mathrm{f2,bind}})/2} = t^{\mathrm{P}_2} & t_1 < t < t_2 = t_\mathrm{c} \\ t^{1/2} & (t > t_\mathrm{c}) \end{cases} \tag{10.1c}$$

In some cases, as mentioned above, a triple-fractal analysis with six parameters (k_1, k_2, k_3, D_{f1}, D_{f2}, and D_{f3}) may be required to adequately model the binding kinetics. This is when the binding curve exhibits convolutions and complexities in its shape perhaps due to the very dilute nature of the analyte (in some of the cases to be presented) or for some other reasons. Also, in some cases, a dual-fractal analysis may be required to describe the dissociation kinetics.

10.3 RESULTS

In this chapter, we analyze the binding (and dissociation) kinetics if applicable of different analytes in solution to receptors enclosed in sol–gel biosensors. We analyze the binding kinetics of (a) different toxic chemicals to recombinant *E. coli* present in thick silicate films (Premkumar *et al.*, 2002), (b) superoxide radical to superoxide dismutase (SOD) plus horseradish peroxidase (HRP) plus probe (Amplex red) enclosed in sol–gels (Pastor *et al.*, 2004), and (c) cholesterol in solution to cholesterol oxidase (COX) enclosed sol–gel based on chiotsan/silica and a multiwalled carbon nanotube (MWCNTs) organic–inorganic hybrid composite material (Tan *et al.*, 2005).

For all practical purposes, even though the receptor is not strictly immobilized on a biosensor surface, as far as the sol–gel is concerned, the receptor is 'confined' to the sol–gel. The analyte has to diffuse from the solution to the receptor that is 'heterogeneously' confined to the sol–gel space or volume. In this way, the receptor exhibits 'heterogeneity', and may be characterized by a fractal dimension. As done for the other biosensor examples presented in the other chapters of this book, the fractal dimension value is obtained from the eqs. (10.1a–c).

The fractal analysis is only one possible explanation for analyzing the diffusion-limited binding and dissociation kinetics assumed to be present in the sol–gel systems analyzed. This method has been used to obtain values for the binding and dissociation rate coefficients as well as affinities for analyte–receptor interactions occurring on sol–gels biosensor surfaces. The parameter values thus obtained would provide an idea of the range of values for the different sol–gel systems analyzed.

Premkumar *et al.* (2002) indicate that the usage of test kits and early-warning systems (biosensors) using recombinant bacteria is increasing (Ames *et al.*, 1975; Bousse, 1996; Daunert *et al.*, 2000). Premkumar *et al.* (2002) emphasize that researchers will, in general, trade-off molecular specificity for an early warning of a physiological stress such as membrane cell deficiency, nutritional deficiency, etc. These authors have developed a luminescence physiological stress sensor based on sol–gel derived silicates in which viable recombinant *E. coli* was enclosed. Figure 10.1a shows the binding of DDT in solution to *E. coli* reporter strain enclosed in a TV1061-silicate sensor (Premkumar *et al.*, 2002).

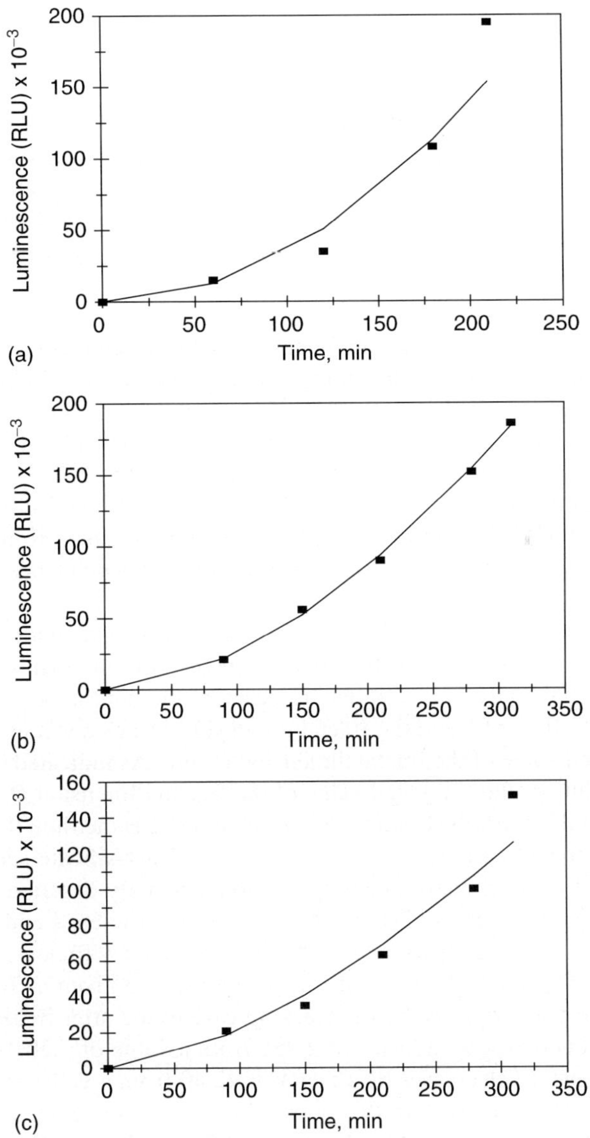

Figure 10.1 Binding of different toxic chemicals in solution to a TV1061-silicate sensor (Premkumar *et al.*, 2002): (a) DDT, (b) Aldicarb, (c) Malathion.

A single-fractal analysis is adequate to describe the binding kinetics. The values of the binding rate coefficient k and the fractal dimension D_f are given in Table 10.1a.

The estimated value of the fractal dimension is zero. This is because the binding curve exhibits a 'convex' shape to the origin. This also indicates that the surface acts as a "Cantor-like' dust. Only the positive value of the error of the fractal dimension is presented since the fractal dimension cannot have a negative value.

The values of the binding rate coefficient and the fractal dimension presented in Table 10.1 were obtained from a regression analysis using Corel Quattro Pro 8.0 (1997) to model the data using eqs. (10.1a and b) wherein the (analyte·receptor) $=k^{(3-D_f)/2}$, for a single-fractal analysis, and (analyte·receptor) $=k_1 t^{(3-D_{f1})/2}$, and $=k_2 t^{(3-D_{f2})/2}$, for a dual-fractal analysis, respectively. The analyte·receptor(t) versus time data is regressed to obtain the values of k and p (single-fractal analysis), and k_1 and k_2 and p_1 and p_2 (dual-fractal analysis). Note that $p_1 = (3-D_{f1})/2$ and $p_2 = (3-D_{f2})/2$ from which relations the fractal dimension values were obtained. Only those values of the binding rate coefficient and the fractal dimension will be analyzed for which the sum of least squares (r^2) obtained regression is equal to or greater than 0.97. In fact, this is one of the criterion used to go from a single-fractal analysis to a dual-fractal analysis.

The values of the binding rate coefficients and the fractal dimensions presented in Table 10.1 are within 95% confidence limits. For example, for the binding of DDT in solution to the *E. coli* reporter enclosed in a TV1061-silicate sensor (Premkumar *et al.*, 2002), the reported binding rate coefficient k value is 0.00367 ± 0.00145. The 95% confidence limit indicates that this value lies between 0.00225 and 0.00512. This indicates this value is precise and significant. The confidence limits were calculated directly by the Corel Quattro Pro 8.0 (1997) software program. Since the data is taken from the literature, no information is provided about the distribution and the independence of the data being analyzed.

Figures 10.1b and c show the binding of Aldicarb and Malathion in solution to the *E. coli* reporters enclosed in TV1061-silicate sensors (Premkumar *et al.*, 2002). Once again, a single-fractal analysis is adequate to describe the binding kinetics. The values of the binding rate coefficient k and the fractal dimension D_f are given in Table 10.1a. Once again, the estimated value of the fractal dimension is zero. As indicated above for the binding of DDT in solution, the estimated value of the fractal dimension D_f is also zero.

Figures 10.2a and b show the binding of 2% ethanol in LB medium (before freezing and after freezing) to the *E. coli* reporters enclosed in TV1061-silicate sensors (Premkumar *et al.*, 2002). A dual-fractal analysis is required to adequately describe the binding kinetics in both the cases. The values of (a) the binding rate coefficient k and the fractal dimension D_f for a single-fractal analysis, and (b) the binding rate coefficients, k_1 and k_2, and the fractal dimension, D_{f1} and D_{f2}, for a dual-fractal analysis are given in Table 10.1b.

Note that as one goes from the before freezing case to the after freezing case, the fractal dimension D_{f1} decreases by a factor of 2.955 from a value of 2.201 to 0.7448, and the binding rate coefficient k decreases by a factor of 42.40 from a value of 7.4258 to 0.1751. Also, and as indicated elsewhere in this book, changes in the degree of heterogeneity or the fractal dimension D_{f1} and in the binding rate coefficient k_1 are in the same direction. In both these cases (before and after freezing), the estimated value of the fractal dimension

Table 10.1

Binding rate coefficients and fractal dimensions for (a) different toxic chemicals (DDT, aldicarb, and malathion) exposed to TV1061-silicate sensors, and (b) ethanol in solution binding to TV1061-silicate sensor before and after freezing by immersion of sensor in liquid nitrogen (Premkumar *et al.*, 2002)

Analyte in solution/receptor on surface	k	k_1	k_2	D_f	D_{f1}	D_{f2}
(a) DDT/*E. coli* reporter on TV1061-silicate sensor	0.00367 ± 0.00145	na	na	0 + 0.6846	na	na
Aldicarb/*E. coli* reporter on TV1061-silicate sensor	0.008868 ± 0.000451	na	na	0 + 0.09841	na	na
Malathion/*E. coli* reporter on TV1061-silicate sensor	0.01752 ± 0.00228	na	na	0 + 0.3410	na	na
(b) 2% ethanol-LB medium/*E. coli* reporter on TV1061-silicate sensor (before freezing)	0.7580 ± 0.2611	7.4258 ± 1.0816	0.001505 ± 0.000007	1.1316 + 0.4106	2.2010 ± 0.3386	0 + 0.03118
2% ethanol-LB medium/*E. coli* reporter on TV1061-silicate sensor (after freezing)	0.05756 ± 0.01735	0.1751 ± 0.0609	0.002068 ± 0.000257	0.2404 + 0.4270	0.7448 ± 0.6828	0 + 0.8770

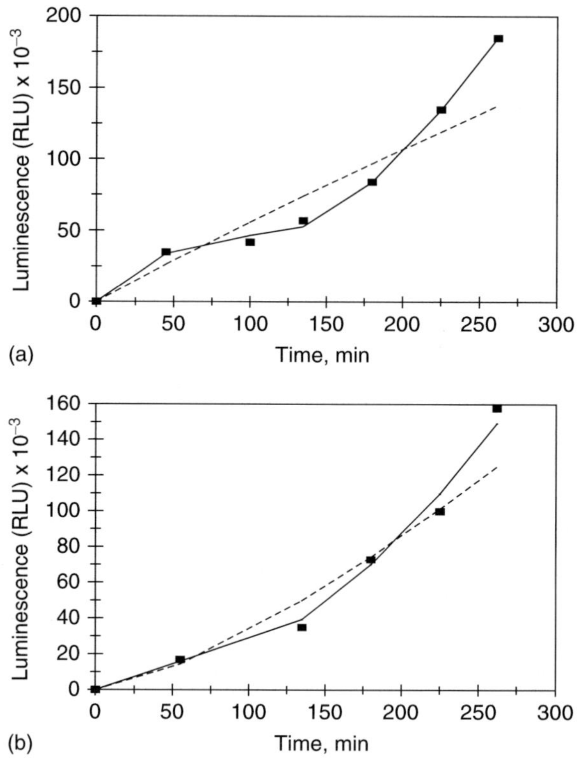

Figure 10.2 Binding of 2% ethanol in soluton to a TV1061-silicate sensor (Premkumar *et al.*, 2002): (a) before freezing, (b) after freezing by immersion of sensor in liquid nitrogen. When only a solid line is used then a single-fractal analysis applies. When both a dashed (------) and a solid (____) line are used, then the dashed line is for a single-fractal analysis and the solid line is for a dual-fractal analysis. In this case, the solid line provides the better fit.

D_{f2} =zero. Only positive error values are given for these two cases since, and as indicated above, the estimated fractal dimension D_{f2} value is zero.

Premkumar *et al.* (2002) have analyzed the binding of different compounds in solution to different silicate (glass or hybrid) sensors. Figure 10.3a shows the binding of 0.43 M ethanol in solution to a TV1061 hybrid sensor. A single-fractal analysis is adequate to describe the binding kinetics. The values of the binding rate coefficient k and the fractal dimension D_f are given in Table 10.2.

Figure 10.3b shows the binding of 0.43 M ethanol in solution to a TV1061 glass sensor. A dual-fractal analysis is required to adequately describe the binding kinetics. The values of (a) the binding rate coefficient k and the fractal dimension D_f for a single-fractal analysis, and (b) the binding rate coefficients, k_1 and k_2, and the fractal dimensions, D_{f1} and D_{f2}, for a dual-fractal analysis are given in Table 10.2. It is of interest to note the change in the binding mechanism as one goes from the hybrid sensor to the glass sensor for the detection of 0.43 M ethanol since a single-fractal analysis is adequate to describe the binding

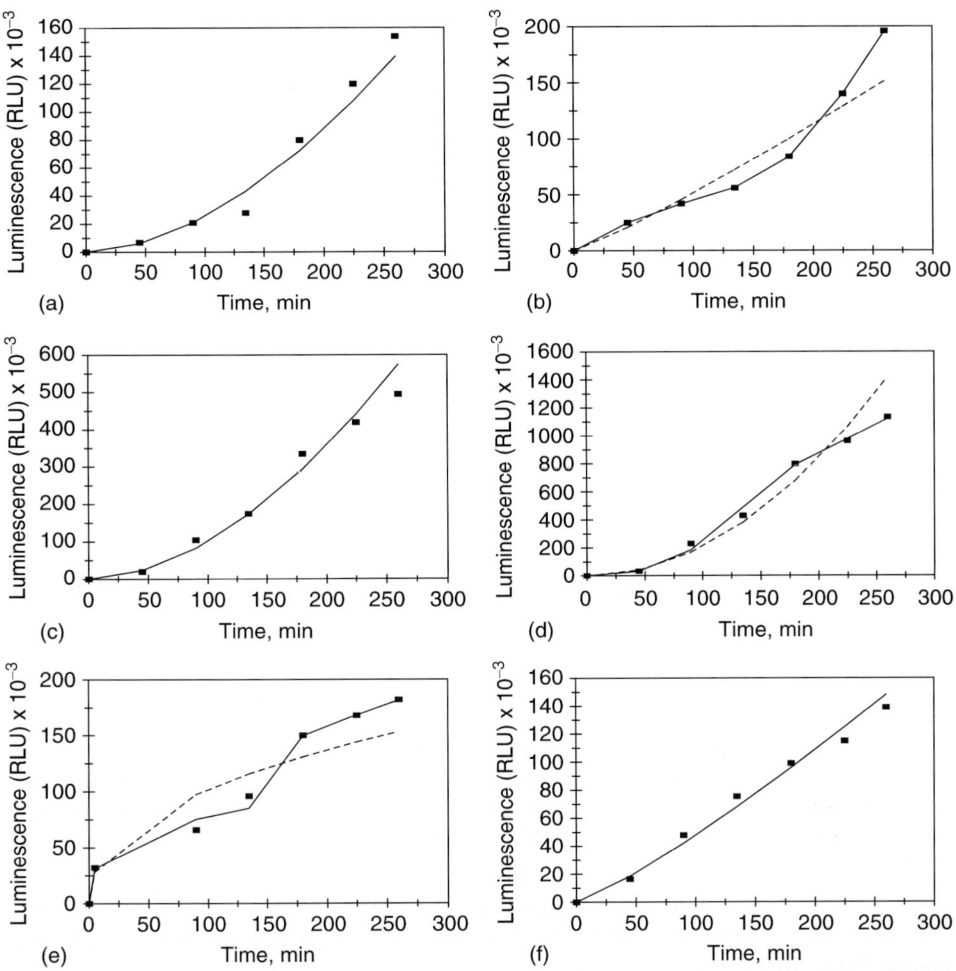

Figure 10.3 Binding of different compounds in solution to different silicate (glass or hybrid) sensors (Premkumar *et al.*, 2002): (a) 0.43 M ethanol to *E. coli* encapsulated on a TV1061 hybrid sensor, (b) 0.43 M ethanol to *E. coli* encapsulated on a TV1061 glass sensor, (c) 6 μM nalidixic acid to *E. coli* encapsulated on a DPD2794 hybrid sensor, (d) 6 μM nalidixic acid to *E. coli* encapsulated on a DPD2794 glass sensor, (e) 0.5 μM phenol to *E. coli* encapsulated on a DPD2544 glass sensor, (f) 0.5 μM phenol to *E. coli* encapsulated on a DPD2544 hybrid sensor, (g) 3.9 mM methyl viologen to *E. coli* encapsulated on a DPD2515 glass sensor, (h) 3.9 mM methyl viologen to *E. coli* encapsulated on a DPD2515 hybrid sensor, (i) 3.9 mM hydrogen peroxide to *E. coli* encapsulated on a DPD2515 glass sensor, (j) 3.9 mM hydrogen peroxide to *E. coli* encapsulated on a DPD2515 hybrid sensor. When only a solid line is used then a single-fractal analysis applies. When both a dashed (------) and a solid (___) line are used, then the dashed line is for a single-fractal analysis and the solid line is for a dual-fractal analysis. In this case, the solid line provides the better fit.

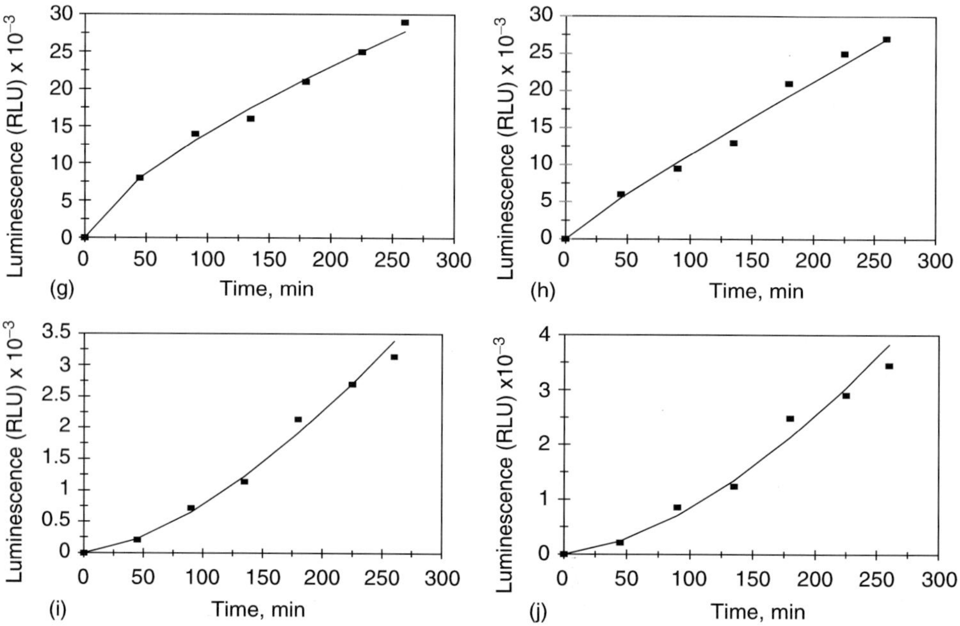

Figure 10.3 Continued.

kinetics using a hybrid sensor, and a dual-fractal analysis is required when the glass sensor is used. Apparently in this case also, the binding kinetics is 'more complex' when the glass sensor is used compared to the use of a hybrid sensor.

Figure 10.3c shows the binding of 6 μM nalidixic acid in solution to a DPD2794 hybrid sensor. A single-fractal analysis is adequate to describe the binding kinetics. The values of the binding rate coefficient k and the fractal dimension D_f for a single-fractal analysis are given in Table 10.2.

Figure 10.3d shows the binding of 6 μM nalidixic acid in solution to a DPD2794 glass sensor. A dual-fractal analysis is required to adequately describe the binding kinetics. The values of (a) the binding rate coefficient k and the fractal dimension D_f for a single-fractal analysis, and (b) the binding rate coefficients, k_1 and k_2, and the fractal dimensions, D_{f1} and D_{f2}, for a dual-fractal analysis are given in Table 10.2. It is of interest to note, once again and as noted previously, the change in the binding mechanism as one goes from the hybrid sensor to the glass sensor for the detection of 6 μM nalidixic acid since a single-fractal analysis is adequate to describe the binding kinetics using a hybrid sensor, and a dual-fractal analysis is required when the glass sensor is used. Apparently in this case also, the binding kinetics is 'more complex' when the glass sensor is used compared to the use of a hybrid sensor.

Figure 10.3e shows the binding of 0.5 mM phenol in solution to a DPD2544 glass sensor. A dual-fractal analysis is required to adequately describe the binding kinetics. The values of (a) the binding rate coefficient k and the fractal dimension D_f for a single-fractal

Table 10.2

Binding rate coefficients and fractal dimensions for the interactions of different compounds in solution to different silicate (glass or hybrid) sensors (Premkumar *et al.*, 2002)

Analyte in solution/receptor on sensor surface and sensor type	k	k_1	k_2	D_f	D_{f1}	D_{f2}
0.43 M ethanol/*E. coli* onTV1061 hybrid sensor	0.006892 ± 0.001916	na	na	0 + 0.3350	na	na
0.43 M ethanol/*E. coli* onTV1061 glass sensor	0.2846 ± 0.0729	1.5234 ± 0.0123	0.00053 ± 0.000002	2.7420 + 0.3116	1.5288 ± 0.0204	0 + 0.0206
6 μM nalidixic acid/*E. coli* on DPD2794-hybrid sensor	0.02301 ± 0.00440	na	na	0 + 0.2388	na	na
6 μM nalidixic acid/*E. coli* on DPD2794 glass sensor	0.02088 ± 0.00602	0.004063 ± 0.001230	5.8316 ± 0.1207	0 + 0.3458	0 + 0.6732	1.1082 ± 0.1565
0.5 mM phenol/*E. coli* on DPD2544-glass sensor	14.3241 ± 4.2771	19.3399 ± 3.7986	9.8432 ± 0.0321	2.1476 ± 0.1579	2.3948 ± 0.1411	1.9514 ± 0.02482
0.5 mM phenol/*E. coli* on DPD2544 hybrid sensor	0.2034 ± 0.0248	na	na	0.6292 ± 0.1570	na	na

(*Continued*)

Table 10.2 *(Continued)*

Analyte in solution/receptor on sensor surface and sensor type	k	k_1	k_2	D_f	D_{f1}	D_{f2}
3.9 mM methyl viologen/*E. coli* on DPD2515 glass sensor	0.5445 ± 0.03350	na	na	1.5862 ± 0.8156	na	na
3.9 mM methyl viologen/*E. coli* on DPD2515-hybrid sensor	0.1786 ± 0.03350	na	na	1.1960 ± 0.1409	na	na
3.9 mM hydrogen peroxide/*E. coli* on DPD2515 glass sensor	0.000597 ± 0.000060	na	na	$0 + 0.1307$	na	na
3.9 mM hydrogen peroxide/ *E. coli* on DPD2515 hybrid sensor	0.000553 ± 0.000092	na	na	$0 + 0.2099$	na	na

analysis, and (b) the binding rate coefficients, k_1 and k_2, and the fractal dimensions, D_{f1} and D_{f2}, for a dual-fractal analysis are given in Table 10.2.

Figure 10.3f shows the binding of 0.5 mM phenol in solution to a DPD2544 hybrid sensor. A single-fractal analysis is adequate to describe the binding kinetics. The values of the binding rate coefficient k and the fractal dimension D_f for a single-fractal analysis are given in Table 10.2. It is of interest to note, once again and as noted previously, the change in the binding mechanism as one goes from the hybrid sensor to the glass sensor for the detection of 0.5 mM phenol in solution since a single-fractal analysis is adequate to describe the binding kinetics using a hybrid sensor, and a dual-fractal analysis is required when the glass sensor is used. Apparently in this case also, the binding kinetics is 'more complex' when the glass sensor is used compared to the use of a hybrid sensor.

Figure 10.3g shows the binding of 3.9 mM methyl viologen in solution to a DPD2515 glass sensor. A single-fractal analysis is adequate to describe the binding kinetics. The values of the binding rate coefficient k and the fractal dimension D_f are given in Table 10.2. The estimated value of the fractal dimension D_f is zero. This indicates that the DPD2515 glass sensor surface acts as a Cantor-like dust.

Figure 10.3h shows the binding of 3.9 mM methyl viologen in solution to a DPD2515 hybrid sensor. Once again, a single-fractal analysis is adequate to describe the binding kinetics. The values of the binding rate coefficient k and the fractal dimension D_f are given in Table 10.2. The estimated value of the fractal dimension D_f is zero. As indicated above, in this case too, the DPD2515 hybrid sensor surface acts as a Cantor-like dust.

It is of interest to note that as one goes from the DPD2515 hybrid sensor to the DPD2515 glass sensor, (a) the fractal dimension D_f increases by 32.6% from a value of 1.1960 to 1.5862, and (b) the binding rate coefficient k increases by a factor of 3.049 from a value of 0.1786 to 0.5445. Note that changes in the fractal dimension D_f or the degree of heterogeneity on the sensor surface and in the binding rate coefficient k are in the same direction.

Figure 10.3i shows the binding of 3.9 mM hydrogen peroxide in solution to a DPD2515 glass sensor. A single-fractal analysis is adequate to describe the binding kinetics. The values of the binding rate coefficient k and the fractal dimension D_f are given in Table 10.2. The estimated value of the fractal dimension D_f is zero. This indicates that the DPD2515 glass sensor surface acts as a Cantor-like dust.

Figure 10.3j shows the binding of 3.9 mM hydrogen peroxide in solution to a DPD2515 hybrid sensor. Once again, a single-fractal analysis is adequate to describe the binding kinetics. The values of the binding rate coefficient k and the fractal dimension D_f are given in Table 10.2. The estimated value of the fractal dimension D_f is zero. As indicated above, in this case too, the DPD2515 hybrid sensor surface acts as a Cantor-like dust. Note that the values of the binding rate coefficient for both the glass and the hybrid sensor are within 7.37% of each other. As indicated above the estimated values of the fractal dimension in both cases is zero.

For the data presented in Tables 10.1 and 2 and for a dual-fractal analysis, Figure 10.4a shows the increase in the binding rate coefficient k_1 with an increase in the fractal dimension D_{f1}. The binding rate coefficient k_1 is given by

$$k_1 = (0.4532 \pm 0.263) D_{f1}^{3.800 \pm 0.4966} \tag{10.2a}$$

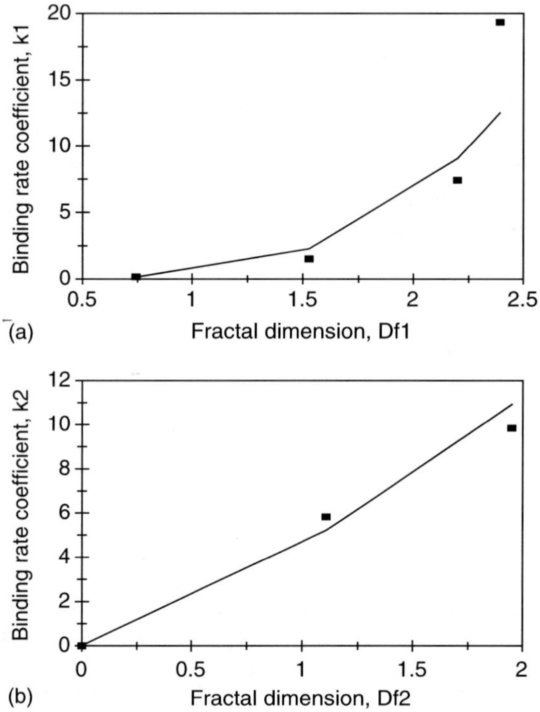

Figure 10.4 (a) Increase in the binding rate coefficient k_1 with an increase in the fractal dimension D_{f1}, and (b) Increase in the binding rate coefficient k_2 with an increase in the fractal dimension D_{f2}.

The fit is good. Only four data points are available. The availability of more data points would lead to a more reliable fit. The binding rate coefficient k_1 is very sensitive to the fractal dimension D_{f1} or the degree of heterogeneity that exists on the sensor surface as noted by the order between three and one-half and four (=3.8) exhibited.

For the data presented in Table 10.2 and for a dual-fractal analysis, Figure 10.4b shows the increase in the binding rate coefficient k_2 with an increase in the fractal dimension D_{f2}. The binding rate coefficient k_2 is given by

$$k_2 = (4.552 \pm 0.759) D_{f2}^{\,1.3101 \pm 0.02585} \tag{10.2b}$$

The fit is good. Only three data points are available. The availability of more data points would lead to a more reliable fit. The binding rate coefficient k_2 exhibits an order of dependence between first and one and one-half (=1.3101) on the fractal dimension D_{f2} or the degree of heterogeneity that exists on the sensor surface.

Diffusional limitations for analyte in solution binding to receptors enclosed in sol–gel sensors do occur. Premkumar *et al.* (2002) prepared different sol–gel TV1061-silicate film thickness sensors. Different volumes of the sol–gel precursors containing the same concentration of cells (10^8 cells/ml) were deposited on the glass slides. Figure 10.5a–f shows the binding of 2% ethanol in solution to 1, 0.9, 0.8, 0.7, 0.6, and 0.5 mm sol–gel TV1061-silicate film sensor. In all the cases, a single-fractal analysis is adequate to describe the binding kinetics. The values of the binding rate coefficient k and the fractal dimension D_f along with the film thickness is given in Table 10.3.

An attempt was made to plot the binding rate coefficient k with the fractal dimension D_f. No clear trend is observable, and the regression coefficient is very low. Therefore, the plot is not given. As noted elsewhere in this chapter and throughout the book, the fractal

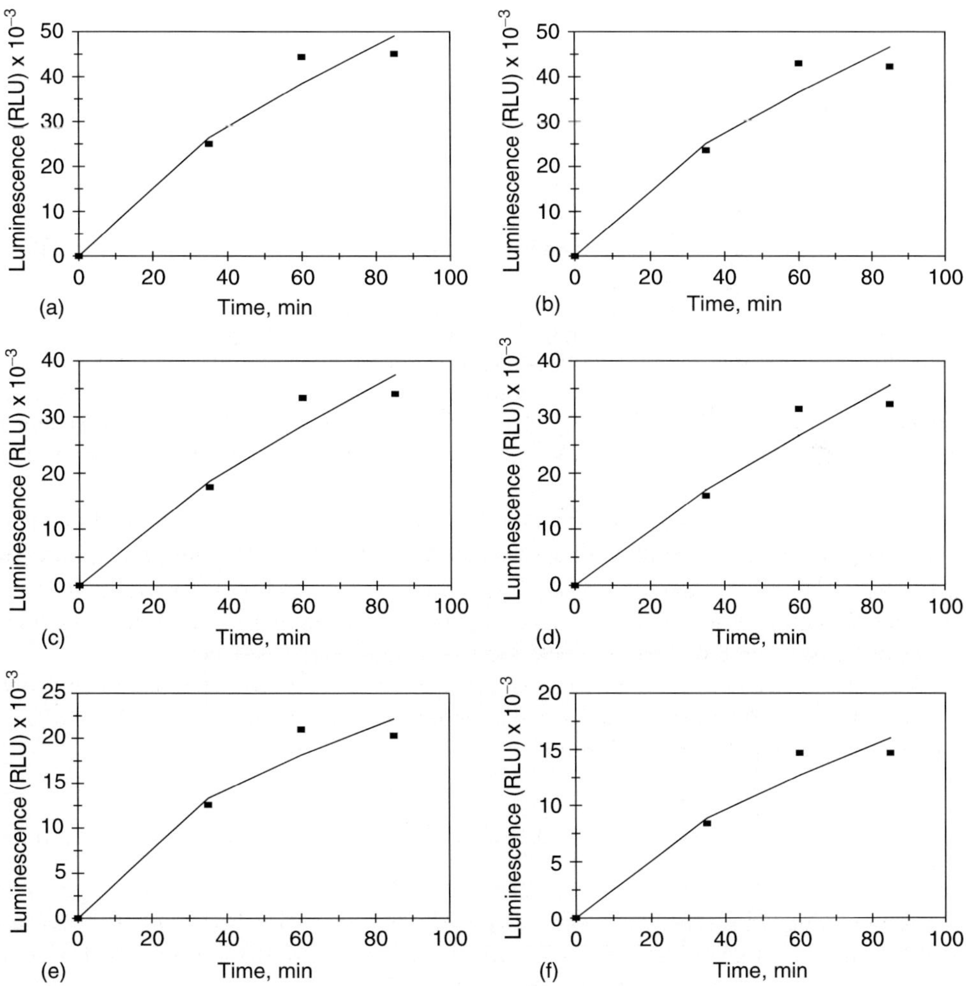

Figure 10.5 Binding of 2% ethanol in solution to different sol–gel TV1061-silicate film thickness (in mm) containing 10^8 cells/ml *E. coli* (Premkumar *et al.*, 2002): (a) 1, (b) 0.9, (c) 0.8, (d) 0.7, (e) 0.6, (f) 0.5.

Table 10.3

Binding rate coefficients and fractal dimensions for the binding of 2% ethanol in solution to different sol–gel TV1061-silicate film thickness containing 10^8 cells/ml *E. coli* (Premkumar *et al.*, 2002)

Silicate film thickness (mm)	D_f	k
1	1.6014±0.5536	2.1997±0.4207
0.9	1.6066±0.6292	2.1117±0.4648
0.8	1.4196±0.6180	1.1219±0.242
0.7	1.3380±0.6344	0.8884±0.1973
0.6	1.8546±0.5668	1.7414±0.3417
0.5	1.6688±0.5630	0.8339±0.1624

dimension D_f or the degree of heterogeneity on the sensor surface does influence the binding rate coefficient.

Table 10.3 and Figure 10.6 show the increase in the binding rate coefficient k with an increase in the silicate film thickness (in mm). The binding rate coefficient k is given by

$$k = (1.9647 \pm 0.8439)\,(\text{film thickness})^{1.1363 \pm 0.6177} \tag{10.3}$$

There is scatter in the fit of the data. Heterogeneity of receptors in the sol–gel sensor would influence the binding rate coefficient k as observed generally for other types of sensors. However, that aspect is not included in the analysis, and may contribute to the scatter in the fit of the data observed. Eq. (10.3) indicates that the binding rate coefficient k exhibits a slightly higher than first-order (=1.1363) dependence on the film thickness. The non-integer order of dependence exhibited lends support to the fractal nature of the system. The increase in the binding rate coefficient k with an increase in the film thickness lends support to the kinetically controlled response of the TV1061 sensors for the

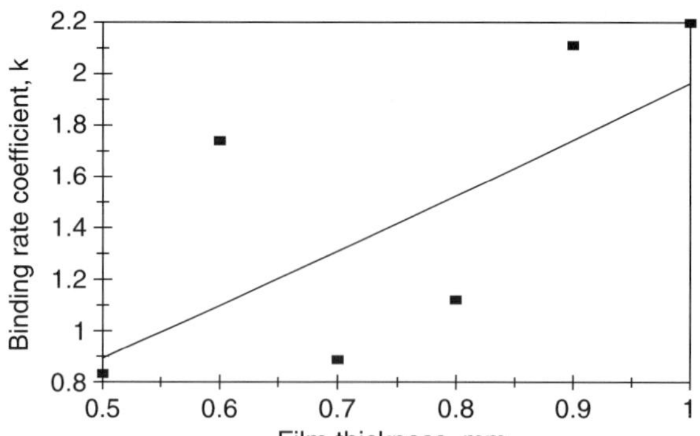

Figure 10.6 Increase in the binding rate coefficient k with an increase in the film thickness (in mm) on the sol–gel biosensor surface.

detection of 2% ethanol in solution. This result is consistent with the analysis and conclusion of Premkumar *et al.* (2002).

Pastor *et al.* (2004) indicate that the detection of superoxide anion O_2^- is an important area of investigation since it is produced in living systems either by autooxidation processes or during the normal metabolism process by enzymes (Fridovich, 1995). If uncontrolled this anion contributes to pathological diseases such as diabetes, atherosclerosis, aging, hypertension, and cancer (Floyd, 1990; Cadenas and Davies, 2000; Diaz *et al.*, 1997). Pastor *et al.* (2004) have summarized the detection methods used for superoxide anion radicals in biological systems (Roubaud *et al.*, 1997; Barbacanne *et al.*, 2000; Yao *et al.*, 2004; Benov *et al.*, 1998; Wang *et al.*, 1998). Pastor *et al.* (2004) have developed a selective, sensitive, and ready-to-use fluorescent biosensor for the detection of superoxide anions. The sensor uses the coupled reaction of superoxide dismutase (SOD) and horseradish peroxidase (HRP) along with the use of the probe *N*-acetyl-3,7-dihydroxyphenoxazine (Amplex red). Pastor *et al.* (2004) indicate the following reaction scheme that generates the red-fluorescent oxidation product resorufin (Zhou *et al.*, 1997; Martinez-Perez *et al.*, 2003):

$$2H^+ + 2O_2^- \xrightarrow{\text{SOD}} H_2O_2 + O_2$$

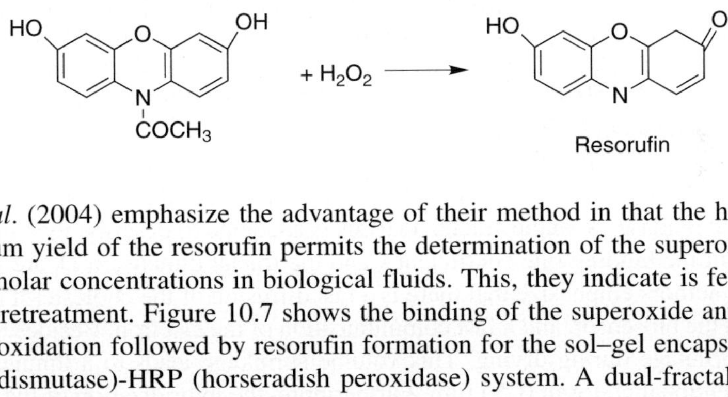

Pastor *et al.* (2004) emphasize the advantage of their method in that the high fluorescence quantum yield of the resorufin permits the determination of the superoxide radical at submicromolar concentrations in biological fluids. This, they indicate is feasible without sample pretreatment. Figure 10.7 shows the binding of the superoxide anion O_2^- and Amplex red oxidation followed by resorufin formation for the sol–gel encapsulated SOD (superoxide dismutase)-HRP (horseradish peroxidase) system. A dual-fractal analysis is required to adequately describe the binding kinetics. The values of (a) the binding rate coefficient k and the fractal dimension D_f and (b) the binding rate coefficients, k_1 and k_2, and the fractal dimensions, D_{f1} and D_{f2}, for a dual-fractal analysis are given in Table 10.4. The authors (Pastor *et al.*, 2004) indicate that complete oxidation was achieved in approximately 60 min. They noted the augmentation of the fluorescence emission at 590 nm as a function of the incubation time.

It is of interest to note that for a dual-fractal analysis, as the fractal dimension increases by a factor of 17.89 from a value of D_{f1}= 0.1522 to D_{f2}= 2.7232, the binding rate coefficient increases by more than an order of magnitude from a value of k_1= 2361.08 to D_{f2}= 24825.27.

Tan *et al.* (2005) have very recently developed an amperometric cholesterol sol–gel biosensor. This is based on a hybrid inorganic–organic composite material that consists of chitosan/silica and multiwalled carbon nanotubes (MWCNTs). These authors used this

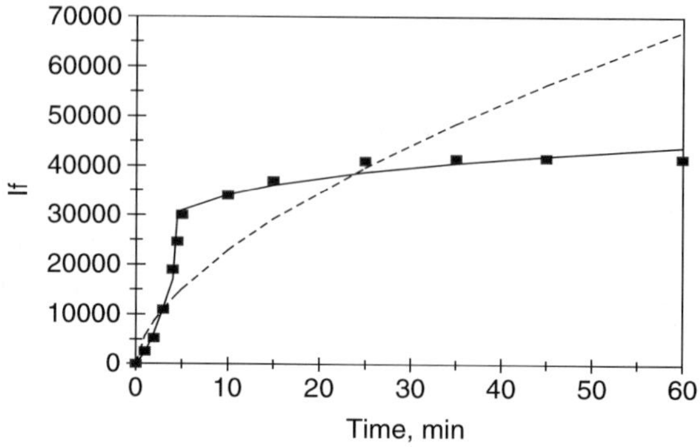

Figure 10.7 Binding of superoxide anion O_2^- in solution to superoxide dismutase (SOD) plus horseradish peroxidase (HRP) plus Amplex probe on a sol–gel biosensor (Pastor *et al.*, 2004). When both a dashed (--------) and a solid (________) line are used, then the dashed line is for a single-fractal analysis and the solid line is for a dual-fractal analysis. In this case the solid line provides the better fit.

composite material to immobilize cholesterol oxidase. They emphasize the importance of cholesterol detection in clinical analysis since its concentration in blood is a primary determinant in the prevention and diagnosis of diseases such as hypertension, arteriosclerosis, cerebral thrombosis, and heart disease (Nauck *et al.*, 2000; Singh *et al.*, 2004; Shmyantseva *et al.*, 2004). Tan *et al.* (2005) indicate that there are quite a few cholesterol oxidase (COX)-based (cholesterol oxidase and cholesterol-esterase-based) electrochemical biosensors (Brahim *et al.*, 2001; Li and Peng, 2002, 2003; Vidal *et al.*, 1999; Wang and Mu, 1999; Ram *et al.*, 2001; Garcya *et al.*, 2004).

Figure 10.8a shows the binding of 0.5 mM cholesterol in solution to a GC/PB/CS-SiO_2-COX-MWCNT electrode (Tan *et al.*, 2005). PB is Prussian blue, CS is chitosan, and COX is cholesterol oxidase. A single-fractal analysis is adequate to describe the binding kinetics. The values of the binding rate coefficient k and the fractal dimension D_f are given in Table 10.4. These authors emphasize that there is a fast diffusion of the cholesterol in their hybrid film used in the biosensor, and a fast communication of the electron. Besides, there is a considerable shrinkage during drying. This volume shrinkage leads to a thinner sol–gel film than a conventional chitosan (CS) film. Furthermore, the hybridization of the chitosan with the SiO_2 leads to an increased pore size of the silica glass. Both of these factors help minimize the diffusional limitations apparently inherently present in sol–gel biosensor applications. Furthermore, Tan *et al.* (2005) emphasize that the electrocatalytic properties of the MWCNTs promote electron exchange.

Figure 10.8b shows the binding of 0.5 mM cholesterol in solution to a GC/PB/CS-SiO_2-COX electrode. A single-fractal analysis is adequate to describe the binding kinetics. The values of (a) the binding rate coefficient k and the fractal dimension D_f for a single-fractal analysis and (b) the dissociation rate coefficient k_d and the fractal dimension D_{fd} for the dissociation phase are given in Table 10.4. Note that in this case, unlike when MWCNT is used, there is a dissociation phase. Also, as one goes from the GC/PB/CS-SiO_2 –COX-MWCNT electrode to the GC/PB/CS-SiO_2-COX electrode for a single-fractal analysis there is an increase in the fractal dimension D_f by a factor of 1.493 from a value of 0.8950 to 1.3360,

Table 10.4

Binding and dissociation rate coefficients and fractal dimensions for the binding and dissociation phase for different analytes in solution to receptors immobilized in sol–gels

Analyte in solution/receptor on surface	k	k_1	k_2	k_d	D_f	D_{f1}	D_{f2}	D_{fd}	Reference
Oxygen radical/SOD + HRP+ probe (Amplex red)	5752.5 ±3589.2	2361.08 ±428.11	24825.27 ±1029.78	na	1.7988 ±0.2218	0.1522 +0.32	2.7232 ±0.0375	na	Pastor *et al.* (2004)
Cholesterol/GC/PB/CS-SiO_2-COX-MWCNT electrode	0.000885 ±0.00005	na	na	na	0.8950 ±0.0381	na	na	na	Tan *et al.* (2005)
Cholesterol/GC/PB/CS-SiO_2-COX	0.001611 ±0.00009	na	na	0.000679 ±0.00003	1.3360 ±0.0422	na	na	1.6398 ±0.7524	Tan *et al.* (2005)

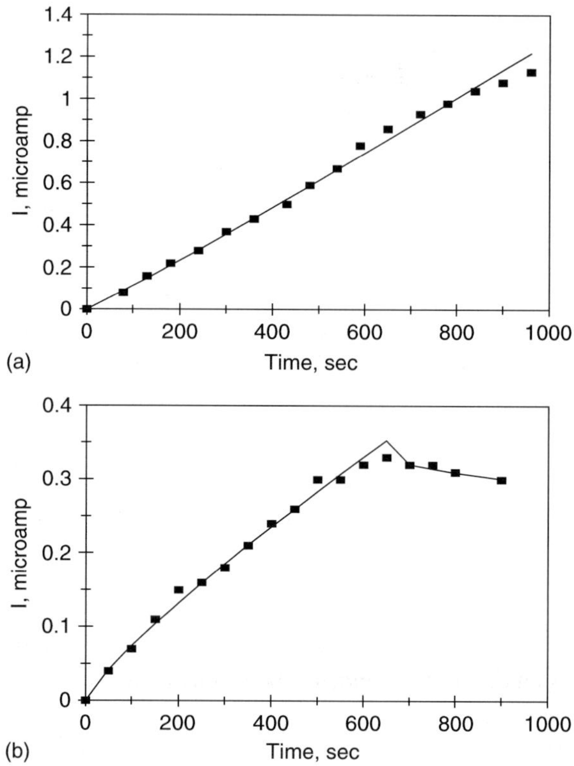

Figure 10.8 Binding of 0.5 mM cholesterol in solution to (a) a GC/PB/CS-SiO_2-COX-MWCNT electrode, and (b) a GC/PB/CS-SiO_2-COX electrode (Tan *et al.*, 2005).

and a corresponding increase in the binding rate coefficient k by a factor of 1.820 from a value of 0.000885 to 0.001611. Changes in the degree of heterogeneity or the fractal dimension D_f and in the binding rate coefficient k are in the same direction.

10.4 CONCLUSIONS

A fractal analysis is presented for the binding of different analytes in solution to receptors encapsulated in sol–gel biosensors. The binding (and the dissociation, if applicable) kinetics of (a) different toxic chemicals to recombinant *E. coli* present in thick silica films (Premkumar *et al.*, 2002), (b) superoxide radical O_2^- to superoxide dismutase (SOD) plus horseradish peroxidase (HRP) plus probe (Amplex red) enclosed in sol–gels (Pastor *et al.*, 2004), and (c) cholesterol in solution to cholesterol oxidase (COX) encapsulated on a sol–gel based organic–inorganic hybrid composite material (Tan *et al.*, 2005) are analyzed. The binding kinetics may be adequately described by either a single- or a dual-fractal analysis. The isolated case of dissociation kinetics presented is adequately described by a single-fractal analysis. A double-fractal analysis was used only if a single-fractal analysis did not provide an adequate fit. This was determined by the regression analysis provided by Corel Quattro Pro 8.0 (1997).

As indicated elsewhere in the book, the fractal dimension values provide a quantitative indication of the degree of heterogeneity present in the different sol–gel biosensors used. In this case, the heterogeneity may not be actually of the surface as is presented throughout the different chapters in the book, but the heterogeneity of the 'receptors' enclosed in the volume of the sol–gel film thickness. One aspect of this heterogeneity could be the way in which the active site on the receptor is presented to the analyte after it diffuses into the sol–gel and comes in the vicinity of the receptor. Quite possibly, the receptors may or may not be actually attached to the sol–gel film. Binding rate coefficient values for the single- and the dual-fractal analysis cases are provided. The fractal dimension, D_f, D_{f1}, and D_{f2} values presented are not a typical independent variable, such as analyte concentration, that may be directly manipulated. It is estimated from eqs. (10.1a–c), and one may consider them to as a derived variable.

An increase in the fractal dimension value or the degree of heterogeneity on the sol–gel biosensor leads to an increase in the binding rate coefficient. For the binding of different analytes in solution to a TV1061 or to a different glass or hybrid sol–gel biosensor (Premkumar *et al.*, 2002), the binding rate coefficients, k_1 and k_2, for a dual-fractal analysis exhibit an increase as the fractal dimensions, D_{f1} and D_{f2}, respectively, exhibit increases. The binding rate coefficient k_1 is very sensitive to the degree of heterogeneity since it exhibits a 3.8 order of dependence on the fractal dimension D_{f1}. The binding rate coefficient k_2 exhibits an order of dependence between one and one and one-half (=1.310) on the fractal dimension D_{f2}.

Sol–gel biosensors are, in general, constrained by diffusional limitations. Premkumar *et al.* (2002) analyzed the influence of film thickness on the diffusional limitations of the binding kinetics of 2% ethanol in solution to sol–gel TV1061 biosensors with different film thicknesses (in mm). Our fractal analysis indicates that the binding rate coefficient k increases with an increase in the film thickness. This indicates that diffusional limitations are not present in these sol–gel biosensor systems. This is consistent with the analysis of Premkumar *et al.* (2002).

The binding kinetics of (a) supeoxide anion, O_2^- in solution to superoxide dismutase (SOD) plus horseradish peroxidase (HRP) plus Amplex red probe on a sol–gel biosensor (Pastor *et al.*, 2004), and (b) cholesterol in solution to either a GC/PB/CS-SiO_2-COX-MWCNT electrode or a GC/PB/CS-SiO_2-COX electrode is presented. A dual-fractal analysis is required to adequately describe the binding kinetics of superoxide anion O_2^- in solution to superoxide dismutase (SOD) plus horseradish peroxidase (HRP) plus Amplex red probe sol–gel biosensor. A single-fractal analysis is adequate to describe the binding kinetics of cholesterol in solution to the amperometric cholesterol biosensor (Tan *et al.*, 2005) in the presence and in the absence of the multiwalled carbon nanotube. Both of the examples presented above have biomedical implications.

The fractal (and kinetic) analysis of the different examples of analytes that may be detected by sol–gel biosensors presented in this chapter should hopefully encourage experimentalists to pay further attention, and to explore the potential of sol–gels for biosensor applications in different areas. Biomedical applications are of particular interest, as emphasized in the different chapters in the book, and sol–gels do seem to exhibit the potential where they may be used to advantage. As with all biosensor applications, the sol–gel method too has advantages and disadvantages. Hopefully, the advances in nanotechnology and nanobiotechnology, and the development of materials particularly of hybrid materials, should help alleviate some of these disadvantages and help extend the applications of sol–gel biosensors to novel areas. For example, Daunert *et al.* (2005) have recently developed a novel stimuli-responsive hydrogel with genetically engineered proteins. These authors claim that this form of the hydrogel may have applications in microfluidic devices and in drug-delivery systems.

REFERENCES

Akbarian, F., Lin, A., Dunn, B., Valentine, J.S., and Zink, J.I. (1997). Spectroscopic determination of cholinesterase activity and inhibition in sol-gel media. *Journal of Sol–Gel Science and Technology*, **8**, 1067–1070.

Ames, B.N., McCann, J., and Yamasaki, E. (1975). *Mutation Research*, **31**, 347.

Avnir, D., Braun, S., Lev, O., and Ottolenghi, M. (1994). *Chemical Materials,* **9**, 2632.

Aylott, J.W., Richardson, D.J., and Russell, D.A. (1997a). Optical biosensing of nitrate ions using a sol–gel immobilized nitrate reductase. *Analyst*, **122**, 77–80.

Aylott, J.W., Richardson, D.J., and Russell, D.A. (1997b). *Chemical Materials*, **9**, 2261.

Barbacanne, M.A., Souchard, J.P., Darblade, B., Iliou, J.P., Nepveu, F., Pipy, B., Bayard, F., and Arnal, J.F. (2000). Detection of superoxide anion released extracellularly by endothelial cells using cytochrome *c* reduction, ESR, fluorescence and luicgenin-enhanced chemiluminescence techniques. *Free Radicals in Biology and Medicine*, **29**, 388–396.

Benov, L., Sztejnberg, L., and Fridovich, I. (1998). Critical evaluation of the use of hydroethidine as a measure of superoxide anion radical. *Free Radicals in Biology and Medicine*, **25**, 826–831.

Bousse, L. (1996). Whole cell biosensors. *Sensors & Actuators B: Chemical*, **34**, 270–275.

Brahim, S., Narinesingh, D., and Guiseppi-Elie, A. (2001). Amperometric determination of cholesterol in serum using a biosensor of cholesterol oxidase contained within a polypyrrole-hydrogel membrane. *Analytica Chimica Acta*, **448**, 27–36.

Bronshtein, A., Aharonson, N., Avnir, D., Turniansky, A., and Altstein, M. (1997). *Chemical Materials*, **9**, 2261.

Cadenas, E., and Davies, K.J.A. (2000). Mitochondrial free radical generation, oxidative stress, and aging. *Free Radicals in Biology and Medicine*, **29**, 222–230.

Corel Quattro Pro, 8.0., Corel Corporation, Ottawa, Canada, 1997.

Daunert, S., Barrett, G., Feliciano, J.S., Shetty, R.S., Shrestha, S. and Smith-Spencer, W. (2000). Genetically engineered whole-cell sensing systems coupling biological recognition with reporter genes. *Chemical Reviews*, **100**, 2705.

Diaz, M.N., Frei, B., Vita, J.A., and Keaney Jr. J.F. (1997). Antioxidants and atherosclerotic heart disease. *New England Journal of Medicine*, **337**, 408–416.

Dickey, F.H. (1955). Specific adsorption. *Journal of Physical Chemistry*, **58**, 695–707.

Floyd, R.A. (1990). Role of oxygen free radicals in carcinogenesis and brain ischemia. *FASEB*, **4**, 2587–2597.

Fridovich, I. (1995). Superoxide radical and superoxide dismutases. *Annual Reviews in Biochemistry*, **64**, 97–112.

Garcya, E., Vidal, J.C., Aramendia, M.T., and Castillo, J.R. (2004). Design of an interference-free cholesterol amperometric biosensor based on electrosynthesis of polymeric films of diaminonaphthalene isomers. *Electroanalysis*, **16**, 497–504.

Havlin, S. (1989). Molecular diffusion and reaction. In: *The Fractal Approach to Heterogeneous Chemistry: Surfaces, Colloids, Polymers* (ed. D. Avnir), Wiley, New York, pp. 251–269.

Hench , L.L., and West, J.K. (1990). The sol–gel process. *Chemical Reviews*, **90**, 33–72.

Ignarro, L.J. (2000). *Nitric Oxide: Biology and Pathobiology*, Academic Press, New York.

Ignarro, L.J., Buga, G.M., Wood, K.S., Byrns, R.E., and Chaudhari, G. (1987). Endothelium-derived relaxing factor produced and released from artery and vein is nitric oxide. *Proceedings of the National Academy of Sciences USA*, **84**, 9265–9269.

Jin, W., and Brennan, J.D. (2002). Properties and applications of proteins encapsulated within sol–gel derived materials. *Analytica Chimica Acta*, **461**, 1–36.

Jordan, J.D., Dunbar, R.A., and Bright, F.V. (1996). Aerosol generated sol–gel-derived thin films as biosensing platforms. *Analytica Chimica Acta*, **332**, 83.

Li, J.P., and Peng, T.Z. (2002). Polyaniline/Prussian blue composite film electrochemical biosensor for cholesterol detection. *Chinese Journal of Chemistry*, **20**, 1038–1043.

Li, J.P., and Peng, T.Z. (2003). An amperometric cholesterol biosensor based on cholesterol oxidase immobilized in polypyrrole on the Prussian blue modified electrode. *Chinese Journal of Analytical Chemistry*, **31**, 669–673.

Lincoln, J., Hoyle, C.H., and Burnstock, G. (1997). *Nitric Oxide in Health and Disease*, Cambridge University Press: Cambridge, UK.

Lundgren, J.S. and Bright, F.V. (1996). Biosensor for the nonspecific determination of ionic surfactants. *Analytical Chemistry*, **68**, 3377.

Martinez-Perez, D., Ferrer, M.L., and Mateo, C.R. (2003). A reagentless fluorescent sol–gel biosensor for uric acid detection in biological fluids. *Analytical Biochemistry*, **322**, 238–242.

Narang, U., Prasad, P.N., Bright, F.V., Ramanathan, K., Kumar, N.D., Malhotra, B.D., Kamalasanan, M.N., and Chandra, S. (1994). *Analytical Chemistry*, **66**, 3139.

Nauck, M., Marz, W., and Weiland, H. (2000). Is lipoprotein (a) cholesterol a significant indicator of cardiovascular risk? *Clinical Chemistry*, **46**, 436–437.

Nishida, F., McKiernan, J.M., Dunn, B., Zink, J.I., Brinker, C.J., and Hurd, A.J. (1995). *Journal of the American Chemical Society*, **78**, 1640.

Pastor, I., Esquembre, R., Micol, V., Mallavia, R., and Mateo, C.R. (2004). A ready-to-use fluorimeter biosensor for superoxide radical using superoxide dismutase and peroxidase immobilized in sol–gel glasses. *Analytical Biochemistry*, **334**, 335–343.

Premkumar, J.R., Rosen, R., Belkin, S., and Lev, O. (2002). Sol-gel luminescence biosensors: Encapsulation of recombinant *E. coli* reporters in thick silicate films. *Analytica Chimica Acta*, **462**, 11–23.

Ram, M.K., Bertoncello, P., Ding, H., Paddeu, S., and Nicolini, C. (2001). Cholesterol biosensor prepared by layer-by-layer technique. *Biosensors & Bioelectronics*, **16**, 849–856.

Ramakrishnan, A., and Sadana, A. (2001). A single-fractal analysis of cellular analyte-receptor binding kinetics utilizing biosensor. *Biosystems*, **59**(1), 35–45.

Roubaud, V., Sankarapandi, S., Kuppusamy, P., Tordo, P., and Zweier, J.L. (1997). Quantitative measurement of superoxide generation using the spin trap 5-(diethoxyphosphoryl)-5-methyl-1-pyrroline-*N*-oxide. *Analytical Biochemsitry*, **247**, 404–411.

Sadana, A. (2001). A fractal analysis approach for the evaluation of hybridization kinetics in biosensors. *Journal of Colloid and Interface Science*, **234**, 9–18.

Shin, J.H., Weinmann, S.W., and Schoenfisch, M.H. (2005). Sol–gel derived amperometric nitric oxide microsensor. *Analytical Chemistry*, **77**, 3494–3501.

Shmyantseva, V., Deluca, G., Bulko, T., Carrara, S., Nicolini, C., Usanov, S.A., and Archakov, A. (2004). Cholesterol amperometric biosensor based on cytochrome P450. *Biosensors & Bioelectronics*, **19**, 971–976.

Singh, S., Chaubey, A., and Malhotra, B.D. (2004). Amperometric cholesterol biosensor based on immobilized cholesterol esterase and cholesterol oxidase on conducting films. *Analytica Chimica Acta*, **502**, 229–234.

Tan, X., Li, M., Cai, P., Luo, L., and Zou, X. (2005). An amperometric cholesterol bionsensor based on multiwalled carbon nanotubes and organically modified sol-gel/chitosan hybrid composite film. *Analytical Biochemistry*, **337**, 111–120.

Vidal, J.C., Garcya, E., and Castillo, J.R. (1999). *In-situ* preparation of a cholesterol biosensor: entrapment of cholesterol oxidase in an overoxidized polypyrrole film electrodeposited in a flow system and determination of total cholesterol in serum. *Analytica Chimica Acta*, **385**, 213–222.

Walcarius, A. (1998). Silica modified carbon paste electrodes for copper determination in ammoniacal medium. *Electroanalysis*, **9**, 707–713.

Walcarius, A. (2001). *Chemical Materials*, **13**, 3351–3372.

Wang, H.D., Pagano, P.J., Du, Y., Cayatte, A.J., Quinn, M.T., Brecher, P., and Cohen, R.A. (1998). Superoxide anion from the advential of the rat thoracic aorta inactivates nitric oxide. *Circulation Research*, **82**, 810–818.

Wang, H.Y., and Mu, S.L. (1999). Bioelectrochemical characteristics of cholesterol oxidase immobilized in a polyaniline film. *Sensors & Actuators B: Chemical*, **57**, 219–226.

Yamanaka, S.A., Dunn, S.B., Valentine, J.S., and Zink, J.I. (1995). Nicotinamide adenine dinucleotide phosphate fluorescence and adsorption monitoring of enzymic activity in silicate sol–gels for chemical sensing applications. *Journal of the American Chemical Society,* **117**, 9095–9096.

Yao, D., Vlessidis, A.G., Gou, Y., Zhou, X., Zhou, Y., and Evmiridis, N.P. (2004). Chemiluminescence detection of superoxide anion release and superoxide dismutase activity: modulation effect of *Pulsatilla chinensis. Analytical and Bioanalytical Chemistry,* **379**(1), 171–177.

Zhou, M.J., Diwu, Z.J., PanchukVoloshina, N., and Haugland, R.P. (1997). A stable nonfluorescent derivative of resorufin for the fluorometric determination of trace hydrogen peroxide: applications in detecting the activity of phagocyte NADPH oxidase and other oxidases. *Analytical Biochemistry*, **253**, 162–168.

– 11 –

Novel Biosensing Methods

11.1 INTRODUCTION

Biosensors have received increasing attention as a means of detecting different types of analytes in different arenas such as bioterrorism, and in the early detection of the emergence and proliferation of different types of diseases. Though some techniques such as ELISA and the surface plasmon resonance (SPR) biosensor technique have gained considerable attention recently, needless to say these detection techniques are not perfect, and do have flaws, which need to be improved upon. For example, ELISA is time-sensitive, and the surface plasmon resonance (SPR) technique does involve the presence of diffusional limitations and heterogeneities, for example, of the receptors on the biosensor surface. Also, on using the ELISA technique, Quinn *et al.* (2002) indicate that for the detection of human IgG, 19 h are required from analyte immobilization to binding confirmation by chromophore detection. Some of the drawbacks may be minimized by appropriate handling and experimental procedures. Nevertheless, as with all techniques there is room for improvement in these techniques themselves or in the development of other novel biosensing techniques. Some of the novel techniques to be presented in this chapter may be in their infancy, but they do exhibit potential as a useful detection technique. Modifications in the standard SPR biosensor technique, as an example, is also presented. The biosensor techniques to be presented in this chapter may be taken as only model examples. In no way are we indicating that these are in any way representative of all other biosensor techniques recently developed or in the process of development.

In this chapter, we present the fractal analysis of (a) the binding of different concentrations of mouse monoclonal anti-rabbit IgG in solution to rabbit patterned cells using a diffraction-based sandwich immunoassay (Goh *et al.*, 2003), (b) the binding and hybridization of DNA using a double-wavelength technique for SPR measurements (Zybin *et al.*, 2005), (c) binding and dissociation of different concentrations of GSH (Glutathione)-MPC (monolayer protected clusters) in solution to polyclonal antibody plus protein A immobilized on a quartz crystal microbalance (QCM) (Gerdon *et al.*, 2005), and the (d) binding and dissociation of substance P (neuropeptide) (SP) in solution to quencher-labeled mAb (monoclonal antibody) SP31 using the SPIT-FRI (solid-phase immobilized tripod for fluorescent renewable immunoassay) procedure (Volland *et al.*, 2005). Binding and dissociation rate coefficients, and affinity values are provided where ever possible. Fractal dimension values for the binding and the dissociation phase are also given. It should be pointed out again, and as indicated elsewhere in this book that the fractal analysis is one possible way to analyze the binding and the dissociation kinetics present in these types of analyte–receptor biosensor systems. The fractal analysis method does, however, include the involvement of external diffusional limitations and the presence of heterogeneities present in these types of analyte–receptor biosensor systems.

11.2 THEORY

Havlin (1989) has reviewed and analyzed the diffusion of reactants toward fractal surfaces. The details of the theory and the equations involved for the binding and the dissociation phases for analyte–receptor binding are available (Sadana, 2001). The details are not repeated here, except that just the equations are given to permit an easier reading. These equations have been applied to other biosensor systems (Ramakrishnan and Sadana, 2001; Sadana, 2001). For most applications, a single- or a dual-fractal analysis is often adequate to describe the binding and the dissociation kinetics. Peculiarities in the values of the binding and the dissociation rate coefficients as well as in the values of the fractal dimensions with regard to the dilute analyte systems being analyzed will be carefully noted, if applicable.

11.2.1 Single-fractal analysis

Binding rate coefficient

Havlin (1989) indicates that the diffusion of a particle (analyte [Ag]) from a homogeneous solution to a solid surface (e.g., receptor [Ab]-coated surface) on which it reacts to form a product (analyte–receptor complex; (Ab·Ag)) is given by

$$(\text{Analyte}\cdot\text{Receptor}) \sim \begin{cases} t^{(3-D_{\text{f,bind}})/2} = t^{p} & (t < t_c) \\ t^{1/2} & (t > t_c) \end{cases} \tag{11.1a}$$

Here $D_{\text{f,bind}}$ or D_f (used later on in the chapter) is the fractal dimension of the surface during the binding step; t_c is the cross-over value. Havlin (1989) indicates that the cross-over value may be determined by using $r_c^2 \sim t_c$. Above the characteristic length r_c, the self-similarity of the surface is lost and the surface may be considered homogeneous. Above time t_c, the surface may be considered homogeneous, since the self-similarity property disappears and 'regular' diffusion is now present. For a homogeneous surface where $D_\text{f} = 2$, and when only diffusional limitations are present, $p = ½$ as it should be. Another way of looking at the $p = ½$ case (where $D_{\text{f,bind}} = 2$) is that the analyte in solution views the fractal object; in our case, the receptor-coated biosensor surface from a 'large distance.' In essence, in the association process, the diffusion of the analyte from the solution to the receptor surface creates a depletion layer of width $(Đt)^{1/2}$ where Đ is the diffusion constant. This gives rise to the fractal power law

$$(\text{Analyte}\cdot\text{Receptor}) \sim t^{(3-D_{\text{f,bind}})/2}$$

For the present analysis, t_c is arbitrarily chosen and we assume that the value of the t_c is not reached. One may consider the approach as an intermediate 'heuristic' approach that may be used in the future to develop an autonomous (and not time-dependent) model for diffusion-controlled kinetics.

Dissociation rate coefficient

The diffusion of the dissociated particle (receptor [Ab] or analyte [Ag]) from the solid surface (e.g., analyte [Ag]–receptor [Ab]) complex coated surface) into solution may be given, as a first approximation by

$$\begin{aligned}(\text{Analyte}\cdot\text{Receptor}) &\sim -t^{(3-D_{\text{f,diss}})/2} \qquad (t > t_{\text{diss}}) \\ &= -k_{\text{diss}} t^{(3-D_{\text{f,diss}})/2}\end{aligned} \tag{11.1b}$$

Here $D_{\text{f,diss}}$ is the fractal dimension of the surface for the dissociation step. This corresponds to the highest concentration of the analyte–receptor complex on the surface. Henceforth, its concentration only decreases. The dissociation kinetics may be analyzed in a manner 'similar' to the binding kinetics.

11.2.2 Dual-fractal analysis

Binding rate coefficient

Sometimes, the binding curve exhibits complexities and two parameters (k, D_f) are not sufficient to adequately describe the binding kinetics. This is further corroborated by low values of r^2 factor (goodness-of-fit). In that case, one resorts to a dual-fractal analysis (four parameters; k_1, k_2, D_{f1}, and D_{f2}) to adequately describe the binding kinetics. The single-fractal analysis presented above is thus extended to include two fractal dimensions. At present, the time ($t = t_1$) at which the 'first' fractal dimension 'changes' to the 'second' fractal dimension is arbitrary and empirical. For the most part, it is dictated by the data analyzed and experience gained by handling a single-fractal analysis. A smoother curve is obtained in the 'transition' region if care is taken to select the correct number of points for the two regions. In this case, the product (antibody–antigen, or analyte–receptor complex; Ab·Ag, or analyte·receptor) is given by

$$(\text{Ab}\cdot\text{Ag}) \approx \begin{cases} t^{(3-D_{\text{f1,bind}})/2} = t^{p1} & (t < t_1) \\ t^{(3-D_{\text{f2,bind}})/2} = t^{p2} & t_1 < t < t_2 = t_c \\ t^{1/2} & (t > t_c) \end{cases} \tag{11.1c}$$

In some cases, as mentioned above, a triple-fractal analysis with six parameters (k_1, k_2, k_3, D_{f1}, D_{f2}, and D_{f3}) may be required to adequately model the binding kinetics. This is when the binding curve exhibits convolutions and complexities in its shape perhaps due to the very dilute nature of the analyte (in some of the cases to be presented) or for some other reasons. Also, in some cases, a dual-fractal analysis may be required to describe the dissociation kinetics.

11.3 RESULTS

At the outset it is appropriate to indicate that a fractal analysis will be applied to the data obtained for analyte–receptor binding and dissociation (if applicable) data for different types of either novel biosensor systems or biosensor systems with significant modifications. This is one possible explanation for analyzing the diffusion-limited binding and dissociation kinetics assumed to be present in all of the systems analyzed. The parameters thus obtained would provide a useful comparison of different situations. Alternate expressions involving saturation, first-order reaction, and no diffusion limitations are possible, but they are apparently deficient in describing the heterogeneity that inherently exists on the surface. The analyte–receptor binding as well as the dissociation reaction is a complex reaction, and the fractal analysis via the fractal dimension and the rate coefficient for binding or dissociation provide a useful lumped parameter(s) analysis of the diffusion-limited situation.

Also, we do not present any independent proof or physical evidence for the existence of fractals in the analysis of these analyte–receptor binding/dissociation systems except by indicating that it has been applied in other areas and that it is a convenient means of making more quantitative the degree of heterogeneity that exists on the surface. Thus, in all fairness, this is one possible way by which to analyze this analyte–receptor binding/dissociation data for novel biosensor systems. One might justifiably argue that appropriate modeling may be achieved by using a Langmuirian or other approach. The Langmuirian approach has a major drawback because it does not allow for or accommodate the heterogeneity that exists on the surface.

The Langmuir approach was originally developed for gases (Thomson and Webb, 1968). Researchers in the past have successfully modeled the adsorption behavior of analytes in solution to a solid surface using the Langmuir model even though it does not conform to the theory. Rudzinski *et al.* (1983) indicate that other appropriate "liquid" counterparts of the empirical isotherms have been developed. These include counterparts of the Freundlich (Dabrowski and Jaroniec, 1979), (Oscik *et al.*, 1976), and Toth (Jaroniec and Derylo, 1981) empirical equations. These studies with their known constraints, have provided some "restricted" physical insights into the adsorption of adsorbates on different surfaces. The Langmuir approach may be used to model the data presented, if one assumes the presence of discrete classes of sites (for example, double exponential analysis as compared to a single exponential analysis). Lee and Lee (1995) indicate that the fractal approach has been applied to surface science, for example, adsorption and reaction processes. These authors emphasize that the fractal approach provides a convenient means of representing the different structures and morphology at the reaction surface. These authors emphasize using the fractal approach to develop optimal structures and as a predictive approach.

Goh *et al.* (2003) indicate that diffraction-based sensing may be used to analyze interactions between two or more molecules. In order that one may obtain kinetic and equilibrium rate coefficients these authors emphasize that specific molecular interactions may be continuously monitored by observing intensity changes of the diffraction image from a patterned substrate. They further indicate that diffraction-based sensing has been used as a diagnostic tool to assay pH (Nakajima *et al.*, 1999), small molecules (Mines *et al.*, 2002; Bailey and Hupp, 2002), proteins (Goh *et al.*, 2002; Tsay *et al.*, 1991), and whole cells (St. John *et al.*, 1998; Morhard *et al.*, 2000). In spite of the disadvantages of using secondary labels such as high cost and longer sample preparation time, Goh *et al.* (2003) used diffraction-based sensing through the use of a secondary label to reduce the limit of detection of anti-rabbit IgG by a factor of 40 on using a gold-conjugated secondary antibody.

Figure 11.1a shows the binding of 25 ng/ml mouse monoclonal anti-rabbit IgG in solution to rabbit IgG patterned cells (Goh *et al.*, 2003). A single-fractal analysis is adequate to describe the binding kinetics. Table 11.1 shows the values of the binding rate coefficient k and the fractal dimension D_f, obtained using Corel Quattro Pro (1997) to fit the data. Eq. (11.1a) was used to fit the data, and the equation was used to obtain the value of k and D_f for a single-fractal analysis. The values of the parameters presented in Table 11.1 are within 95% confidence limits. For example, for the binding of 25 ng/ml mouse monoclonal anti-rabbit IgG in solution to rabbit IgG patterned cells, the binding rate coefficient k for a single-fractal analysis reported is 5.8179±0.0540. The 95% confidence limit indicates that 95% of the k values will lie between 5.7639 and 5.8179. This indicates that the values are precise and significant.

Figures 11.1b and c show the binding of (1) 50 and (2) 100 ng/ml mouse monoclonal anti-rabbit IgG in solution to rabbit IgG patterned cells (Goh *et al.*, 2003). In both these cases, a dual-fractal analysis is required to adequately describe the binding kinetics. The values of (a) the binding rate coefficient k and the fractal dimension D_f for a single-fractal analysis, and (b) the binding rate coefficients, k_1 and k_2, and the fractal dimensions, D_{f1} and D_{f2}, for a dual-fractal analysis are given in Table 11.1. In case (1), it is of interest to note that as the fractal dimension (or the degree of heterogeneity on the surface) increases by 43.3% from a value of D_{f1}=1.7750 to D_{f2}=2.5434, the binding rate coefficient increases by a factor of 2.24 from a value of k_1=24.427 to k_2 =54.595. As indicated elsewhere in this book, in this case too an increase in the degree of heterogeneity leads to an increase in the binding rate coefficient.

Also in case (2) it is of interest to note that as the fractal dimension (or the degree of heterogeneity on the surface) increases by 68.5% from a value of D_{f1}=1.4836 to D_{f2}=2.5004, the binding rate coefficient increases by a factor of 2.71 from a value of k_1=38.7559 to k_2=105.0776. Once again, an increase in the degree of heterogeneity leads to an increase in the binding rate coefficient. Also, note that an increase in the mouse monoclonal anti-rabbit IgG in solution by a factor of 2 from 50 to 100 ng/ml in solution leads to an increase in (a) the binding rate coefficient k_1 by a factor of 1.587 from a value of 24.427 to 38.756, and in (b) the binding rate coefficient k_2 by a factor of 1.924 from a value of 54.5952 to 105.0776.

Figures 11.1d and e show the binding of (1) 250 and (2) 500 ng/ml mouse monoclonal anti-rabbit IgG in solution to rabbit IgG patterned cells (Goh *et al.*, 2003). In both these

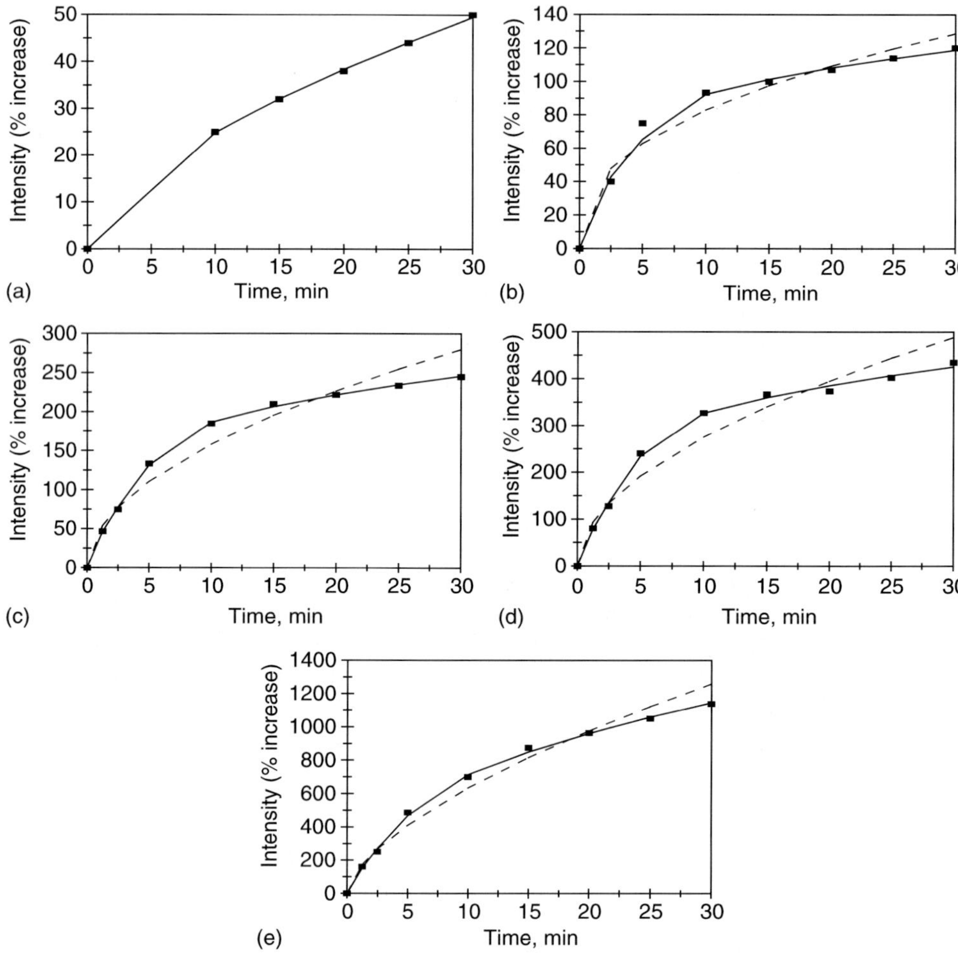

Figure 11.1 Binding of different concentrations (in ng/ml) of mouse monoclonal anti-rabbit IgG in solution to rabbit IgG patterned cells (Goh *et al.*, 2003): (a) 25, (b) 50, (c) 100, (d) 250, (e) 500. When only a solid line is used then a single-fractal analysis applies. When both dashed (-----) and solid (____) lines are used, then the dashed line is for a single-fractal analysis and the solid line is for a dual-fractal analysis. In this case, the solid line provides a better fit.

cases, a dual-fractal analysis is required to adequately describe the binding kinetics. The values of (a) the binding rate coefficient k and the fractal dimension D_f for a single-fractal analysis, and (b) the binding rate coefficients, k_1 and k_2, and the fractal dimensions, D_{f1} and D_{f2}, for a dual-fractal analysis are given in Table 11.1.

In both the cases presented in Figures 11.1d and e the trends are similar to the ones observed in Figures 11.1b and c. They are not repeated here to avoid repetition.

For a dual-fractal analysis, Table 11.1 and Figure 11.2a indicate the increase in the binding rate coefficient k_1 with an increase in the anti-rabbit IgG concentration in solution. In

Table 11.1

Binding rate coefficients and fractal dimensions for (a) the binding of different concentrations of mouse monoclonal anti-rabbit IgG in solution to rabbit IgG patterned cells, and (b) binding of different 5 μg/ml monoclonal anti-rabbit IgG in solution to rabbit IgG patterned cells, and (c) followed by 1 and 5 μg/ml anti-mouse gold conjugate (Goh *et al.*, 2003)

Mouse monoclonal anti-rabbit IgG in solution (ng/ml)/rabbit IgG patterned cells	k	k_1	k_2	D_f	D_{f1}	D_{f2}
(a) 25	5.8719±0.0540	NA	NA	1.7464±0.0212	NA	NA
50	33.1758±4.6043	24.4270±4.4288	54.5952±0.7562	2.2032±0.1169	1.7750±0.3398	2.5434±0.03172
100	48.0623±7.0313	38.7559±1.6747	105.0776±1.1673	1.9638±0.08964	1.4836±0.08630	2.5004±0.02548
250	83.0593±13.5538	65.2400±4.4233	187.3647±4.8296	1.9578±0.09924	1.4122±0.1338	2.5176±0.05868
500	151.3012±16.990	NA	NA	1.7542±0.06986	NA	NA
(b) 5 μg/ml monoclonal anti-rabbit in solution/ rabbit IgG patterned cells	19.7865±1.7548	NA	NA	2.2290±0.06618	NA	NA
(c) 1 μg/ml anti-mouse gold conjugate/rabbit IgG patterned cells	21.4603±2.2370	17.5609±1.3825	32.2395±2.2430	1.4368±0.0821	1.1494±0.2212	1.7336±0.10898
5 μg/ml anti-mouse gold conjugate/rabbit IgG patterned cells	60.0433±11.0704	31.0439±3.8759	149.086±6.881	1.7206±0.1349	0.5810±0.4778	2.3736±0.1055

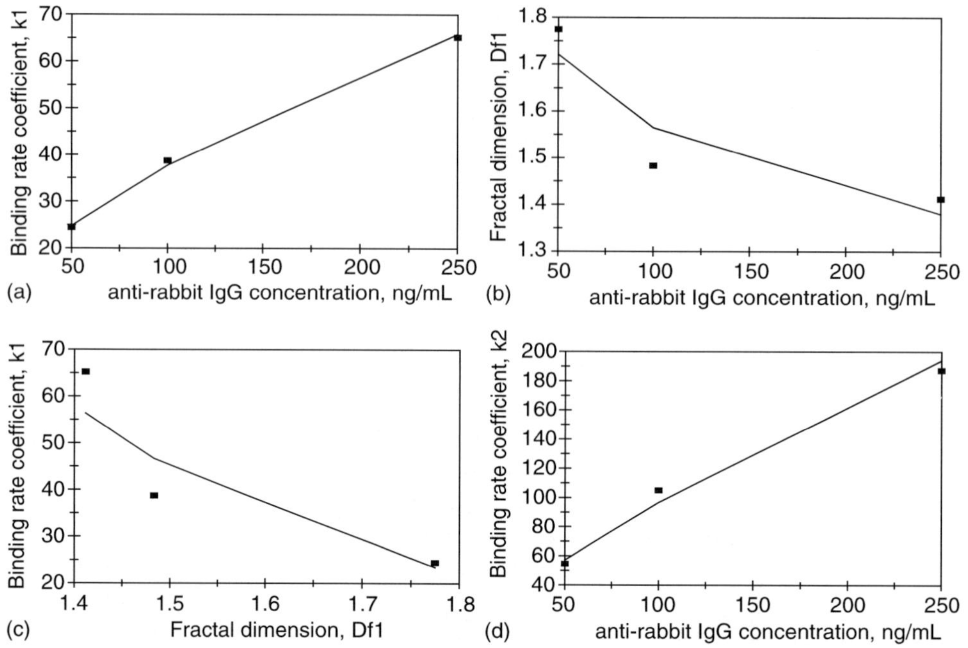

Figure 11.2 (a) Increase in the binding rate coefficient k_1 with an increase in the anti-rabbit IgG concentration (in ng/ml) in solution. (b) Decrease in the fractal dimension D_{f1} with an increase in the anti-rabbit IgG concentration (in ng/ml) in solution. (c) Decrease in the binding rate coefficient k_1 with an increase in the fractal dimension D_{f1}. (d) Increase in the binding rate coefficient k_2 with an increase in the anti-rabbit IgG concentration (in ng/ml) in solution.

the 50–250 ng/ml anti-rabbit IgG concentration in solution the binding rate coefficient k_1 is given by

$$k_1 = (2.2955 \pm 0.0731)\,[\text{anti - rabbit IgG}]^{0.6082 \pm 0.0274} \tag{11.2a}$$

The fit is good. Only three data points are available. The availability of more data points would lead to a more reliable fit. The binding rate coefficient k_1 is only mildly sensitive to the anti-rabbit IgG concentration in solution as noted by the 0.6082 order of dependence exhibited. The non-integer order of dependence exhibited lends support to the fractal nature of the system.

For a dual-fractal analysis, Table 11.1 and Figure 11.2b indicate the decrease in the fractal dimension D_{f1} with an increase in the anti-rabbit IgG concentration in the solution. In the 50–250 ng/ml anti-rabbit IgG concentration in solution the fractal dimension D_{f1} is given by

$$D_{f1} = (2.9477 \pm 0.2005)\,[\text{anti - rabbit IgG}]^{-0.1375 \pm 0.05764} \tag{11.2b}$$

The fit is reasonable. Only three data points are available. The availability of more data points would lead to a more reliable fit. The fractal dimension D_{f1} is only very slightly sensitive to the anti-rabbit IgG concentration in solution as noted by the negative 0.1375 order of dependence exhibited. It should be pointed out that the fractal dimension is based on a log scale, and that even small changes in the fractal dimension exhibited indicate significant changes in the degree of heterogeneity on the sensing surface.

For a dual-fractal analysis, Table 11.1 and Figure 11.2c indicate the decrease in the binding rate coefficient k_1 with an increase in the fractal dimension D_{f1}. In the 50–250 ng/ml anti-rabbit IgG concentration in solution the binding rate coefficient k_1 is given by

$$k_1 = (211.801 \pm 57.348)\, D_{f1}^{-3.8434 \pm 1.4081} \tag{11.2c}$$

The fit is good. Only three data points are available. The availability of more data points would lead to a more reliable fit. The binding rate coefficient k_1 is very sensitive to the fractal dimension D_{f1} or the degree of heterogeneity that exists on the surface as noted by the negative 3.8434 order of dependence exhibited.

For a dual-fractal analysis, Table 11.1 and Figure 11.2d indicate the increase in the binding rate coefficient k_2 with an increase in the anti-rabbit IgG concentration in solution. In the 50–250 ng/ml anti-rabbit IgG concentration in solution the binding rate coefficient k_2 is given by

$$k_2 = (2.9356 \pm 0.3109)\,[\text{anti-rabbit IgG}]^{0.7591 \pm 0.08817} \tag{11.2d}$$

The fit is good. Only three data points are available. The availability of more data points would lead to a more reliable fit. The binding rate coefficient k_1 is only mildly sensitive to the anti-rabbit IgG concentration in solution as noted by the 0.7591 order of dependence exhibited. The non-integer order of dependence exhibited lends support to the fractal nature of the system. The binding rate coefficient k_2 exhibits an order of dependence slightly higher (= 0.7591) than k_1 (= 0.6082) on the anti-rabbit IgG concentration in solution.

Figure 11.3a shows the increase in the ratio of the fractal dimensions, D_{f2}/D_{f1}, with an increase in the anti-rabbit IgG concentration in solution. In the 50–250 ng/ml anti-rabbit IgG concentration range in solution, the ratio of the fractal dimensions, D_{f2}/D_{f1}, is given by

$$D_{f2}/D_{f1} = (0.8777 \pm 0.0501)\,[\text{anti-rabbit IgG}]^{0.1319 \pm 0.0486} \tag{11.3a}$$

Only three data points are available. The fit is reasonable. The availability of more data points would lead to a better fit. The equation provides one with an idea of the change in the degree of heterogeneity on the sensing surface as one goes from the first- to the second-phase of binding. The order of dependence exhibited by the ratio of the fractal dimensions, D_{f2}/D_{f1}, on the anti-rabbit IgG concentration in solution is rather small (= 0.1319). But,

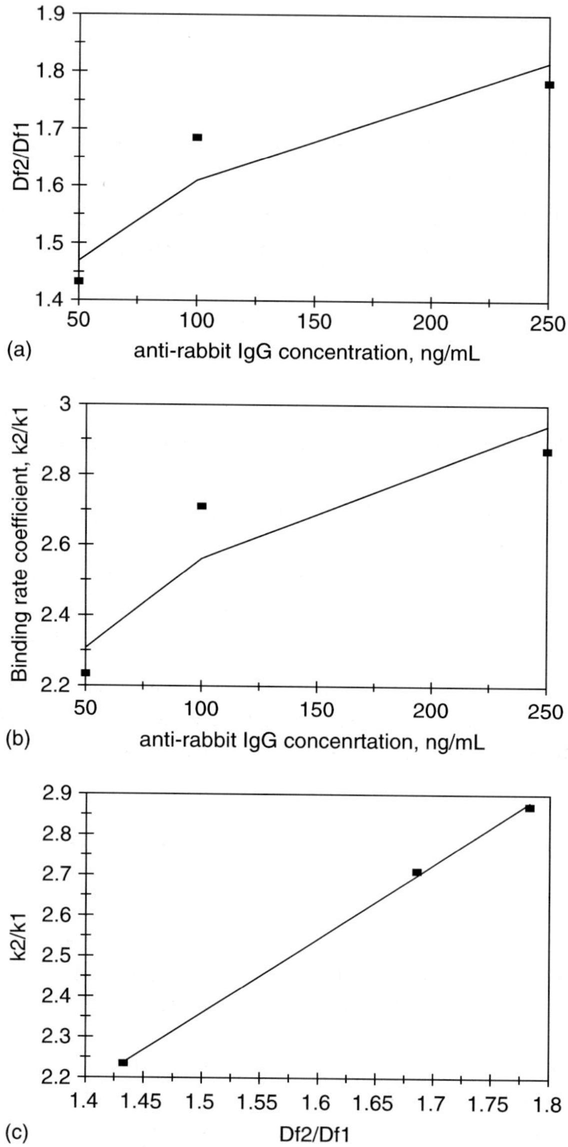

Figure 11.3 (a) Increase in the fractal dimension ratio, D_{f2}/D_{f1}, with an increase in the anti-rabbit IgG concentration (in ng/ml) in solution. (b) Increase in the binding rate coefficient ratio, k_2/k_1, with an increase in the anti-rabbit IgG concentration (in ng/ml) in solution. (c) Increase in the binding rate coefficient ratio, k_2/k_1, with an increase in the ratio of fractal dimensions, D_{f2}/D_{f1}.

once again, one needs to be reminded that the fractal dimension is based on a log scale, and even small changes in the value of the fractal dimension indicate significant changes in the degree of heterogeneity on the sensing surface. Figure 11.3b shows the increase in the ratio of the binding rate coefficients, k_2/k_1, with an increase in the anti-rabbit IgG concentration in solution. In the 50–250 ng/ml anti-rabbit IgG concentration in solution, the ratio of the binding rate coefficients, k_2/k_1, is given by

$$k_2/k_1 = (1.2788 \pm 0.0918)\,[\text{anti-rabbit IgG}]^{0.1509\pm0.0607} \tag{11.3b}$$

Once again, only three data points were available. The availability of more data points would lead to a more reliable fit. The equation does provide one with an estimate of the change in the value of the binding rate coefficient as one goes from the first- to the second-phase of binding as the anti-rabbit IgG concentration in solution changes in the 50–250 ng/ml concentration range. The ratio of the binding rate coefficients, k_2/k_1, exhibits a very small order (= 0.1509) of dependence on the anti-rabbit IgG concentration in solution.

Figure 11.3c shows the increase in the ratio of the binding rate coefficients, k_2/k_1, with an increase in the ratio of fractal dimensions, D_{f2}/D_{f1}. In the 50–250 ng/ml anti-rabbit IgG concentration in solution, and for the data presented in Figure 11.3c the ratio of the binding rate coefficients, k_2/k_1, is given by

$$k_2/k_1 = (1.4757 \pm 0.008)\,(D_{f2}/D_{f1})^{1.1571\pm0.03395} \tag{11.3c}$$

The fit is very good. Once again, only three data points were available. The availability of more data points would lead to a more reliable fit. The ratio of the binding rate coefficients, k_2/k_1, exhibits an order of dependence slightly greater than first (=1.1517) order on the ratio of fractal dimensions.

Figure 11.4a shows the binding of 5 μg/ml monoclonal anti-rabbit antibody in solution to rabbit IgG patterned cells (Goh *et al.*, 2003). A single-fractal analysis is adequate to describe the binding kinetics. The values of the binding rate coefficient k and the fractal dimension D_f are given in Table 11.1b.

Figures 11.4b and c show the binding of 1 and 5 μg/ml mouse monoclonal anti-rabbit IgG to rabbit IgG patterned cells. This was followed by anti-mouse gold conjugate that Goh *et al.* (2003) used as a secondary label to enhance their diffraction-based sensing for diagnostic purposes. In both the cases, a dual-fractal analysis is required to adequately describe the binding kinetics. The values of (a) the binding rate coefficient k and the fractal dimension D_f for a single-fractal analysis, and (b) the binding rate coefficients, k_1 and k_2, and the fractal dimensions, D_{f1} and D_{f2}, for a dual-fractal analysis are given in Table 11.1c.

On using the secondary label (anti-mouse gold conjugate) there is a change in the binding mechanism compared to when it is not used. This is because a dual-fractal analysis is required to adequately describe the binding kinetics when the secondary label is used, and a single-fractal analysis is adequate to describe the binding kinetics when the secondary label is not used. It is not entirely unexpected that the use of the secondary label leads to a more complex binding mechanism.

It is interesting to note that as one increases the mouse monoclonal anti-rabbit IgG concentration in solution (when the secondary label anti-mouse gold conjugate is present) from 1 to 5 μg/ml, both of the binding rate coefficients, k_1 and k_2, exhibit increases.

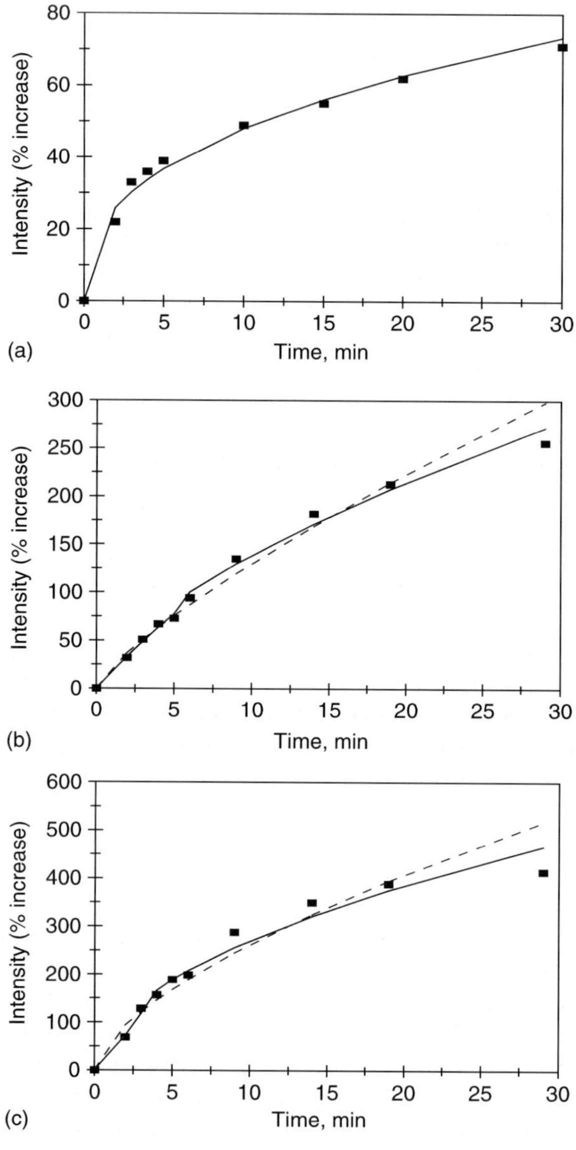

Figure 11.4 (a) Binding of 5 μg/ml monoclonal anti-rabbit antibody in solution to rabbit IgG patterned cells (Goh *et al.*, 2003). (b) Binding of 1 μg/ml anti-mouse gold conjugate antibody in solution to rabbit IgG patterned cells (Goh *et al.*, 2003). (c) Binding of 5 μg/ml anti-mouse gold conjugate antibody in solution to rabbit IgG patterned cells (Goh *et al.*, 2003). When only a solid line is used then a single-fractal analysis applies. When both dashed (----) and solid (_____) lines are used, then the dashed line is for a single-fractal analysis and the solid line is for a dual-fractal analysis. In this case, the solid line provides a better fit.

Zybin *et al.* (2005) have used the double-wavelength technique for the on-line monitoring of analyte binding to sensor surfaces using the SPR biosensor. These authors have used their recently developed system to analyze and to detect short nucleotides of the sequences used for genotyping human hepatitis C viruses. They were able to obtain a selective response to complementary oligonucleotides.

Zybin *et al.* (2005) provide an excellent history of the initial development of the SPR biosensor starting with Leidberg *et al.* (1995), and followed by subsequent modifications of the technique. They emphasize that the improvement in the detection sensitivity has been one of the main goals in improving the SPR sensing technique. For example, immobilization procedures have been optimized (Hermanson, 1995; Henke and Krull, 1999; Albers *et al.*, 2001; Wrobel *et al.*, 2002, 2003). Also, the plasmon layer has been optimized by using bimetallic Ag/Au layers (Zynio *et al.*, 2002; Alieva and Konopsky, 2004). Zybin *et al.* (2005) indicate that the SPR technique is a highly sensitive technique that may be affected by changes in the surface roughness that influence possible shape changes in the resonance. These authors emphasize that their double-wavelength measurement technique does help alleviate these problems. In essence, in this technique the surface is irradiated by a parallel beam that is composed of two laser beams with different wavelengths.

In their hybridization (binding) experiments, Zybin *et al.* (2005) used sequences of the DNA oligonucleotides that were used earlier for their diagnostics of human hepatitis C virus (G), and also two types of virus T_1 and T_2 were used. The following receptor sequences were used: 5′-CCAAGAAAGGACCCG-3′ (G), 5′-CTCCAGGCATTGAGC-3′ (T1), and 5′-CAACCCAACGCTACT-3′ (T2). Zybin *et al.* (2005) indicate that the analytes in solution were synthetic oligonucleotides of complementary sequence without the thymine spacer. They are designated as cG, cT1, and cT2, respectively.

Figure 11.5a shows the binding (hybridization) of G-HCV-c of the sequence 5′-TTTCGGGTCCTTTCTTGG-3′ to the spot coated with the complementary oligonucleotide G-HCV of the sequence 5′-TTTTTCCAAGAAAGGACC-3′. A dual-fractal analysis is required to adequately describe the hybridization (binding) kinetics. The values of (a) the binding rate coefficient k and the fractal dimension D_f for a single-fractal analysis, and (b) the binding rate coefficients, k_1 and k_2, and the fractal dimensions, D_{f1} and D_{f2}, for a dual-fractal analysis are given in Table 11.2a.

Figures 11.5b and c show the selective hybridization (binding) of G-HCV-c and T_2-HCV-c in a mixture of G-HCV-c and T_2-HCV-c in solution to G-HCV and T_2-HCV complementary sequences immobilized at different spots. In both the cases, a dual-fractal analysis is required to adequately describe the hybridization (binding) kinetics. The values of (a) the hybridization (binding) rate coefficient k and the fractal dimension D_f for a single-fractal analysis, and (b) the hybridization (binding) rate coefficients, k_1 and k_2, and the fractal dimensions, D_{f1} and D_{f2}, for a dual-fractal analysis are given in Table 11.2b.

Note that the selective hybridization rate coefficients k_1 for G-HCV-c/G-HCV is lower than that for T_2-HCV-c/T_2-HCV. However, the selective hybridization rate coefficient k_2 for G-HCV-c/G-HCV is higher than that for T_2-HCV-c/T_2-HCV. It is of interest to note that the corresponding fractal dimensions also follow similar trends. In this case, higher fractal dimensions or higher degrees of heterogeneity on the spot surface lead to higher hybridization (binding) rate coefficients and vice versa. For example, the fractal dimension D_{f1} values for G-HCV-c/G-HCV and T_2-HCV-c/T_2-HCV are 0.7706 and 1.9236, respectively.

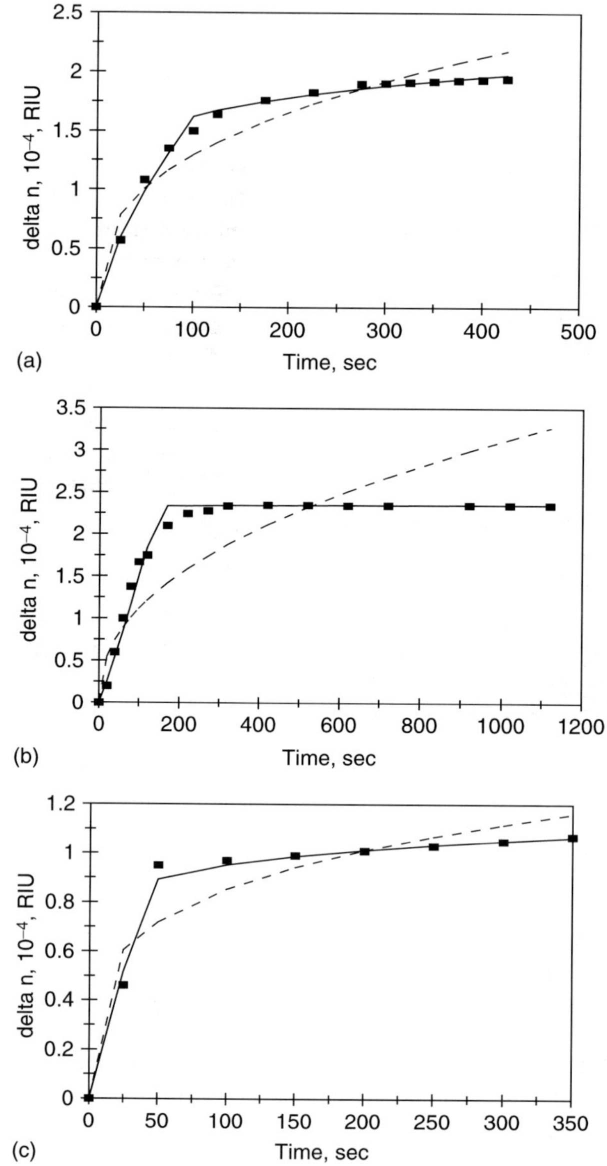

Figure 11.5 (a) Binding and hybridization of G-HCV-c of the sequence 5′-TTTCGGGTC-CTTTCTTGG3′ to the spot coated with the complementary oligonucleotide of the sequence 5′-TTTTTCCAAGAAAGGACC-3′ (Zybin *et al.*, 2005). (b) Selective hybridization and binding of G-HCV-c in solution to G-HCV immobilized on a spot surface (Zybin *et al.*, 2005). (c) Selective hybridization and binding of T_2-HCV-c in solution to T_2-HCV immobilized on a spot surface (Zybin *et al.*, 2005). When only a solid line is used then a single-fractal analysis applies. When both dashed (----) and solid (____) lines are used, then the dashed line is for a single-fractal analysis and the solid line is for a dual-fractal analysis. In this case, the solid line provides a better fit.

Table 11.2

Binding rate coefficients and fractal dimensions for the hybridization of (a) G-HCV-c of the sequence 5′-TTTCGGGTCCTTTCTTGG3′ to the spot coated with complementary oligonucleotide G-HCV of the sequence 5′-TTTTTCCAAGAAAGGACC-3′, (b) G-HCV-c to the spot coated with G-HCV, and T_2-HCV-c to the spot coated with T_2 -HCV, and (c) (G-HCV-c) + (T_1 -HCV-c) + (T_2 -HCV-c) to the spot coated with G-HCV, T_1 -HCV, and T_2 -HCV (Zybin *et al.*, 2005)

Analyte in solution/receptor on spot surface	k	k_1	k_2	D_f	D_{f1}	D_{f2}
(a) G-HCV-c/G-HCV	0.2463±0.0359	0.0611±0.0060	0.8676±0.0121	2.2796±0.0860	1.5796±0.1788	2.7282±0.02338
(b) Selective hybridization of G-HCV-c and T_2-HCV-c/G-HCV and T_2-HCV (complementary sequences immobilized at different spots)						
G-HCV-c/G-HCV	0.1451±0.0652	0.00886±0.00192	2.2926±0.0009	2.1126±0.1483	0.7706±0.221	2.9928±0.001598
T_2-HCV-c/T_2-HCV	0.2734±0.0524	0.09150±0.0305	0.6247±0.0025	2.5062±0.1431	1.9236±0.5866	2.8176±0.01196
(c) (G-HCV-c) + (T_1-HCV-c) + (T_2-HCV-c)/T_1-HCV	0.05855±0.01554	0.005446±0.000724	0.6989±0.0130	1.7330±0.1424	0.5770±0.1963	2.6158±0.0304
(G-HCV-c) + (T_1-HCV-c) + (T_2-HCV-c)/T1-HCV	0.2297±0.0258	0.06072±0.00582	0.3233±0.0107	2.5530±0.06450	1.8792±0.1756	2.6810±0.06454
(G-HCV-c) + (T_1-HCV-c) + (T_2-HCV-c)/T1-HCV	0.01484±0.00312	0.002039±0.000074	0.1249±0.0050	1.4862±0.1157	0.5186±0.06816	2.2442±0.05452

Correspondingly, the hybridization (binding) rate coefficients k_1 are 0.0086 and 0.09150, respectively. Similarly, the fractal dimension D_{f2} values for G-HCV-c/G-HCV and T_2-HCV-c/ T_2-HCV are 2.9928 and 2.8176, respectively. Correspondingly, the hybridization (binding) rate coefficients k_2 are 2.2926 and 0.6247, respectively.

Figures 11.6a–c show the hybridization (binding) of (G-HCV-c) + (T_1-HCV-c) + (T_2-HCV-c) in solution to G-HCV, T_1-HCV, and T_2-HCV immobilized on a spot surface (Zybin *et al.*, 2005). In all the cases, a dual-fractal analysis is required to adequately describe the hybridization (binding) kinetics. The values of (a) the binding rate coefficient k and the fractal dimension D_f for a single-fractal analysis, and (b) the binding rate coefficients, k_1 and k_2, and the fractal dimensions, D_{f1} and D_{f2}, for a dual-fractal analysis are given in Table 11.2c.

It is of interest to note that the fractal dimension D_{f1} increases in the sequence T_2-HCV-c, G-HCV-c, and T_1-HCV-c immobilized on the spot surface, and the corresponding binding rate coefficients k_1 also increase in the same sequence T_2-HCV-c, G-HCV-c, and T_1-HCV-c.

Tables 11.2a and b, and Figure 11.7a show the increase in the (hybridization) binding rate coefficient k_1 with an increase in the fractal dimension D_{f1}. For the binding of G-HCV-c in solution to G-HCV immobilized on the spot surface, and for the selective hybridization of G-HCV-c and T_2-HCV-c in solution to complementary sequences immobilized at different spot surfaces, the binding rate coefficient k_1 is given by

$$k_1 = (0.01762 \pm 0.00141) D_{f1}^{2.5894 \pm 0.1128} \tag{11.4a}$$

Only three data points are available. The fit is good. The availability of more data points would lead to a more reliable fit. The binding rate coefficient k_1 is sensitive to the degree of heterogeneity that exists on the spot surface, as noted by the slightly greater than two and one-half (=2.5894) order of dependence exhibited.

Figure 11.7b shows the increase in the (hybridization) ratio of the binding rate coefficients, k_2/k_1, with an increase in the ratio of the fractal dimensions, D_{f2}/D_{f1}. For the binding of G-HCV-c in solution to G-HCV immobilized on the spot surface, and for the selective hybridization of G-HCV-c and T_2-HCV-c in solution to complementary sequences immobilized at different spot surfaces, the ratio of the binding rate coefficients, k_2/k_1, is given by

$$k_2/k_1 = (1.7717 \pm 0.1682)\,(D_{f2}/D_{f1})^{3.6824 \pm 0.1229} \tag{11.4b}$$

Only three data points are available. The fit is good. The availability of more data points would lead to a more reliable fit. The ratio of the binding rate coefficients, k_2/k_1, is very sensitive to the degree of heterogeneity that exists on the spot surface, as noted by the order between three and one-half and four (=3.6824) exhibited.

Table 11.2c and Figure 11.7c show the increase in the (hybridization) binding rate coefficient k_1 with an increase in the fractal dimension D_{f1}. In this case, the binding rate coefficient k_1 is given by

$$k_1 = (0.01387 \pm 0.00936)\, D_{f1}^{2.3901 \pm 0.5104} \tag{11.4c}$$

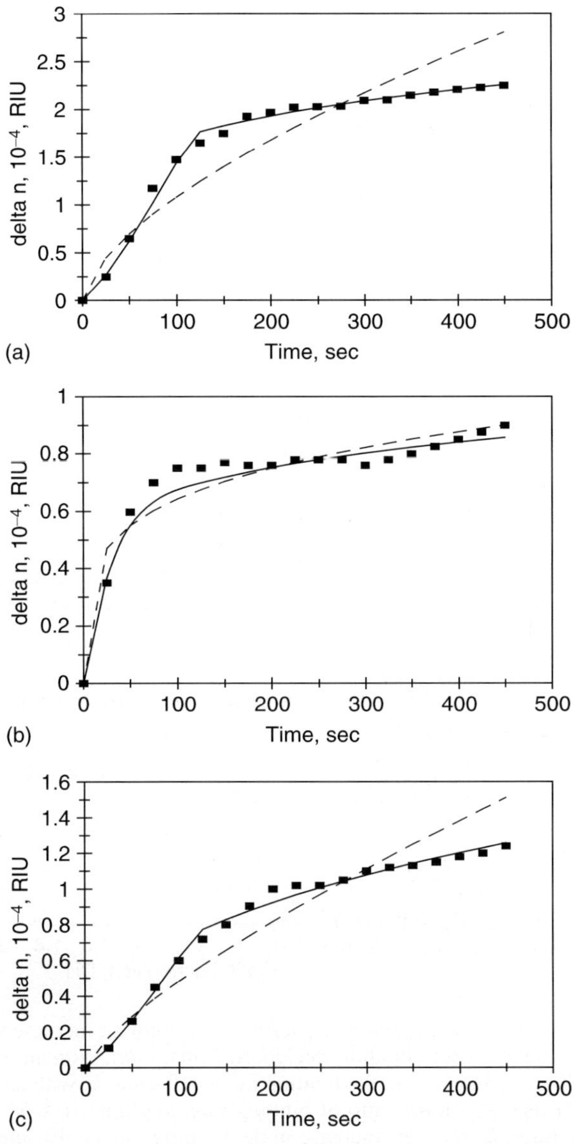

Figure 11.6 (a) Binding and hybridization of (G-HCV-c) + (T_1 -HCV-c) + (T_2 -HCV-c) in solution to G-HCV immobilized on a spot surface (Zybin *et al.*, 2005). (b) Binding and hybridization of (G-HCV-c) + (T_1 -HCV-c) + (T_2 -HCV-c) in solution to T_1-HCV immobilized on a spot surface (Zybin *et al.*, 2005). (c) Binding and hybridization of (G-HCV-c) + (T_1 -HCV-c) + (T_2 -HCV-c) in solution to T_2-HCV immobilized on a spot surface (Zybin *et al.*, 2005). When only a solid line is used then a single-fractal analysis applies. When both dashed (----) and solid (____) lines are used, then the dashed line is for a single-fractal analysis and the solid line is for a dual-fractal analysis. In this case, the solid line provides a better fit.

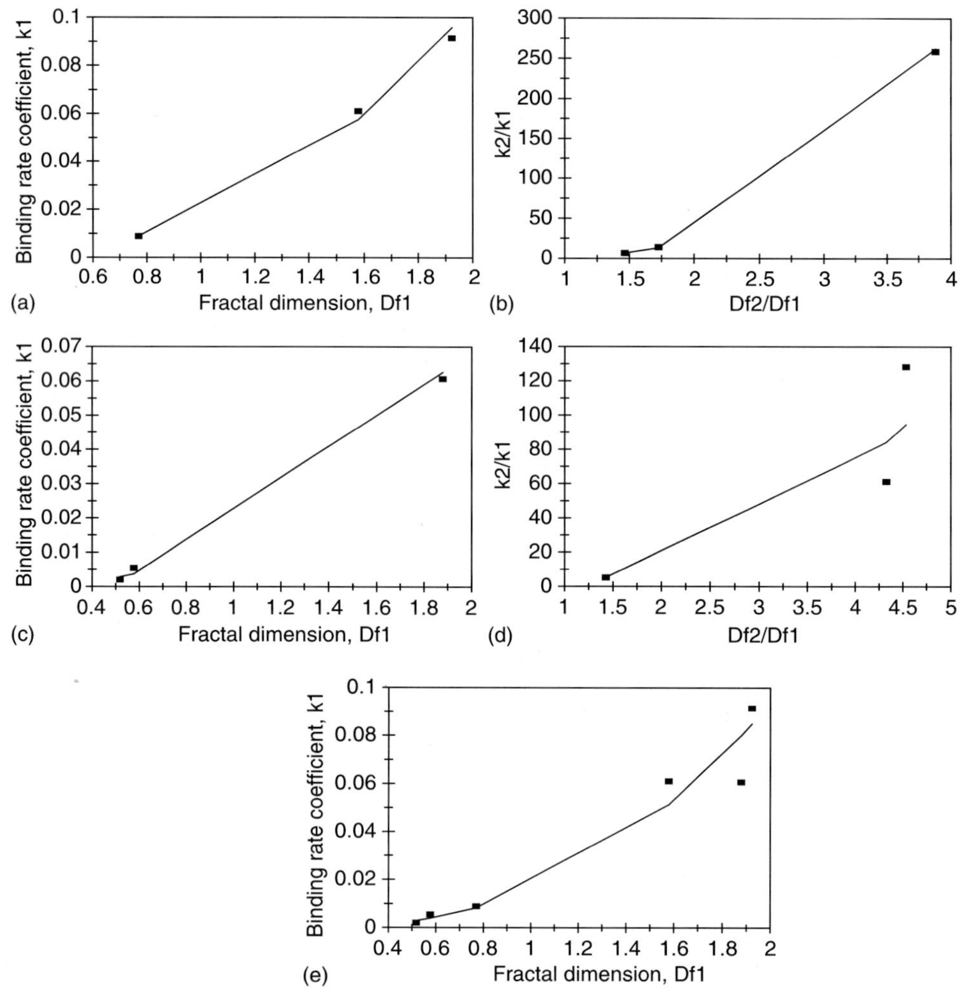

Figure 11.7 (a) Increase in the binding rate coefficient k_1 with an increase in the fractal dimension D_{f1}. (b) Increase in the ratio of binding rate coefficients, k_2/k_1, with an increase in the fractal dimension ratio, D_{f2}/D_{f1}. (c) Increase in the binding rate coefficient k_1 with an increase in the fractal dimension D_{f1}. (d) Increase in the ratio of binding rate coefficients, k_2/k_1, with an increase in the fractal dimension ratio, D_{f2}/D_{f1}. (e) Increase in the binding rate coefficient k_1 with an increase in the fractal dimension D_{f1}.

Only three data points are available. The fit is good. The availability of more data points would lead to a more reliable fit. The binding rate coefficient k_1 is sensitive to the degree of heterogeneity that exists on the spot surface, as noted by the 2.3901 order of dependence exhibited.

Figure 11.7d shows the increase in the ratio of the (hybridization) binding rate coefficients, k_2/k_1, with an increase in the ratio of the fractal dimensions, D_{f2}/D_{f1}. For the data presented in Table 11.2c the ratio of the binding rate coefficients, k_2/k_1, is given by

$$k_2/k_1 = (2.1627 \pm 1.1995)\,(D_{f2}/D_{f1})^{2.4996 \pm 0.4767} \tag{11.4d}$$

Only three data points are available. The fit is good. The availability of more data points would lead to a more reliable fit. The ratio of the binding rate coefficients, k_2/k_1, is sensitive to the ratio of the fractal dimensions, D_{f2}/D_{f1}, that exists on the spot surface, as noted by the very close to two and one-half order of dependence (=2.4996) exhibited.

Tables 11.2a–c and Figure 11.7e show the increase in the (hybridization) binding rate coefficient k_1 with an increase in the fractal dimension D_{f1}. In this figure, we have plotted the results presented in Figures 11.7a and c on one plot. This was to see if it made a lot of difference on plotting both of these types of analyte–receptor systems on one plot. In this case, the binding rate coefficient k_1 is given by

$$k_1 = (0.01596 \pm 0.00568)\,D_{f1}^{2.5564 \pm 0.2256} \tag{11.4e}$$

In this case, since two sets of data are plotted together, six data points are available. Considering that data from two different sets are plotted together, the fit is good. The availability of more data points would nevertheless lead to a more reliable fit. The binding rate coefficient k_1 is sensitive to the degree of heterogeneity that exists on the spot surface, as noted by the 2.5564 order of dependence exhibited.

Gerdon *et al.* (2005) recently indicated that there is a need to develop a label-free, fast-responsive, quantitative assay for the detection of 3D substrates and multilayer adsorptions (Crowther, 2001). They indicated that two techniques that used wave propagation techniques have gained prominence. They are the SPR and the quartz crystal microbalance (QCM) techniques. Gerdon *et al.* (2005) indicate that the QCM technique uses acoustic waves, whereas the SPR technique uses optical waves. The QCM technique measures the acoustic impedance minimum, and the SPR technique measures the angle of reflection minimum. These authors emphasize the ability of the QCM technique to work with 3D substrates and multilayer adsorptions.

Figure 11.8 shows the binding of 4.1 μM Tiop (Tiopronin)-MPC (monolayer protected clusters) in solution to polyclonal anti-glutathione + protein A immobilized on a QCM (Gerdon *et al.*, 2005). The plot of the binding is a straight line. A single-fractal analysis is adequate to describe the binding kinetics. The binding rate coefficient k=0.006, and the fractal dimension D_f=2.0. This D_f=2.0 value is the value of a homogeneous distribution in two-dimensional space.

Figures 11.9a–c show the binding and dissociation kinetics of (1) 2.3, (2) 4.5, and (3) 9.1 μM glutathione (GSH)-MPC in solution to polyclonal antibody (polyclonal anti-GSH) + protein A immobilized on a QCM (Gerdon *et al.*, 2005). In all the cases, a dual-fractal analysis is required to adequately describe the binding kinetics. A single-fractal analysis is adequate to describe the dissociation kinetics. The values of (a) the binding rate

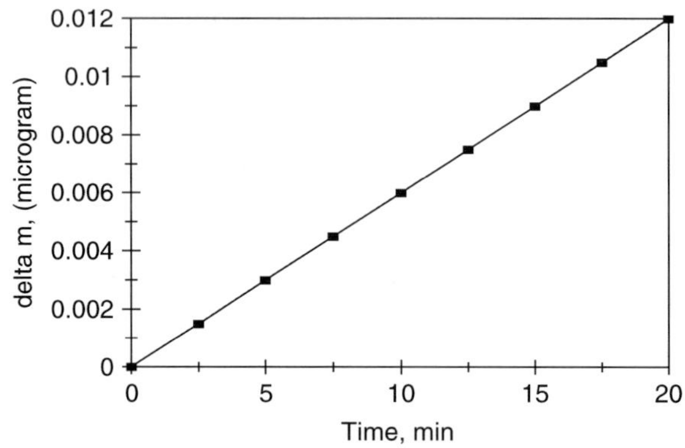

Figure 11.8 Binding of 4.1 μM Tiop (Tiopronin)-MPC (monolayer protected clusters) in solution to polyclonal anti-glutathione + protein A immobilized on a quartz crystal microbalance (QCM) (Gerdon *et al.*, 2005).

coefficient k and the fractal dimension D_f for a single-fractal analysis, and (b) the binding rate coefficients, k_1 and k_2, and the fractal dimensions, D_{f1} and D_{f2}, for a dual-fractal analysis are given in Table 11.1c.

In case (2), it is interesting to note that as the glutathione (GSH)-MPC concentration in solution increases by almost a factor of 2 from 2.3 to 4.5 μM, the binding rate coefficient k_1 value increases by a factor of 1.752 from a value of 0.001145 to 0.002007, and the binding rate coefficient, k_2, value increases by a factor of 1.902 from a value of 0.0256 to 0.04868.

In case (3), it is interesting to note that as the glutathione (GSH)-MPC concentration in solution increases by almost a factor of 2 from 4.5 to 9.1 μM, the binding rate coefficient, k_1, value increases by a factor of 10.36 from a value of 0.002007 to 0.02079, and the binding rate coefficient, k_2, value increases by a factor of 1.064 from a value of 0.04868 to 0.05178.

Figure 11.10a and b and Table 11.3a show the increase in the binding rate coefficients, k_1 and k_2, with an increase in the GSH-MPC concentration in solution. In the 2.3–9.1 μM GSH-MPC concentration range, the binding rate coefficients, k_1 and k_2, are given by

$$k_1 = (0.000147 \pm 0.000147)\,[\text{GSH - MPC}]^{2.1178 \pm 0.7164} \quad (11.5a)$$

$$k_2 = (0.01857 \pm 0.00513)\,[\text{GSH - MPC}]^{0.5084 \pm 0.2507} \quad (11.5b)$$

The fit is quite good. Only three data points are available. The availability of more data points would lead to a more reliable fit. The binding rate coefficient k_1 is sensitive to the [GSH–MPC] concentration in solution as noted by the slightly higher than second order (=2.1177) order of dependence exhibited. The binding rate coefficient k_2 is only mildly sensitive to the [GSH-MPC] concentration in solution as noted by the close to half order (=0.5084) order of dependence exhibited.

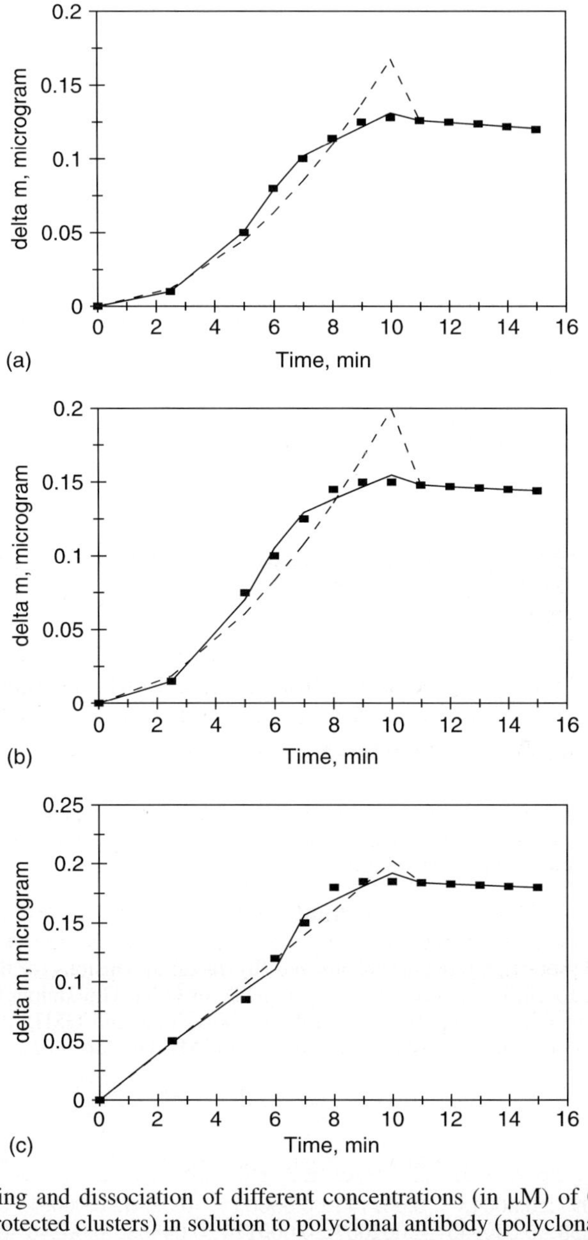

Figure 11.9 Binding and dissociation of different concentrations (in μM) of GSH (Glutathione)-MPC (monolayer protected clusters) in solution to polyclonal antibody (polyclonal anti-GSH) + protein A immobilized on a quartz crystal microbalance (QCM) (Gerdon *et al.*, 2005): (a) 2.3, (b) 4.5, (c) 9.1. When only a solid line is used then a single-fractal analysis applies. When both dashed (----) and solid (____) lines are used, then the dashed line is for a single-fractal analysis and the solid line is for dual-fractal analysis. In this case, the solid line provides a better fit.

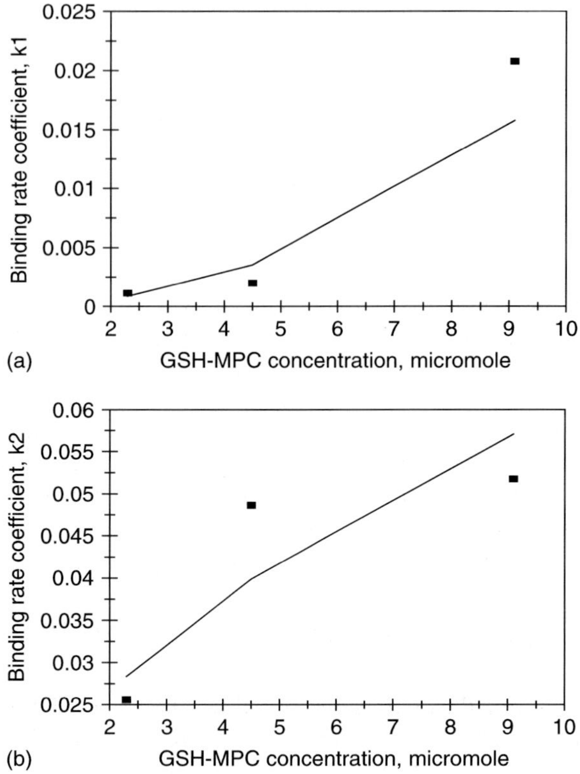

Figure 11.10 (a) Increase in the binding rate coefficient k_1 with an increase in the GSH-MPC concentration (in μM) in solution. (b) Increase in the binding rate coefficient k_2 with an increase in the GSH-MPC concentration (in μM) in solution.

Table 11.3

(a) Binding and dissociation rate coefficients and (b) fractal dimensions for the binding and the dissociation phase for different concentrations (in μM) for GSH (Glutathione)-MPC (monolayer protected clusters) in solution to polyclonal antibody (polyclonal anti-GSH) + protein A immobilized on a quartz crystal microbalance (QCM) (Gerdon *et al.*, 2005)

(a)

GSH-MPC (μM)	k	k_1	k_2	k_d
2.3	0.002097±0.00045	0.001145±0.00003	0.0256±0.00081	0.001823±0.000254
4.5	0.003858±0.000945	0.002007±0.000174	0.04868±0.00246	0.001943±0.000071
9.1	0.01867±0.00280	0.02079±0.00280	0.05178±0.00331	0.001±0

(b)

	D_f	D_{f1}	D_{f2}	D_{fd}
2.3	0.0+0.3550	0.0+0.08752	1.5806±0.2356	2.9964±0.1027
4.5	0.0+0.3814	0.0+0.2544	1.9948±0.3696	1.6384±0.05694
9.1	0.9294±0.1744	1.1298±0.3850	1.8608±0.4672	1.0±1.4E-16

Table 11.3a and Figure 11.11a and b show the increase in the binding rate coefficients, k_1 and k_2 with an increase in the fractal dimensions, D_{f1} and D_{f2}. In the 2.3–9.1 μM GSH-MPC concentration range, the binding rate coefficients, k_1 and k_2 are given by

$$k_1 = (0.01509 \pm 0.0104) D_{f1}^{1.1611 \pm 0.2899} \tag{11.6a}$$

$$k_2 = (0.006689 \pm 0.001496) D_{f2}^{3.0371 \pm 1.1939} \tag{11.6b}$$

The fit is good. Only three data points are available. The availability of more data points would lead to a more reliable fit. The binding rate coefficient k_1 exhibits slightly more than first order (=1.1611) of dependence on the fractal dimension D_{f1} or the degree of heterogeneity that exists on the quartz crystal microbalance surface. The binding rate coefficient k_2 exhibits slightly more than third order (=3.0371) of dependence on the fractal dimension D_{f2} or the degree of heterogeneity that exists on the quartz crystal microbalance surface.

Figure 11.11c shows the increase in the ratio of the binding rate coefficients, k_2/k_1, with an increase in the fractal dimension ratio, D_{f2}/D_{f1}. In the 2.3–9.1 μM GSH–MPC concentration range, the ratio of the binding rate coefficients, k_2/k_1, is given by

$$k_2/k_1 = (1.3983 \pm 0.1134)(D_{f2}/D_{f1})^{0.8632 \pm 0.04704} \tag{11.6c}$$

The fit is good. Only three data points are available. The availability of more data points would lead to a more reliable fit. The ratio of the binding rate coefficients, k_2/k_1, exhibits less than a first order (=0.8632) of dependence on the ratio of fractal dimensions, (D_{f2}/D_{f1}) that exists on the QCM surface.

Volland *et al.* (2005) have recently suggested some of the desirable characteristics for immunosensors or biosensors. These include: recognition of different analytes present in solution, preferably simultaneously, response in a few minutes, field analysis (autonomy and ruggedness), continuous-flow analysis, and regeneration of the sensor or repeated use. The last characteristic is often difficult to realize in practice, and single-use biosensors have been developed. Volland *et al.* (2005) have recently developed a new concept of an immunosensor that permits a sensitive measurement as well as non-aggressive regeneration (Volland *et al.*, 2002). This new method is termed as solid-phase immobilized tripod for fluorescent renewable immunoassay (SPIT-FRI).

Figures 11.12a–c shows the binding of (1) 1, (2) 0.1, and (3) 0.01 μM substance P (neuropeptide) (SP) in solution to quencher-labeled mAb (monoclonal antibody) SP31 using the SPIT-FRI procedure (Volland *et al.*, 2005). In cases (1) and (3), a single-fractal analysis is adequate to describe the binding kinetics, and a dual-fractal analysis is required to adequately describe the dissociation kinetics. In case (2), a single-fractal analysis is adequate to describe the binding as well as the dissociation kinetics. The values of (a) the binding rate coefficient k and the fractal dimension D_f for a single-fractal analysis, (b) the dissociation rate coefficient k_d and the fractal dimenison D_{fd} for a single-fractal analysis, and (c) the dissociation rate coefficients, k_{d1} and k_{d2}, and the fractal dimensions, D_{fd1} and D_{fd2}, for a dual-fractal analysis are given in Tables 11.4a and b.

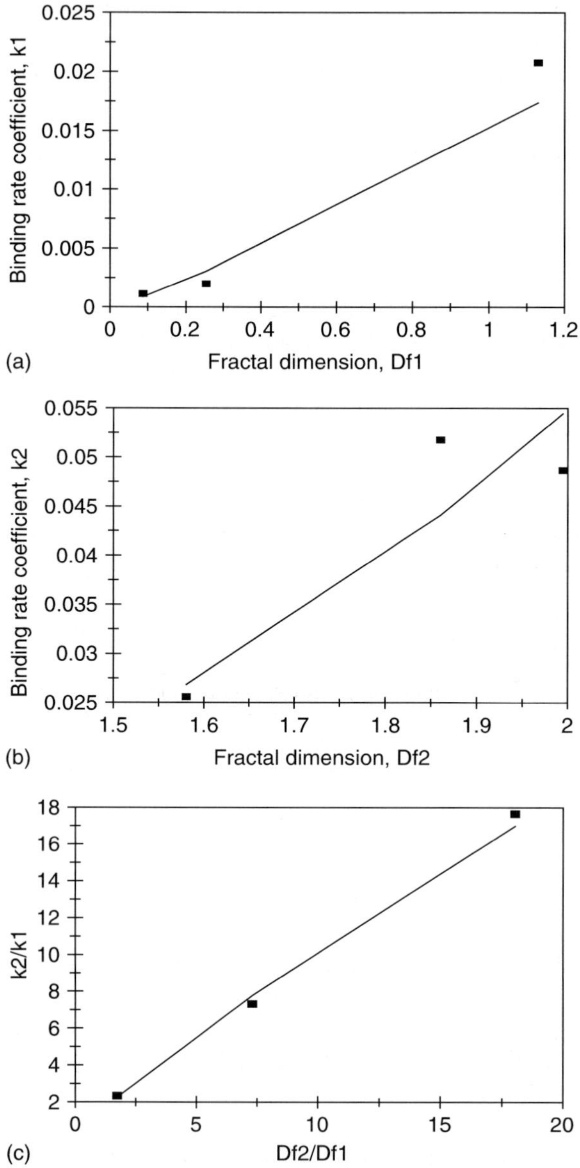

Figure 11.11 a) Increase in the binding rate coefficient k_1 with an increase in the fractal dimension D_{f1}. (b) Increase in the binding rate coefficient k_2 with an increase in the fractal dimension D_{f2}. (c) Increase in the ratio of binding rate coefficients, k_2/k_1, with an increase in the fractal dimension ratio, D_{f2}/D_{f1}.

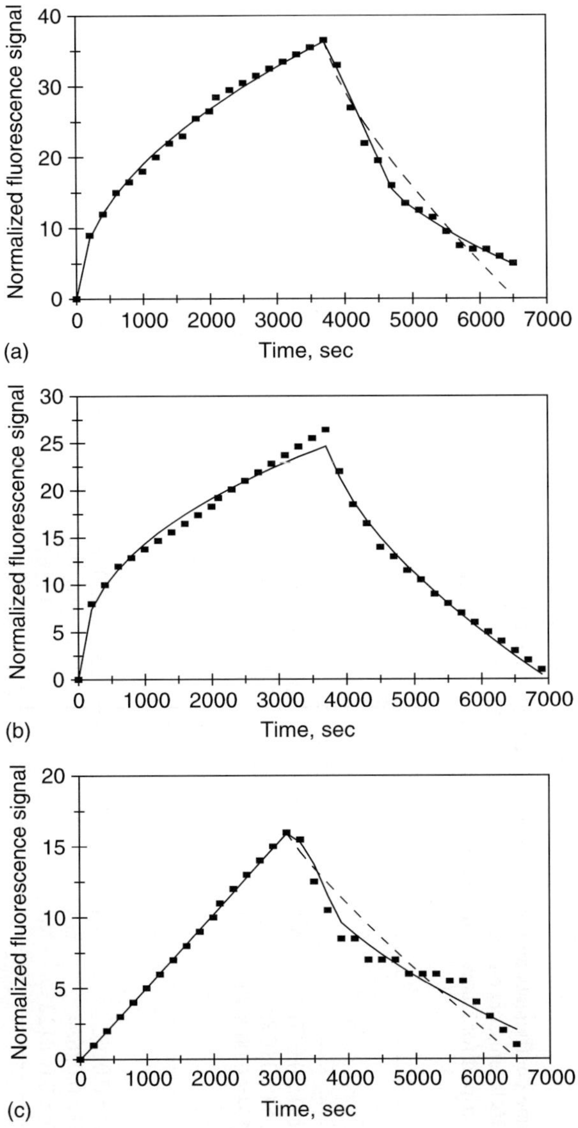

Figure 11.12 (a) Binding and dissociation of different concentrations (in μM) of substance P (neuropeptide) (SP) in solution to quencher-labeled mAb SP 31 using the SPIT-FRI (solid-phase immobilized tripod for fluorescence immunoassay) procedure (Volland *et al.*, 2005): (a) 1, (b) 0.1, (c) 0.01. When only a solid line is used then a single-fractal analysis applies. When both dashed (-----) and solid (____) lines are used, then the dashed line is for a single-fractal analysis and the solid line is for dual-fractal analysis. In this case the solid line provides a better fit.

Table 11.4

(a) Binding and dissociation rate coefficients and (b) fractal dimensions for the binding and the dissociation phase for the interaction of substance P (neuropeptide) (SP) at different concentrations in solution to quencher-labeled mAb SP31 using the SPIT-FRI (solid-phase immobilized tripod for fluorescent renewable immunoassay) procedure (Volland *et al.*, 2005)

(a)

Analyte in solution (μM)	k	k_1	k_2	k_d	k_{d1}	k_{d2}
1	0.6191±0.0176	NA	NA	0.1089±0.020	0.01781±0.00270	1.5674±0.0347
0.1	0.8364±0.0388	NA	NA	0.2123±0.0103	NA	NA
0.01	0.004514±0.000070	NA	NA	0.01499±0.00586	0.000083±0.000041	0.1697±0.0134

(b)

	D_f	D_{f1}	D_{f2}	D_{fd}	D_{fd1}	D_{fd2}
1	2.0088±0.0166	NA	NA	1.5372±0.1158	0.9522±0.1908	2.2430±0.05968
0.1	2.1762±0.02684	NA	NA	1.8084±0.03096	NA	NA
0.01	0.9676±0.010122	NA	NA	1.2856±0.1861	0.0+0.6220	1.9158±0.1078

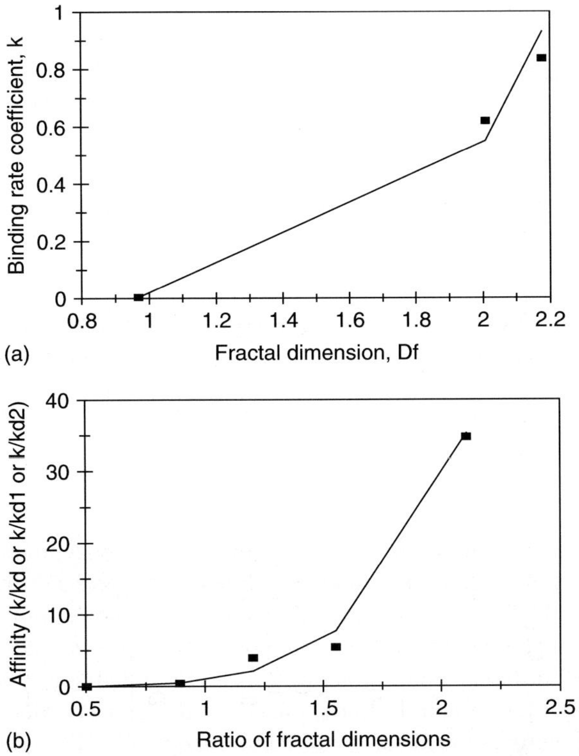

Figure 11.13 (a) Increase in the binding rate coefficient k with an increase in the fractal dimension D_{f}. (b) Increase in the affinity K with an increase in the ratio of fractal dimension present in the binding phase to that present in the dissociation phase.

Tables 11.4a and b and Figure 11.13a show the increase in the binding rate coefficient k with an increase in the fractal dimension D_{f}. In the 0.01–1.0 μM substance P (neuropeptide) (SP) concentration range in solution the binding rate coefficient k is given by

$$k = (0.005668 \pm 0.000978) D_{\mathrm{f}}^{6.5595 \pm 0.2519} \tag{11.7a}$$

The fit is good. Only three data points are available. The availability of more data points would lead to a more reliable fit. The binding rate coefficient k is very sensitive to the degree of heterogeneity or the fractal dimension D_{f} as noted by the order between six and seven (=6.5595) exhibited.

As indicated elsewhere in this book, affinity values are of interest to practicing biosensorists. Affinity may be defined as the ratio of the binding to the dissociation rate coefficient. There are few points available if one were to look at either the single- or the dual-fractal analysis used to model the dissociation kinetics. In order to offset this drawback, we will plot the affinity values that include both the single-fractal analysis as well as the dual-fractal analysis used to model the dissociation kinetics on a single plot. In other words, affinity may be defined as k/k_d, k/k_{d1}, or k/k_{d2} as far as this plot is concerned.

For the data presented in Tables 11.4a and b, Figure 11.13b shows the increase in the affinity K value with an increase in the corresponding ratio of the binding rate coefficient to the dissociation rate coefficient. The affinity K is given by

$$K = (0.8446 \pm 0.4479)\,(\text{ratio of the fractal dimension present in the binding phase to that in the dissociation phase})^{5.00 \pm 0.3887} \quad (11.7b)$$

The fit is very good, considering that two different data sets are used and plotted together. The affinity K is very sensitive to the ratio of fractal dimensions present in the binding phase and in the dissociation phases as noted by the fifth order of dependence exhibited.

11.4 CONCLUSIONS

A fractal analysis is presented for (a) the binding of different concentrations of mouse monoclonal anti-rabbit IgG in solution to rabbit patterned cells using a diffraction-based sandwich immunoassay (Goh *et al.*, 2003), (b) the binding and hybridization of DNA using a double-wavelength technique for SPR measurements (Zybin *et al.*, 2005), (c) the binding and dissociation of different concentrations of GSH-(Glutathione)-MPC in solution to polyclonal antibody plus Protein A immobilized on a QCM (Gerdon *et al.*, 2005), and (d) the binding and dissociation of substance P (neuropeptide) (SP) in solution to quencher-labeled mAb SP31 using the SPIT-FRI procedure (Volland *et al.*, 2005). The fractal analysis is used to provide better insights into the analyte–receptor interactions occurring on these novel types of sensing surfaces or modifications of previously employed sensing surfaces. The fractal analysis helps relate the binding (and the dissociation) rate coefficients with the degree of heterogeneity that exists on these sensing surfaces.

As indicated in the earlier chapters, the fractal analysis provides a quantitative indication of the state of disorder (fractal dimension) and the binding (and dissociation) rate coefficients on the sensor chip surface. Both types of examples are given wherein either a single- or a dual-fractal analysis was used. The dual-fractal analysis was used only when the single-fractal analysis did not provide an adequate fit. This was done by the regression analysis provided by Corel Quattro Pro 8.0 (1997).

In accordance with the prefactor analysis or fractal aggregates (Sorenson and Roberts, 1997), quantitative predictive expressions are developed for (a) the binding rate coefficient k_1 as a function of the fractal dimension D_{f1} for the binding of 50–250 ng/ml anti-rabbit IgG concentration in solution to rabbit IgG patterned cells (Goh *et al.*, 2003), (b) the binding rate coefficient k_2 as a function of the anti-rabbit IgG concentration in solution (Goh *et al.*, 2003), (c) the ratio of the binding rate coefficients, k_2/k_1, as a function of the ratio of fractal dimensions, D_{f2}/D_{f1} (Goh *et al.*, 2003), and (d) the binding rate coefficient k_1 as a function of the fractal dimension D_{f1} for the binding (hybridization) of G-HCV-c in solution to G-HCV immobilized on the spot surface, and for the selective hybridization of G-HCV-c and T_2-HCV-c in solution to complementary sequences immobilized at different spot surfaces (Zybin *et al.*, 2005).

Predictive-type relations are also developed for (a) the binding rate coefficients, k_1 and k_2, for GSH (glutathione)-MPC in solution to polyclonal anti-GSH + protein A immobilized on a QCM as a function of the GSH-MPC concentration in solution (Gerdon *et al.*, 2005), (b) the binding rate coefficient k_1 as a function of the fractal dimension D_{f1} for the binding of substance P (neuropeptide) (SP) in solution to quencher-labeled mAb SP 31 using the SPIT-FRI procedure (Volland *et al.*, 2005), and finally (c) the affinity K (ratio of binding to dissociation rate coefficients) as a function of the ratio of fractal dimensions present in the binding and in the dissociation phase.

In general, the binding rate coefficient is quite sensitive to the fractal dimension or the degree of heterogeneity that exists on the surface as noted by the order of dependence exhibited on the fractal dimension D_f. This is consistent with the examples presented, in general, in other chapters of the book. As indicated in earlier chapters, the fractal dimension is not a classical independent variable such as analyte (antigen, antibody, or other biological molecule) concentration in solution. Nevertheless, the high orders mentioned above for the dependence of the binding rate coefficient on the fractal dimension of these novel biosensor methods or modifications of existing biosensors being employed emphasize the importance of the extent of heterogeneity on these sensing surfaces, and their impact on the binding rate coefficient. This would also significantly impact biosensor performance parameters such as sensitivity, regeneration, stability, and response time. Hopefully, the emphasis on the surface of these novel biosensors would further facilitate in help improving these biosensors, and further help extending their applicability to other novel areas of usage.

REFERENCES

Albers, W.M., Vikholm, I., Viitala, T., and Peltonen, J. (2001). *Handbook of Surfaces and Materials*, Vol. 5, Academic Press, San Diego, pp. 1–31.

Alieva, E., and Konopsky, V.N. (2004). Biosensor based on surface plasmon resonance interferometry independent on variations of liquid's refractive index. *Sensors & Actuators B: Chemical*, **99**, 90–97.

Bailey, R.C., and Hupp, J.T. (2002). Large-scale resonance amplification of optical sensing of volatile compounds with chemoresponsive visible-region diffraction gratings. *Journal of the American Chemical Society,* **124**, 154–157.

Corel Quattro Pro 8.0 (1997). Corel Corporation, Ottawa, Canada.

Crowther, J.R. (2001). *The ELISA Guidebook, Vol. 149*, Humana Press, Totowa, New Jersey, USA.

Dabrowski, A., and Jaroniec, M. (1979). Effects of surface heterogeneities in adsorption from binary liquid mixtures. III. Analysis of experimental data using Langmuir-Freundlich type equations. *Journal of Colloid & Interface Science*, **73**, 475–482.

Gerdon, A.E., Wright, D.W., and Cliffel, D.E. (2005). Quartz crystal microbalance detection of glutathione-protected nanoclusters using antibody recognition. *Analytical Chemistry*, **77**, 304–310.

Goh, J.B., Loo, R.W., McAloney, R.A., and Goh, M.C. (2002). Diffraction-based assay for detecting multiple analytes. *Analytical and Bioanalytical Chemistry*, **374**, 54–56.

Goh, J.B., Tam, P.L., Loo, R.W., and Goh, M.C. (2003). A quantitative diffraction-based sandwich immunoassay. *Analytical Biochemistry*, **313**, 262–266.

Havlin, S. (1989). Molecular diffusion and reaction. In: *The Fractal Approach to Heterogeneous Chemistry: Surfaces, Colloids, Polymers* (ed. D. Avnir), Wiley, New York, pp. 251–269.

Hermanson, G.T. (1995). *Bioconjugate Technique*, Academic Press, San Diego.

Henke, L., and Krull, U.J. (1999). *Canadian Journal of Analytical Science and Spectroscopy*, **44**, 61.

Jaroniec, M., and Derylo, A. (1981). Simple relationships for predicting multisolute adsorption from dilute aqueous solutions. *Chemical Engineering Science*, **36**, 1017–1019.

Lee, C.K., and Lee, S.L. (1995). Multi-fractal scaling analysis of reactions over fractal surfaces. *Surface Science*, **325**, 294–310.

Leidberg, B., Nylander, C., and Lundstrom, I. (1995). Biosensing with surface plasmon resonance – how it all began. *Biosensors & Bioelectronics,* **10**, i–ix.

Livache, T., Fouque, B., Roget, A., Marchand, J., and Bidan, G. (1998). Polypyrrole DNA chip on a silicon device: example of Hepatitis C virus genotyping. *Analytical Biochemistry*, **255**, 188–194.

Mines, G.A., Tzeng, B.C., Stevenson, K.J., Li, J.L., and Hupp, J.T. (2002). Microporous supramolecular coordination compounds as chemosensory photonic lattices. *Angew Chemistry International Edition*, **41**, 154–157.

Morhard, F., Pipper, J., Dahint, R., and Grunze, M. (2000). Immobilization of antibodies in micropatterns for cell detection by optical diffraction. *Sensors & Actuators B: Chemical*, **70**, 232–242.

Nakajima, F., Hirakawa, Y., Kaneta, T., and Imasaka, T. (1999). Diffractive optical chemical sensor based on light absorption. *Analytical Chemistry*, **71**, 2262–2265.

Oscik, J., Dabrowski, A., Jaroniec, M., and Rudzinski, W. (1976). Effects of surface heterogeneity in adsorption from binary liquid mixtures. I. Adsorption from ideal solutions. *Journal of Colloid & Interface Science*, **56**, 403–412.

Quinn, C.P., Semenova, V.A., Elie, C.M., Romero-Steiner, S., Greene, C., Li, H., Stamey, K., Steward-Clark, E., Schmidt, D.S., and Mothershed et al. (2002). *Emerging Infectious Diseases*, **8**, 1103–1110.

Ramakrishnan, A., and Sadana, A. (2001). A single-fractal analysis of cellular analyte-receptor binding kinetics utilizing biosensors. *Biosystems*, **59**, 35–51.

Rudzinski, W., Latter, L., Zajac, J., Wofram, E., and Paszli, J. (1983). Ideal adsorption from binary liquid mixtures on a heterogeneous solid surface: Equations for excess isotherms and heats of immersion. *Journal of Colloid & Interface Science*, **96**, 339–359.

Sadana, A. (2001). A kinetic study of analyte–receptor binding and dissociation, and dissociation alone for biosensor applications: a fractal analysis. *Analytical Biochemistry*, **291**, 34–47.

Sorenson, C.M., and Roberts, G.C. (1997). The prefactor of fractal aggregates. *Journal of Colloid and Interface Science*, **186**, 447–452.

St. John, P.M., Davis, R., Cady, N., Czaka, J., Batt, C.A., and Craighead, H.G. (1998). Diffraction-based cell detection using a microcontact printed antibody grating. *Analytical Chemistry*, **70**, 1108–1111.

Thomson, S.J., and Webb, G. (1968). *Heterogeneous Catalysis*, Wiley, New York, p. 23.

Tsay, Y.G., Lin, C.I., Lee, J., Gustafson, E.K., Appelqvist, R., Magginetti, P., Norton, R., Teng, N., and Charlton, D. (1991). Optical biosensor assay (OBA^{TM}). *Clinical Chemistry*, **37**(9), 1502–1505.

Volland, H., Creminon, C., Neuberger, L.M., and Grassi, J. (2002). French patent, 0214959.

Volland, H., Neuberger, L.M., Schultz, E., Grassi, J., Perrant, F., and Creminion, C. (2005). Solid-phase immobilized tripod for fluorescent renewable immunoassay. A concept for continuous monitoring of an immunoassay. *Analytical Chemistry*, **77**, 1896–1904.

Wrobel, N., Deininger, W., Hegemann, P., and Mirsky, V.M. (2003). Covalent immobilization of oligonucleotides on electrodes. *Colloids and Surfaces B: Biointerfaces*, **32**, 157–162.

Wrobel, N., Schinkinger, M., and Mirsky, V.M. (2002). A novel ultraviolet assay for testing side reactions of carbodiimides. *Analytical Biochemistry*, **305**, 135–138.

Zybin, A., Granwald, C., Mirsky, V.M., Kuhlmann, J., Wolfbeis, O.S., and Niemax, K. (2005). Double-wavelength technique for surface plasmon resonance measurements: basic concept and applications for single sensors and two-dimensional sensor arrays. *Analytical Chemistry*, **77**(8), 2393–2399.

Zynio, S.A., Samoylov, A.V., Surovtseva, E.R., Mirsky, V.M., and Shirshov, Y.M. (2002). Bimetallic layers increase sensitivity of affinity sensors based on surface plasmon resonance. *Sensors*, **2**, 62.

– 12 –

Fractal Analysis of the Binding and Dissociation of Different Compounds on Biosensor Surfaces

12.1 INTRODUCTION

Sensor research and development is impacting a wide range of applications that include energy, the environment, food safety, manufacturing, and national security. A very recent National Science Foundation Program solicitation document, NSF 05-526 (2005), emphasizes the multidisciplinary activity required to (a) engineer materials, and concepts and designs for new sensors and sensory materials and (b) the interpretation of data obtained for sensors and sensor networks. The emphasis is on a synergistic blend of expertise to make relevant and significant contributions. As expected, nanotechnology and biomicronanotechnologies will significantly impact biosensor applications. For example, Gourley (2005) in a recent review of biomicronanotechnology indicates that the development of novel biosensors that incorporates a reduced-size laser (size of a cell or organelle (Gourley, 1998)) has the potential to assess cell structure to interrogate the human immune system (Gourley, 2003), and be able to distinguish cancerous cells from normal cells (Gourley and Sasaki, 2001).

In order to emphasize the different applications that biosensors may be effectively used, we will analyze the kinetics of binding and dissociation of analytes in solution to receptors on biosensor surfaces with no commonality among them unlike the common theme presented in Chapters 3–11. The four cases presented in this chapter should be treated as just examples, with no specific predisposition toward their selection and subsequent kinetic analysis presented in this chapter. The intent is to provide a 'flavor' of the wide range of biosensor applications available in the literature. Surely, the inclusion of more carefully selected examples in this chapter would further widen the applications of biosensors to different areas.

In this chapter, we use fractal analysis to analyze the binding and dissociation (if applicable) kinetics of (a) adenosine triphosphate (ATP) in solution (with and without the interferent, ascorbate) to a platinum (Pt) microelectrode coated with a layer containing glycerol kinase plus glycol-3-phosphatase (Liaudet *et al.*, 2005), (b) human and mouse leptin (Ob) in solution to leptin receptor (ObR) immobilized on a sensor chip surface (Mistrik *et al.*, 2004),

(c) human myeloid protein 14 (MRP 14) in solution to rabbit polyclonal antibody against MRP 14 immobilized on a sensor chip surface, and human fatty acid binding protein (FABP) in solution to rabbit polyclonal antibody against human FABP immobilized on a sensor chip surface (Grote *et al.*, 2005), and (d) the influence of pH on alkaline phosphatase (ALP) in solution to a graphite–Teflon composite biosensor using phenyl phosphatase as a substrate (Serra *et al.*, 2005).

Values of the binding and dissociation (if applicable) rate coefficients along with affinity values (if applicable) are provided. As mentioned in the previous chapters, the fractal analysis may be considered as an alternate analysis to the kinetic analysis presented in the above-mentioned references. Note that the fractal analysis, as indicated earlier, includes the effects of diffusion and the heterogeneity of the receptors present on the biosensor surface.

12.2 THEORY

Havlin (1989) has reviewed and analyzed the diffusion of reactants toward fractal surfaces. The details of the theory and the equations involved for the binding and the dissociation phases for analyte–receptor binding are available (Sadana, 2001). The details are not repeated here, except that just the equations are given to permit an easier reading. These equations have been applied to other biosensor systems (Ramakrishnan and Sadana, 2001; Sadana, 2001).

For most applications, a single- or a dual-fractal analysis is often adequate to describe the binding and dissociation kinetics. Peculiarities in the values of the binding and dissociation rate coefficients, as well as in the values of the fractal dimensions with regard to the dilute analyte systems being analyzed, will be carefully noted, if applicable.

12.2.1 Single-fractal analysis

Binding rate coefficient

Havlin (1989) indicates that the diffusion of a particle (analyte [Ag]) from a homogeneous solution to a solid surface (e.g., receptor [Ab]-coated surface) on which it reacts to form a product (analyte–receptor complex, (Ab.Ag)) is given by

$$(\mathrm{Ab.Ag}) \approx \begin{cases} t^{(3-D_{\mathrm{f,bind}})/2} = t^{p}, & t < t_{\mathrm{c}} \\ t^{1/2}, & t > t_{\mathrm{c}} \end{cases} \tag{12.1a}$$

Here $D_{\mathrm{f,bind}}$ or D_{f} (used later on in the chapter) is the fractal dimension of the surface during the binding step and t_{c} is the cross-over value. Havlin (1989) indicates that the cross-over value may be determined by $r_{\mathrm{c}}^2 \sim t_{\mathrm{c}}$. Above the characteristic length r_{c}, the self-similarity of the surface is lost and the surface may be considered homogeneous. Above time t_{c}, the surface may be considered homogeneous, since the self-similarity property disappears, and 'regular' diffusion is now present. For a homogeneous surface where $D_f = 2$, and when only diffusional limitations are present, $p = ½$ as it should be. Another way of looking at the $p = ½$ case (where $D_{f,bind} = 2$) is that the analyte in solution views the fractal object, in our case, the receptor-coated biosensor surface, from a 'large distance.' In essence, in the association process, the diffusion of the analyte from the solution to the receptor surface creates a depletion layer of width $(Ðt)^{½}$ where $Ð$ is the diffusion constant. This gives rise to the fractal power law, (Analyte·Receptor)~$t^{(3-D_{f,bind})/2}$. For the present analysis, t_c is arbitrarily chosen and we assume that the value of t_c is not reached. One may consider the approach as an intermediate 'heuristic' approach that may be used in future to develop an autonomous (and not time-dependent) model for diffusion-controlled kinetics.

Dissociation rate coefficient

The diffusion of the dissociated particle (receptor [Ab] or analyte [Ag]) from the solid surface (e.g., analyte [Ag]–receptor [Ab] complex coated surface) into solution may be given, as a first approximation by

$$(\text{Ab.Ag}) \approx -k' t^{(3-D_{f,diss})/2}, \qquad t > t_{diss} \tag{12.1b}$$

Henceforth, its concentration only decreases. The dissociation kinetics may be analyzed in a manner 'similar' to the binding kinetics.

12.2.2 Dual-fractal analysis

Binding rate coefficient

Sometimes, the binding curve exhibits complexities and two parameters (k, D_f) are not sufficient to adequately describe the binding kinetics. This is further corroborated by low values of r^2 factor (goodness-of-fit). In that case, one resorts to a dual-fractal analysis (four parameters; k_1, k_2, D_{f1}, and D_{f2}) to adequately describe the binding kinetics. The single-fractal analysis presented above is thus extended to include two fractal dimensions. At present, the time ($t = t_1$) at which the 'first' fractal dimension 'changes' to the 'second' fractal dimension is arbitrary and empirical. For the most part, it is dictated by the data analyzed and experience gained by handling a single-fractal analysis. A smoother curve is obtained in the 'transition' region, if care is taken to select the correct number of points for the two regions. In this case, the product (antibody–antigen or analyte–receptor complex, Ab.Ag or analyte.receptor) is given by

$$(\text{Ab.Ag}) \approx \begin{cases} t^{(3-D_{f1,bind})/2} = t^{p1}, & t < t_1 \\ t^{(3-D_{f2,bind})/2} = t^{p2}, & t_1 < t < t_2 = t_c \\ t^{1/2}, & t > t_c \end{cases} \tag{12.1c}$$

In some cases, as mentioned above, a triple-fractal analysis with six parameters (k_1, k_2, k_3, D_{f1}, D_{f2}, and D_{f3}) may be required to adequately model the binding kinetics. This is when the binding curve exhibits convolutions and complexities in its shape due to perhaps the very dilute nature of the analyte (in some of the cases to be presented) or for some other reasons. Note that for dilute systems, the signal/noise ratio is small. Thus, (a) there is scatter in the data and (b) errors in the estimated values of the binding and dissociation rate coefficients, and in the fractal dimension values for the binding and the dissociation phase may be anticipated. Also, in some cases, a dual-fractal analysis may be required to describe the dissociation kinetics.

12.3 RESULTS

The fractal analysis will be applied to the binding and dissociation (if applicable) kinetics of (a) ATP in solution (with and without the interferent, ascorbate) to a Pt microelectrode coated with a layer containing glycerol kinase plus glycol-3-phosphatase (Liaudet *et al.*, 2005), (b) human and mouse leptin (Ob) in solution to leptin receptor (ObR) immobilized on a sensor chip surface (Mistrik *et al.*, 2004), (c) MRP 14 in solution to rabbit polyclonal antibody against MRP 14 immobilized on a sensor chip surface, and human FABP in solution to rabbit polyclonal antibody against human FABP immobilized on a sensor chip surface (Grote *et al.*, 2005), and (d) the influence of pH on ALP in solution to a graphite–Teflon composite biosensor using phenyl phosphatase as a substrate (Serra *et al.*, 2005).

Understandably, alternate expressions for fitting the data that include saturation, first-order reaction, and no diffusion limitations are available, but these expressions are apparently deficient in describing the heterogeneity that inherently exists on the surface. One might justifiably argue that appropriate modeling may be achieved by using a Langmuirian approach. The Langmuirian approach may be used to model the data presented if one assumes the presence of discrete classes of sites (e.g., double-exponential analysis as compared with single-fractal analysis). Lee and Lee (1995) indicate that the fractal approach has been applied to surface science, e.g. adsorption and reaction processes. These authors emphasize that the fractal approach provides a convenient means to represent the different fractal structures and morphology at the reaction surface. These authors also emphasize using the fractal approach to develop optimal structures as a predictive tool. Another advantage of the fractal technique is that the analyte–receptor association and dissociation reactions are complex reactions, and fractal analysis via the fractal dimension and the rate coefficient provides a useful lumped parameter(s) analysis of the diffusion-limited reaction occurring on a heterogeneous surface.

Fractals have been used to analyze the morphology of different processes. Sarkar *et al.* (2005) have recently analyzed the fractal structure of polycation–DNA complexes. They used fractal concepts to explain aggregation processes for viral vector particles. They emphasize that the fractal dimension may provide a useful means for monitoring the physical state of nonviral delivery-vector particles for gene-delivery systems. This is during the preparation process and during storage. These authors emphasize that fractals have been used to describe the morphology of quite a few structures resulting from naturally occurring growth processes. Furthermore, they indicate that many growth processes in biotechnology such as the growth of filamentous microorganisms and microbial colonies demonstrate fractal properties (Grosskinsky *et al.*, 2002). Sarkar *et al.* (2005) further indicate that fractal aggregation of particles has been observed for adsorbed antibody fragments, coagulation of wastewater sludge, and other nonbiological systems (Amal and Raper, 1993; Shamlou *et al.*, 1996; Bohr *et al.*, 1997; Magazu *et al.*, 1989; Molina-Bolivar *et al.*, 1998).

In the classical situation, to demonstrate fractality, one should make a log–log plot, and one should definitely have a large amount of data. It may also be useful to compare the fit to some other forms, such as exponential or one involving saturation, etc. At present, we do not offer any independent demonstration or physical evidence of fractals in the examples presented. It is a convenient means, as indicated in the previous chapters (since it provides a lumped parameter) to make the degree of heterogeneity that exists on the surface more quantitative. Thus, there is some arbitrariness in the fractal model to be presented. The fractal approach provides additional information about interactions that may not be obtained by conventional analysis of biosensor data.

There is no nonselective adsorption of an analyte. The analysis presented in this chapter, as in other chapters, does not include nonselective adsorption. We do recognize that, in some cases, this may not be a significant component of the adsorbed material and that this rate of association (binding), which is of a temporal nature, would depend on surface availability. If we were to accommodate the nonselective adsorption into the model, there would be an increase in the heterogeneity on the surface, since, by its very nature, nonspecific adsorption is more heterogeneous than specific adsorption. This would lead to higher fractal dimension values since the fractal dimension is a direct measure of the degree of heterogeneity that exists on the surface.

For a first-order reaction, as expected, an increase in the degree of heterogeneity on the surface due to nonspecific association would lead to lower values of the (specific) association rate coefficient. The deletion of this nonspecific association in the analysis would lead to (artificially) higher values of the association rate coefficient for first-order reactions. The reactions to be presented in this chapter are, in general, higher than first-order. Sadana and Chen (1996) have shown that for reaction orders higher than one a certain amount of heterogeneity is beneficial for the association rate coefficient. There is apparently an optimum range. This is due to steric factors. Thus, depending on whether one is inside or outside of this optimum range, the deletion of nonspecific association in the analysis would lead to either an increase or a decrease in the association rate coefficient. In other words, if one is within the optimum range for a particular reaction order, then the presence of nonspecific association would lead to higher values of the (specific) association rate coefficient. In this case, the deletion of the nonspecific association leads to lower than real-life values of the association rate coefficient.

As indicated previously, four examples of analyte–receptor binding and association will be analyzed. In three of the examples to be presented (Serra *et al.*, 2005; Grote *et al.*, 2005; Liaudet *et al.*, 2005) no kinetic analysis was presented. In one of the examples, that is the binding of leptin–leptin interactions, the surface plasmon resonance (SPR) biosensor was used by the authors (Mistrik *et al.*, 2004) to obtain the values of the binding and dissociation rate coefficients.

The software provided by the SPR manufacturers (2002) is used to obtain the values of the rate coefficient(s) and affinities. They do not include the presence of heterogeneity of the receptors on the surface. This, as indicated in the different chapters in the book, is an important aspect in the kinetic analysis. This is especially true if the degree of heterogeneity on the surface significantly affects the association and dissociation rate coefficients, as well as the affinity values. Also, diffusion effects are also neglected in the single kinetic analysis that is to be re-analyzed (Mistrik *et al.*, 2004). The noninclusion of heterogeneities on the biosensor surface and the effect of diffusional limitations may lead to incorrect choices, for example, when the SPR biosensor is used to screen for potential drugs from a selection of potential drug candidates in a drug-screening process.

Liaudet *et al.* (2005) indicate that the purines, ATP, ADP (adenosine diphosphate), and adenosine are important extracellular signaling agents. They are involved in signaling in blood (Kunapuli *et al.*, 2003), kidney (Schwiebert and Kishore, 2001), lungs (Adriaensen and Timmermans, 2004), brain (Sim *et al.*, 2004; Bowser and Khakh, 2004), and heart (Kitakaze and Hori, 2000). Liaudet *et al.* (2005) indicate that ATP plays a significant role in cellular metabolism. Furthermore, ATP acts as a link between non-neural and neural cells, and is involved in sensory transduction, e.g. pain (Cook *et al.*, 1997; Cook and McCleskey, 2002). Liaudet *et al.* (2005) developed an ATP biosensor by coating a Pt electrode with an ultrathin layer containing glycerol kinase and glycerol-3-phosphate oxidase. Their detection scheme depends on glycerol kinase and glycerol-3-phosphate oxidase to detect ATP in the presence of the co-substrate, glycerol (Murphy and Galley, 1994; Katsu *et al.*, 1994).

Figure 12.1 shows the binding of ATP in solution to an ATP sensor (Liaudet *et al.*, 2005). A dual-fractal analysis is required to adequately describe the binding kinetics. The values of (a) the binding (association) rate coefficient k and the fractal dimension D_f for a single-fractal analysis, and (b) the binding rate coefficients, k_1 and k_2, and the fractal dimensions, D_{f1} and D_{f2}, for a dual-fractal analysis are given in Table 12.1a. The values of the binding rate coefficients and the fractal dimensions presented in Table 12.1 were obtained from a regression analysis using Corel Quattro Pro 8.0 (1997) to model the experimental data using, e.g., Eq. 12.1a wherein (Ab.Ag) $= kt^p$ for the binding step, and (Ab.Ag) $= -kt^p$ for the dissociation step.

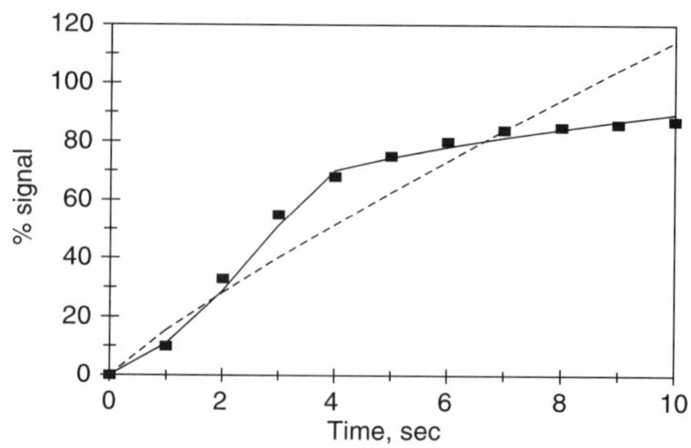

Figure 12.1 Binding of ATP in solution to an ATP sensor (Liaudet *et al.*, 2005). When both a dashed line (-------) and a solid (____) line are used, then the dashed line is for a single-fractal analysis and the solid line is for a dual-fractal analysis. In this case, the solid line provides the better fit.

Table 12.1

(A) Binding and dissociation rate coefficients and (B) Fractal dimensions for the binding and the dissociation phase for ATP in solution to a Pt microelectrode coated with an ultrathin layer containing glycerol kinase and glycerol-3-phosphate oxidase (Liaudet *et al.*, 2005)

(A) Analyte in solution/ receptor on surface	k	k_1	k_2	k_d	k_{d1}	k_{d2}
(a) ATP/ATP sensor	15.4665 ± 4.647	10.848 ± 1.747	48.640 ± 1.313	0.4368 ± 0.0382	NA	NA
(b) 1 μM ATP/ATP sensor	0.3836 ± 0.0589	0.2069 ± 0.0139	1.5719 ± 0.0162	NA	NA	NA
5 μM ATP/ATP sensor	2.5634 ± 0.1156	NA	NA	0.03379 ± 0.00916	0.01604 ± 0.00159	0.8276 ± 0.0175
10 μM ATP/ATP sensor	8.9649 ± 0.8489	6.1165 ± 0.3072	18.7437 ± 0.0316	1.01217 ± 0.3067	0.3408 ± 0.0676	10.1401 ± 0.0367
(c) ATP, pH 6.5/ATP sensor	0.4198 ± 0.0554	0.3069 ± 0.0414	0.8548 ± 0.0049	0.3303 ±0.046	NA	NA
ATP, pH 7.0/ATP sensor	0.5794 ± 0.0918	0.4092 ± 0.0668	1.5002 ± 0.0034	0.7485 ± 0.1160	NA	NA
ATP, pH 8.0/ATP sensor	0.5746 ± 0.1104	0.3740 ± 0.0396	1.8512 ± 0.0014	0.006223 ± 0.00417	NA	NA
(d) ATP + PC/ATP sensor	12.8443 ± 2.0014	6.0951 ± 1.4516	25.1676 ± 0.3265	NA	NA	NA
ATP/ATP sensor	18.813 ± 2.2759	14.3266 ± 1.6951	63.7562 ± 0.3665	NA	NA	NA

(B) Analyte in solution/ receptor on surface	D_f	D_{f1}	D_{f2}	D_{fd}	D_{fd1}	D_{fd2}
(a) ATP/ATP sensor	1.2612 ± 0.2390	0.1744 + 0.2868	2.4710 ± 0.06656	1.7764 ± 0.1321	NA	NA
(b) 1 μM ATP/ATP sensor	1.2276 ± 0.1406	1.2756 ± 0.1246	2.6468 ± 0.0441	NA	NA	NA
5 μM ATP/ATP sensor	2.5082 ± 0.1406	NA	NA	0 + 0.2859	0 + 0.2878	1.9294 ± 0.1765
10 μM ATP/ATP sensor	2.3802 ± 0.08894	2.0362 ± 0.1238	2.8115 ± 0.00726	1.4544 ± 0.1600	0.6346 ± 0.1944	2.5982 ± 0.008566
(c) ATP, pH 6.5/ATP sensor	2.2108 ± 0.07708	1.8968 ± 0.1493	2.5982 ± 0.01235	1.9294 ± 0.1155	NA	NA
ATP, pH 7.0/ATP sensor	2.2968 ± 0.1105	1.9162 ± 0.2212	2.8526 ± 0.00807	2.1380 ± 0.1859	NA	NA
ATP, pH 8.0/ATP sensor	2.1822 ± 0.1315	1.6590 ± 0.2050	2.90989 ± 0.005048	0 + 0.9342	NA	NA
(d) ATP + PC/ATP sensor	2.0817 ± 0.09522	1.4150 ± 0.4358	2.4136 ± 0.02346	NA	NA	NA
ATP/ATP sensor	1.7514 ± 0.1486	1.3466 ± 0.2211	2.6943 ± 0.04814	NA	NA	NA

The binding and the dissociation rate coefficients presented in Table 12.1 are within 95% confidence limits. For example, for the binding of ATP in solution to the ATP biosensor the reported value of the binding rate coefficient, k_1 for a dual-fractal analysis is 10.848 ± 1.747. The 95% confidence limit indicates that the k_1 value lies between 9.101 and 12.795. This indicates that the values are precise and significant. In general, the r^2 value is greater than 0.98. No F value for the significance of the model was obtained using Corel Quattro Pro, 8.0 (1997). Thus, no value is reported.

Figure 12.2a shows the binding and dissociation of 1 μM ATP in solution to the ATP biosensor. A dual-fractal analysis is required to adequately describe the binding kinetics. The dissociation kinetics is adequately described by a single-fractal analysis. The values of (a) the binding (association) rate coefficient k and the fractal dimension D_f for a single-fractal analysis, (b) the binding rate coefficients, k_1 and k_2, and the fractal dimensions, D_{f1} and D_{f2}, for a dual-fractal analysis, and (c) the dissociation rate coefficient k_d and the fractal dimension for the dissociation phase D_{fd} for a single-fractal analysis are given in Table 12.1b.

Figure 12.2b shows the binding and dissociation of 5 μM ATP in solution to the ATP biosensor. A single-fractal analysis is required to adequately describe the binding kinetics. A dual-fractal analysis is required to adequately describe the dissociation kinetics. The values of the binding (association) and dissociation rate coefficients and the fractal dimensions are given in Table 12.1b.

Figure 12.2c shows the binding and dissociation of 10 μM ATP in solution to the ATP biosensor. A dual-fractal analysis is required to adequately describe the binding and dissociation kinetics. The values of the binding (association) and dissociation rate coefficients and the fractal dimensions are given in Table 12.1b. No explanation is offered at present for the change from a single- to a dual-fractal analysis and then back to a single-fractal analysis for describing the binding kinetics as one goes from 1 to 5 to 10 μM ATP in solution. Also, at the lower concentration of ATP used (1 μM) a single-fractal analysis is required to describe the dissociation kinetics, whereas at the higher ATP concentrations used (5 and 10 μM), a dual-fractal analysis is required to describe the dissociation kinetics.

Figure 12.3 and Table 12.1 show the increase in the affinity K_i $(= k_i/k_{di})$ with an increase in the fractal dimension ratio (D_{fi}/D_{fdi}). k_i can be k_1, k_2, or just k. Similarly, k_{di} can be k_{d1}, k_{d2}, or just k_d. The corresponding D_{fi} can be D_{f1}, D_{f2}, or D_f. Similarly, D_{fdi} can be D_{fd1}, D_{fd2}, or D_{fd}. For the data presented in Table 12.1, the affinity K_i is given by

$$K_i = (1.0959 \pm 0.4761)(D_{fi}/D_{fdi})^{2.5042 \pm 0.3283} \tag{12.2}$$

The fit is good. Only five data points are available. The availability of more data points would lead to a more reliable fit. The affinity K_i exhibits very close to two and one-half order of dependence on the ratio of the fractal dimensions in the binding and dissociation phase D_{fi}/D_{fdi}.

Liaudet *et al.* (2005) analyzed the pH sensitivity of the ATP biosensor in the pH range 6.5 to 8. Figures 12.4a–c show the binding and dissociation of ATP in solution at (1) pH 6.5,

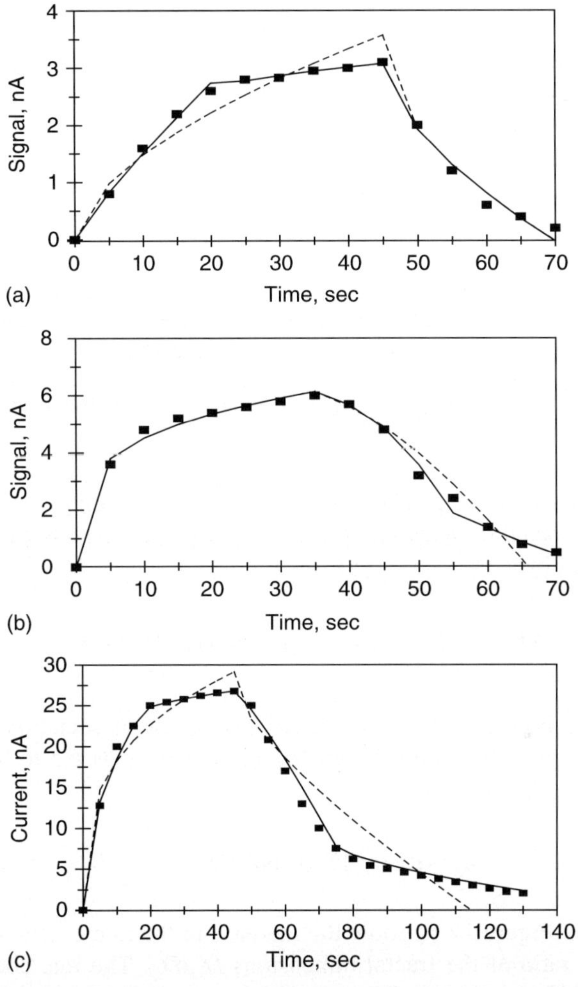

Figure 12.2 Binding and dissociation of different concentrations (in μM) of ATP in solution to the ATP biosensor (Liaudet *et al.,* 2005): (a)1, (b)5, (c)10. When both a dashed line (-------) and a solid (——) line are used, then the dashed line is for a single-fractal analysis and the solid line is for a dual-fractal analysis. In this case, the solid line provides the better fit.

(2) pH 7, and (3) pH 8 to the ATP biosensor. A dual-fractal analysis is required to describe the binding kinetics, and a single-fractal analysis is adequate to describe the dissociation kinetics. The values of (a) the binding rate coefficients and the fractal dimensions D_f for a single-fractal analysis, (b) the binding rate coefficients, k_1 and k_2, and the fractal dimensions, D_{f1} and D_{f2}, for a dual-fractal analysis, and (c) the dissociation rate coefficient k_d and the fractal dimension for the dissociation phase D_{fd} for a single-fractal analysis are given in Table 12.1c.

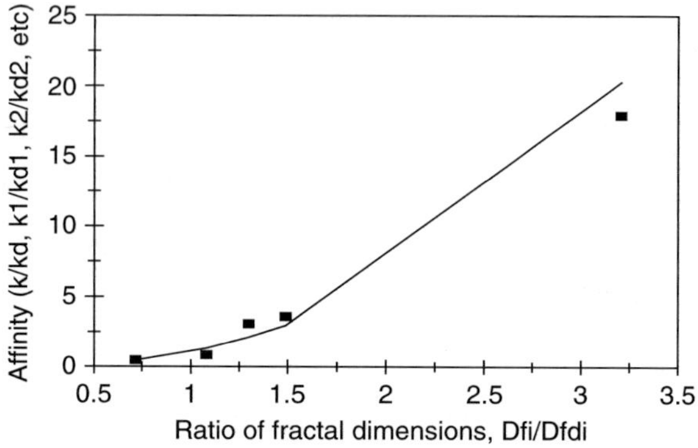

Figure 12.3 Increase in the affinity K_i (k_i/k_{di}) with an increase in the fractal dimension ratio D_{fi}/D_{fdi}.

(1) Table 12.1c and Figure 12.5a show the increase in the binding rate coefficient, k_2 with an increase in the fractal dimension, D_{f2}. In the pH range 6.5 to 8.0 the binding rate coefficient, k_2 is given by:

$$k_2 = (0.001581 \pm 0.000095) D_{f2}^{6.582 \pm 0.6848} \tag{12.3a}$$

(2) Table 12.1c and Figure 12.5b show the increase in the dissociation rate coefficient, k_d with an increase in the fractal dimension, D_{fd}. In the pH range 6.5 to 8.0 the dissociation rate coefficient, k_d is given by:

$$k_d = (0.3641 \pm 0.2647) D_{fd}^{0.4422 \pm 0.0675} \tag{12.3b}$$

(3) Table 12.1c and Figure 12.5c show the increase in the rate coefficient ratio k_d/k_2 with an increase in the ratio of the fractal dimensions D_{fd}/D_{f2}. The rate coefficient ratio k_d/k_2 (reciprocal of the affinity, $K_2 = k_2/k_d$) is given by

$$k_d/k_2 = (0.3410 \pm 0.1715)(D_{fd}/D_{f2})^{0.3786 \pm 0.0508} \tag{12.3c}$$

(4) Table 12.1c and Figure 12.5d show the increase in the binding rate coefficient k_2 with an increase in the pH. The binding rate coefficient k_2 is given by

$$k_2 = (0.00181 \pm 0.000484) \text{pH}^{3.3552 \pm 1.5926} \tag{12.3d}$$

The fit is good. Only three data points are available. The availability of more data points would lead to a more reliable fit. The binding rate coefficient k_2 is very sensitive to the fractal dimension D_{f2} as noted by the order of dependence between 6 and 7 exhibited.

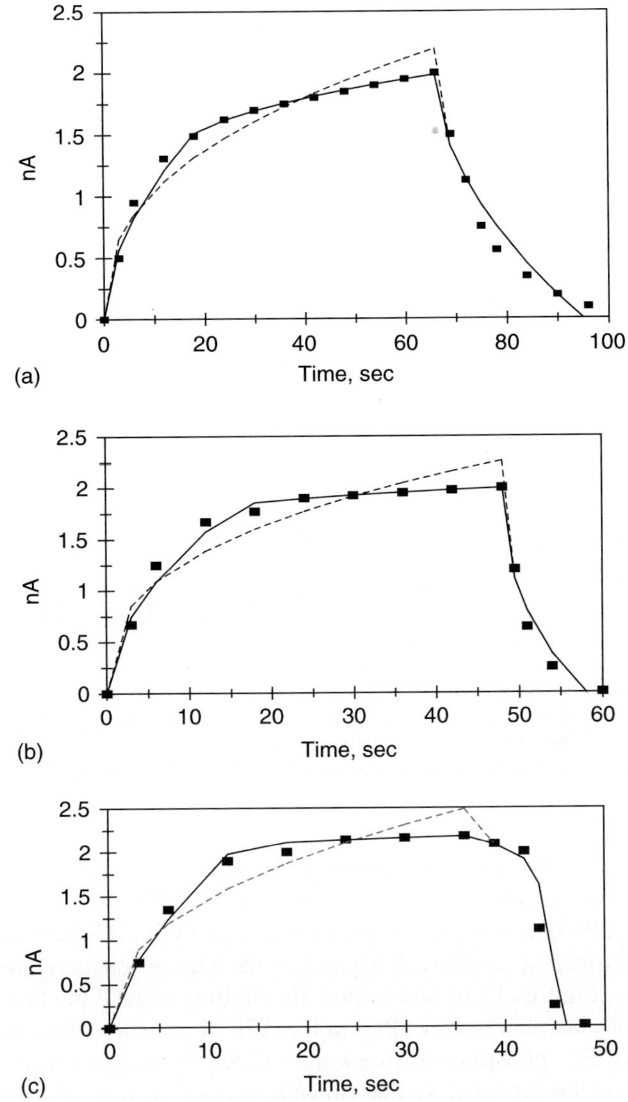

Figure 12.4 Binding and dissociation of ATP in solution at different pH to the ATP biosensor (Liaudet *et al.*, 2005): (a) 6.5, (b) 7.0, (c) 8.0. When both a dashed line (--------) and a solid (_______) line are used, then the dashed line is for a single-fractal analysis and the solid line is for a dual-fractal analysis. In this case, the solid line provides the better fit.

The dissociation rate coefficient k_d is only mildly sensitive to the fractal dimension D_{fd} as noted by the order of dependence less than one-half (= 0.4422) exhibited.

The rate coefficient ratio k_d/k_2 is only mildly sensitive to the fractal dimension ratio D_{fd}/D_{f2} as noted by the order of dependence less than one-half (= 0.3786) exhibited.

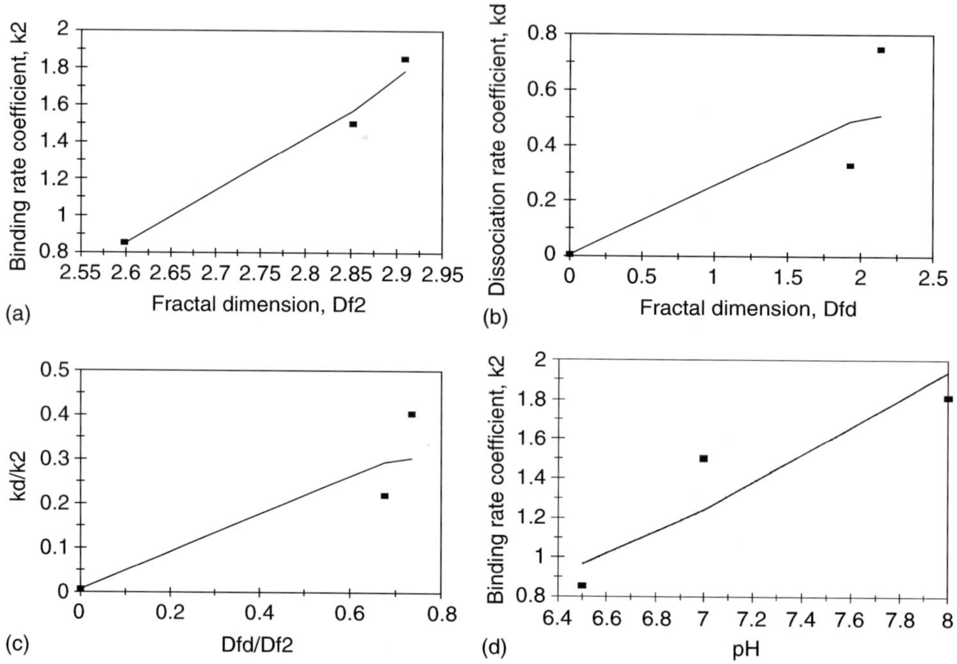

Figure 12.5 (a) Increase in the binding rate coefficient k_2 with an increase in the fractal dimension D_{f2}. (b) Increase in the dissociation rate coefficient k_d with an increase in the fractal dimension D_{fd}. (c) Increase in the rate coefficient ratio k_d/k_2 with an increase in the ratio of fractal dimensions D_{fd}/D_{f2}. (d) Increase in the binding rate coefficient k_2 with an increase in the pH.

The binding rate coefficient k_2 is sensitive to the pH in the range 6.5 to 8.0 as noted by the order of dependence exhibited between 3 and 4 (= 3.3552).

Liaudet *et al.* (2005) compared the performance of their ATP biosensor with regard to sensitivity and speed of response. They noted that a more sensitive biosensor exhibits a slower response. Figures 12.6a and b show the binding of ATP in (1) the presence and (2) the absence of phosphocreatine (PC) to the ATP biosensor. These authors state that in the presence of PC, phosphocreatine kinase (CK) regenerates ATP from ADP. The ADP produced can be recycled in the enzymatic cascade process, thus significantly amplifying the signal. The authors do emphasize that the increase in the sensitivity comes with a price, i.e. a slower response. A dual-fractal analysis is required to describe the binding kinetics. The values of (a) the binding rate coefficients k and the fractal dimension D_f for a single-fractal analysis, and (b) the binding rate coefficients, k_1 and k_2, and the fractal dimensions, D_{f1} and D_{f2}, for a dual-fractal analysis are given in Table 12.1d.

As expected, the corresponding binding rate coefficients when PC is used are lower when compared to when PC is not used. This is consistent with the results of Liaudet *et al.* (2005) who indicate the increase in sensitivity is obtained at the expense of a slower response.

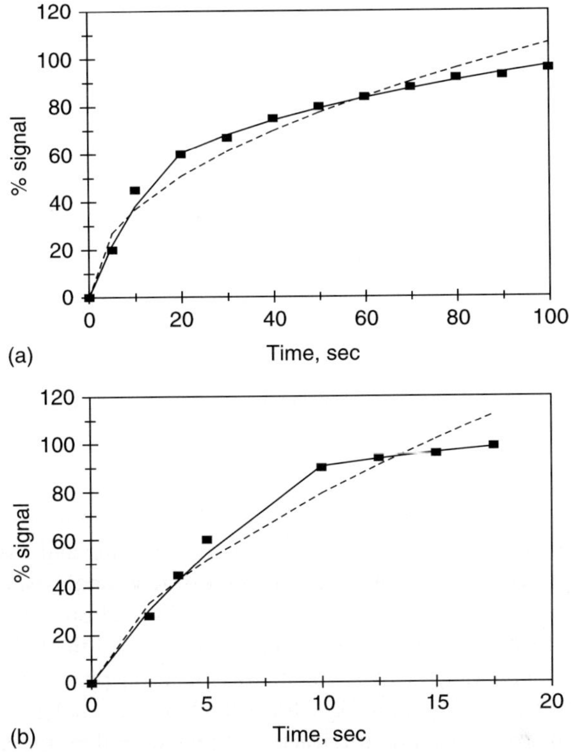

Figure 12.6 Binding of ATP in solution (a) in the presence of and (b) in the absence of PC to the ATP biosensor. When both a dashed line (--------) and a solid (______) line are used, then the dashed line is for a single-fractal analysis and the solid line is for a dual-fractal analysis. In this case, the solid line provides the better fit.

Table 12.1c and Figure 12.7 show the increase in the rate coefficient ratio k_2/k_1 with an increase in the ratio of the fractal dimensions D_{f2}/D_{f1}. For the data presented in Table 12.1, the binding rate coefficient ratio k_2/k_1 for a dual-fractal analysis is given by

$$k_2/k_1 = (1.7509 \pm 0.1988)(D_{f2}/D_{f1})^{1.5380 \pm 0.3612} \qquad (12.3e)$$

The fit is good. Only five data points are available. The availability of more data points would lead to a more reliable fit. The rate coefficient ratio k_2/k_1 is sensitive to the fractal dimension ratio D_{f2}/D_{f1} as noted by the order of dependence on the ratio of the fractal dimensions D_{f2}/D_{f1} slightly more than one and one-half (=1.5380) exhibited.

Liaudet *et al.* (2005) analyzed the influence of PC on the binding and dissociation kinetics of ATP and ADP in solution to the ATP biosensor. Figures 12.8a–c show the binding and dissociation kinetics of (1) 10 μM ATP, (2) 10 μM ADP in the presence of 0.5 mM PC, and (3) 10 μM ATP in solution in the presence of 0.5 mM PC solution to the ATP

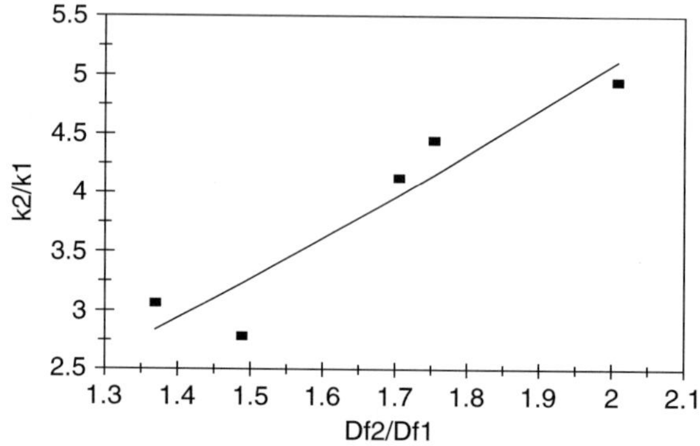

Figure 12.7 Increase in the binding rate coefficient ratio k_2/k_1 with an increase in the fractal dimension ratio D_{f2}/D_{f1}.

biosensor. In case (1) a dual-fractal analysis is required to adequately describe the binding and the dissociation kinetics. The values of the binding and dissociation rate coefficients k and the fractal dimension D_f for a single-fractal analysis, (b) the dissociation rate coefficient k_d and the fractal dimension for the dissociation phase D_{fd}, for a single-fractal analysis, (c) the binding rate coefficients, k_1 and k_2, and the fractal dimensions, D_{f1} and D_{f2}, for a dual-fractal analysis, and (d) the dissociation rate coefficients, k_{d1} and k_{d2}, and the fractal dimensions, D_{fd1} and D_{fd2}, for a dual-fractal analysis are given in Table 12.2. It is of interest to note that as the fractal dimension value increases by a factor of 2.24 from a value of D_{f1}=1.1928 to D_{f2}=2.6750, the binding rate coefficient value increases by more than an order of magnitude (by a factor of 11.54) from a value of k_1=0.1177 to k_2=1.3584. As noted in the different chapters in the book, in general, an increase in the degree of heterogeneity on the biosensor surface leads to an increase in the binding rate coefficient.

In cases (2) and (3) a dual-fractal analysis is required to adequately describe the binding kinetics. The dissociation kinetics is adequately described by a single-fractal analysis. The values of (a) the binding rate coefficient k and the fractal dimension D_f for a single-fractal analysis, (b) the dissociation rate coefficient k_d and the fractal dimension for the dissociation phase D_{fd} for a single-fractal analysis, and the (c) the binding rate coefficients, k_1 and k_2, and the fractal dimensions, D_{f1} and D_{f2}, for a dual-fractal analysis are given in Table 12.2. Once again, it is of interest to note that as the fractal dimension value increases by a factor of 1.66 from a value of D_{f1} = 1.4922 to D_{f2}= 2.4760, the binding rate coefficient value increases by a factor of 8.414 from a value of k_1 =0.1812 to k_2 =1.5246.

Note that in the presence of PC all of the binding and dissociation rate coefficient values are higher when compared with the case when PC is absent in solution. Also note that when PC is used in solution the dissociation phase for ADP may be adequately described by a single-fractal analysis, whereas for ATP a dual-fractal analysis is required to adequately describe the dissociation phase. This suggests that there is a change in the dissociation mechanism as one goes from ADP to ATP in solution.

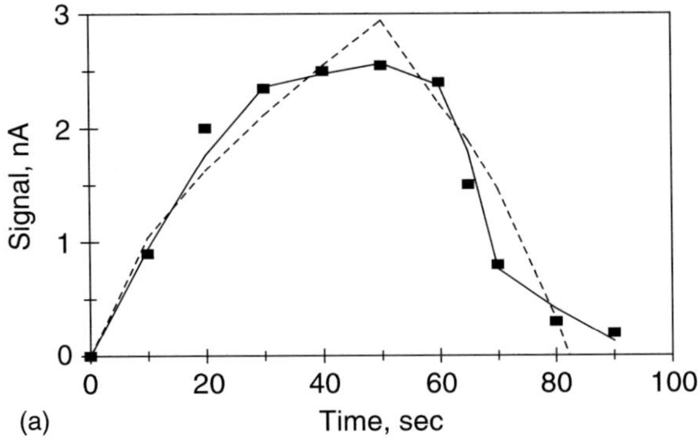

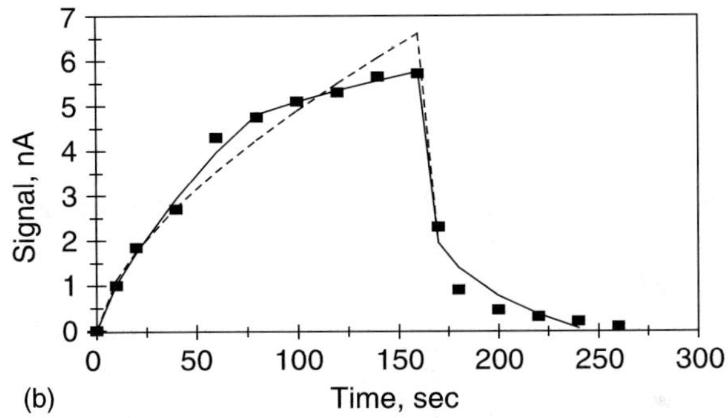

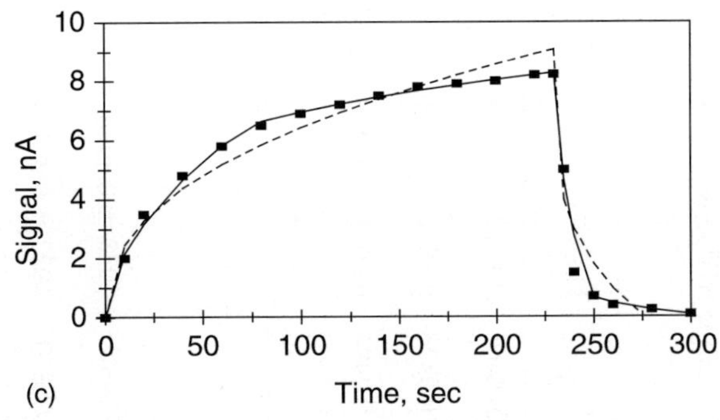

Figure 12.8 (a) Binding and dissociation of 10 μM ATP in solution to the ATP biosensor (Liaudet *et al.*, 2005). (b) Binding and dissociation of 10 μM ADP in solution in the presence of 0.5 mM PC to the ATP biosensor (Liaudet *et al.*, 2005). (c) Binding and dissociation of 10 μM ATP in solution in the presence of 0.5 mM PC to the ATP biosensor (Liaudet *et al.*, 2005). When both a dashed line (---------) and a solid (________) line are used, then the dashed line is for a single-fractal analysis and the solid line is for a dual-fractal analysis. In this case, the solid line provides the better fit.

Table 12.2

(a) Binding and dissociation rate coefficients and (b) Fractal dimensions for the binding and the dissociation phase for the enzymatic amplification of the ATP sensor. Absence and presence of 0.5 mM PC (Liaudet *et al.*, 2005)

(a)

Analyte in solution/ATP sensor	k	k_1	k_2	k_d	k_{d1}	k_{d2}
10 μM ATP	0.2363 ± 0.0461	0.1177 ± 0.0198	1.3584 ± 0.077	0.004217 ± 0.0003742	4.1E-05 ± 2.1E-05	0.4814 ± 0.0318
10 μM ADP + 0.5 mM PC	0.2622 ± 0.0328	0.1812 ± 0.0150	1.5246 ± 0.0205	2.4030 ± 0.2125	NA	NA
10 μM ATP + 0.5 mM PC	0.9414 ± 0.0984	0.6022 ± 0.0501	2.6379 ± 0.0233	2.6160 ± 0.6441	1.3345 ± 0.3866	6.3695 ± 0.0541

(b)

Analyte in solution/ATP sensor	D_f	D_{f1}	D_{f2}	D_{fd}	D_{fd1}	D_{fd2}
10 μM ATP	1.7108 ± 0.2804	1.1928 ± 0.3966	2.6750 ± 0.0714	0 + 1.1570	0 + 1.6790	2.1238 ± 0.2602
10 μM ADP + 0.5 mM PC	1.7284 ± 0.08850	1.4922 ± 0.09450	2.4760 ± 0.07628	0.6074 ± 0.08596	NA	NA
10 μM ATP + 0.5 mM PC	2.1644 ± 0.05906	1.8914 ± 0.9454	2.3994 ± 0.1982	2.3994 ± 0.1982	2.8852 ± 0.5192	6.3695 ± 0.01767

Table 12.2 and Figure 12.9a show the increase in the binding rate coefficient k_1 with an increase in the fractal dimension D_{f1}. For the data presented in Table 12.2 and in Figure 12.9a for the binding of ATP or ADP in solution to the ATP biosensor, the binding rate coefficient k_1 for a dual-fractal analysis is given by

$$k_1 = (0.055552 \pm 0.01907)\, D_{f1}^{3.558 \pm 0.9052} \tag{12.4a}$$

The fit is good. Only three data points are available. The availability of more data points would lead to a more reliable fit. The binding rate coefficient k_1 is very sensitive to the degree of heterogeneity or the fractal dimension D_{f1} as noted by the order of dependence between 3 and 4 (=3.558) exhibited.

Table 12.2 and Figure 12.9b show the increase in the ratio of the binding rate coefficients k_2/k_1 with an increase in the fractal dimension ratio D_{f2}/D_{f1}. For the data presented in Table 12.2 and in Figure 12.9b for the binding of ATP or ADP in solution to the ATP biosensor, the ratio of the binding rate coefficients k_2/k_1 is given by

$$k_2/k_1 = (2.7301 \pm 0.6697)(D_{f2}/D_{f1})^{1.8717 \pm 0.6190} \tag{12.4b}$$

The fit is good. Only three data points are available. The availability of more data points would lead to a more reliable fit. The binding rate coefficient ratio k_2/k_1 is sensitive to the ratio of fractal dimensions D_{f2}/D_{f1} as noted by the order of dependence between 1 and 2 (=1.8717) exhibited.

Liaudet *et al.* (2005) indicate that ascorbate is a known interferent in the detection of ATP and other analytes in the brain. It is present in the extracellular space at concentrations of 200–400 μM (Schenk *et al.*, 1982; Stamford *et al.*, 1984; Miele and Fillenz, 1996; Rice, 2000). Figures 12.10a–f show the binding and dissociation kinetics of (1) 40 μM ATP in solution in the absence of ascorbate, (2) 40 μM ATP in solution in the presence of 40 μM ascorbate, (3) 40 μM ATP in solution in the absence of ascorbate, (4) 40 μM ATP in solution in the presence of 200 μM ascorbate, (5) 40 μM ATP in solution in the absence of ascorbate, and (6) 40 μM ATP in solution in the presence of 400 μM ascorbate, to the ATP biosensor. In cases (1), (2), and (3), a dual-fractal is required to adequately describe the binding kinetics. The dissociation kinetics is adequately described by a single-fractal analysis. The values of the binding rate coefficient k and the fractal dimension D_f for a single-fractal analysis, (b) the dissociation rate coefficient k_d and the fractal dimension for the dissociation phase D_{fd} for a single-fractal analysis, and (c) the binding rate coefficients, k_1 and k_2, and the fractal dimensions, D_{f1} and D_{f2}, for a dual-fractal analysis are given in Table 12.3.

It is of interest to note that when the ascorbate is present in solution, the binding and dissociation rate coefficients for ATP in solution to the ATP biosensor are lower than the corresponding binding and dissociation rate coefficients when it is absent. For example, when only ATP is present in solution (in the absence of ascorbate) the binding rate coefficient k_1 for a dual-fractal analysis is 17.824. When both ATP and 40 μM ascorbate is present in solution, the corresponding binding rate coefficient k_1=8.8816. This is a decrease in the binding rate coefficient k_1 by 50.3%. Similarly, the dissociation rate coefficient

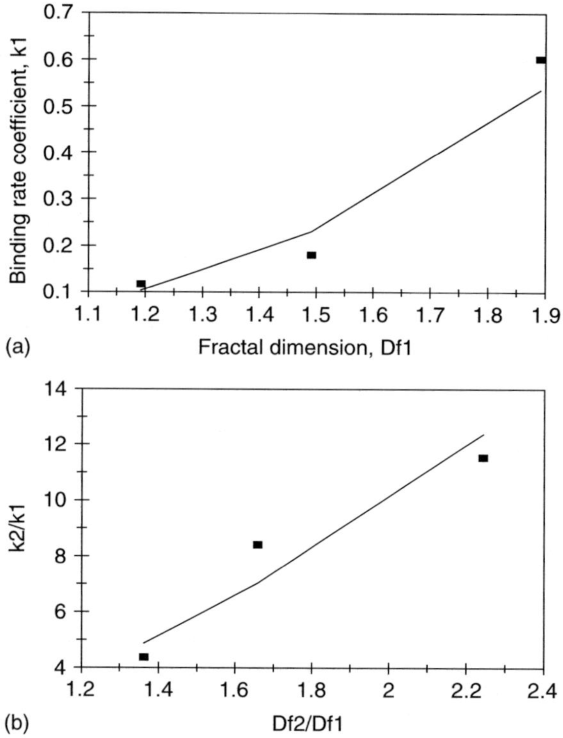

Figure 12.9 (a) Increase in the binding rate coefficient k_1 with an increase in the fractal dimension D_{f1}.(b) Increase in the ratio of the binding rate coefficients k_2/k_1 with an increase in the fractal dimension ratio D_{f2}/D_{f1}.

k_d = 339.21 when 40 μM ATP is present in solution in the absence of ascorbate. When 40 μM ATP is present in solution in the presence of 40 μM ascorbate, there is a decrease in the dissociation rate coefficient k_d by 53.6% from a value of 339.21 to 157.53. Apparently, in this case, the presence of the ascorbate in solution affects the binding and dissociation rate coefficients by approximately the same amount.

In cases (4), (5), and (6) a dual-fractal is required to adequately describe binding and the dissociation kinetics. The values of the binding and dissociation rate coefficients k and the fractal dimensions D_f for a single-fractal analysis, (b) the dissociation rate coefficient k_d and the fractal dimension for the dissociation phase D_{fd} for a single-fractal analysis, (c) the binding rate coefficients, k_1 and k_2, and the fractal dimensions, D_{f1} and D_{f2}, for a dual-fractal analysis, and (d) the dissociation rate coefficients, k_{d1} and k_{d2}, and the fractal dimensions D_{fd1} and D_{fd2}, for a dual-fractal analysis are given in Table 12.3.

It is of interest to note that in this case, the binding rate coefficients when ascorbate is present are higher than the corresponding binding rate coefficients when ascorbate is absent. Also, when ascorbate is absent the dissociation kinetics is adequately described by a single-fractal analysis, whereas a dual-fractal analysis is required to describe the dissociation kinetics when ascorbate is present in solution during the binding of ATP to the ATP

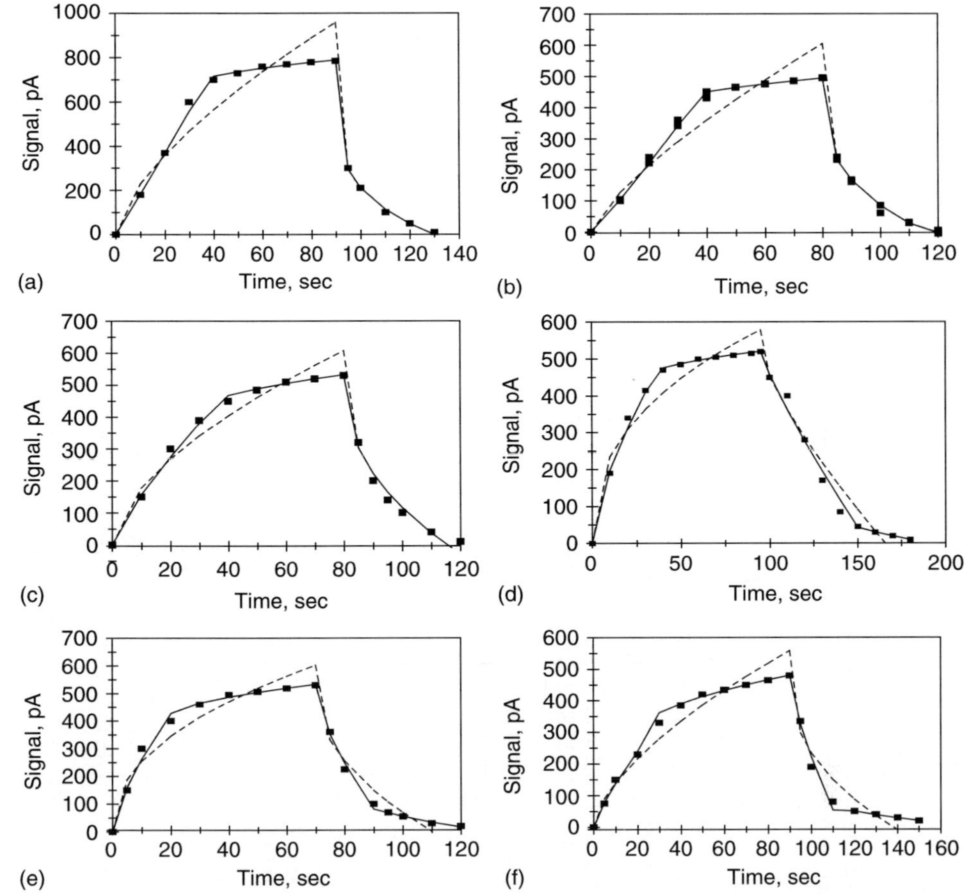

Figure 12.10 Binding and dissociation of 40 μM ATP in solution to the ATP biosensor (Liaudet *et al.*, 2005): (a) in the absence of ascorbate, (b) in the presence of 40 μM ascorbate, (c) in the absence of ascorbate, (d) in the presence of 200 μM ascorbate,(e) in the absence of ascorbate, (f) in the presence of 400 μM ascorbate. When both a dashed line (-------) and a solid (_______) line are used, then the dashed line is for a single-fractal analysis and the solid line is for a dual-fractal analysis. In this case, the solid line provides the better fit.

biosensor. This, once again, indicates a change in the dissociation mechanism of ATP when a comparison is made in the presence and in the absence of ascorbate.

It is of interest to note that except for the dissociation rate coefficient k_{d2}, all the binding and dissociation rate coefficient values are lower when ascorbate is present compared to when it is absent during the binding of ATP in solution to the ATP biosensor. This is consistent with the fact that ascorbate, as indicated by Liaudet *et al.* (2005), is a known interferent in the detection of ATP and other analytes in solution.

Table 12.3 and Figures 12.11a and b show for a dual-fractal analysis the increase in the binding rate coefficients, k_1 and k_2, with an increase in the fractal dimension D_{f1} and D_{f2},

Table 12.3

(a) Binding and dissociation rate coefficients and (b) fractal dimensions for the binding and dissociation phases for the influence of different concentrations (in μM) of a known interferent, ascorbate on ATP in solution to the ATP sensor (Liaudet *et al.*, 2005)

(a)

Analyte in solution/ ATP sensor	k	k_1	k_2	k_d	k_{d1}	k_{d2}
40 μM ATP	51.213 ± 10.386	17.824 ±1.290	461.16 ± 4.079	339.21 ± 4.079	NA	NA
40 μM ATP + 40 μM ascorbate	22.779 ± 4.727	8.8816 ± 0.6867	276.362 ± 0.552	157.526 ± 7.672	NA	NA
40 μM ATP	46.157 ± 6.346	24.803 ± 2.2203	236.82 ± 2.168	112.35 ± 7.982	NA	NA
40 μM ATP + 200 μM ascorbate	92.592 ± 11.127	43.517 ± 3.404	323.87 ± 2.051	19.112 ± 3.217	15.564 ± 2.695	263.046 ± 0.2443
40 μM ATP	92.078 ± 14.875	51.393 ± 9.252	269.48 ± 3.247	92.344 ± 12.415	60.257 ± 6.231	282.49 ± 2.163
40 μM ATP + 400 μM ascorbate	32.914 ± 4.578	21.5547 ± 1.6219	150.047 ± 1.640	91.090 ± 17.330	47.494 ± 7.779	308.827 ± 1.015

(b)

Analyte in solution/ATP sensor	D_f	D_{f1}	D_{f2}	D_{fd}	D_{fd1}	D_{fd2}
40 μM ATP	1.6978 ± 0.1815	0.9768 ± 0.1342	2.7607 ± 0.03792	2.5466 ± 0.01819	NA	NA
40 μM ATP + 40 μM ascorbate	1.5032 ± 0.2026	0.8528 ± 0.1828	2.7346 ± 0.001137	2.3614 ± 0.05628	NA	NA
40 μM ATP	1.8232 ± 0.1384	1.3944 ± 0.1647	2.6302 ± 0.05198	2.1304 ± 0.08114	NA	NA
40 μM ATP + 200 μM ascorbate	2.1948 ± 0.1128	1.6820 ± 0.1449	2.7972 ± 0.03602	1.4482 ± 0.1208	1.2922 ± 0.1616	2.7022 ± 0.008778
40 μM ATP	2.1152 ± 0.1222	1.5850 ± 0.3376	2.6788 ± 0.03578	2.0548 ± 0.1272	1.6612 ± 0.2008	2.6950 ± 0.02872
40 μM ATP +400 μM ascorbate	1.7410 ± 0.09114	1.3930 ± 0.1125	2.4822 ± 0.03214	2.1414 ± 0.1567	1.5360 ± 0.3094	2.8066 ±0.01268

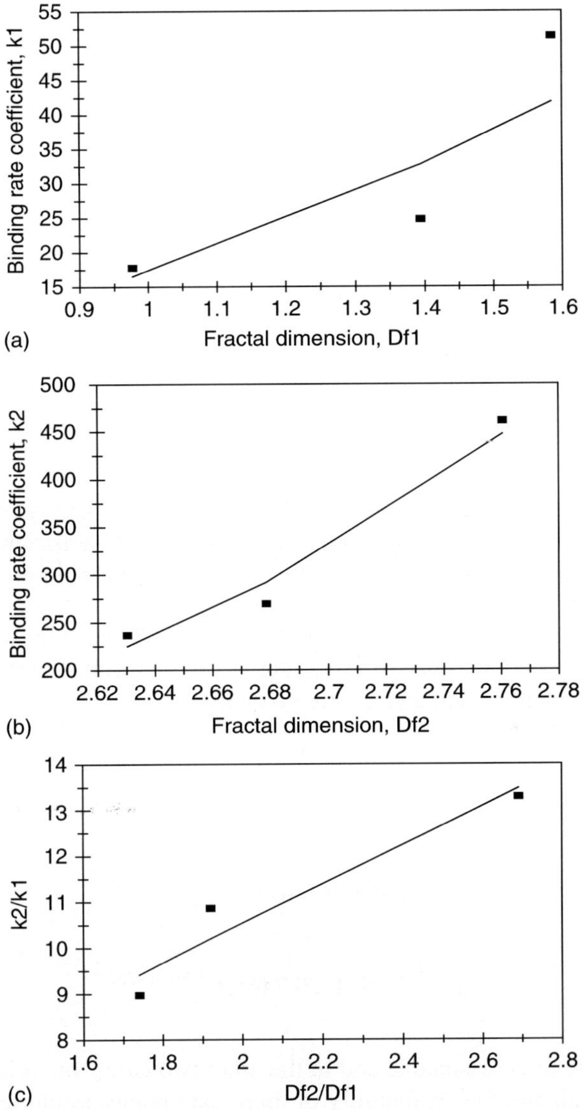

Figure 12.11 (a) Increase in the binding rate coefficient k_1 with an increase in the fractal dimension D_{f1}. (b) Increase in the binding rate coefficient k_2 with an increase in the fractal dimension D_{f2}. (c) Increase in the ratio of the binding rate coefficients k_2/k_1 with an increase in the fractal dimension ratio D_{f2}/D_{f1}.

respectively. For the data presented in Table 12.3 and Figures 12.11a and b, the binding rate coefficients, k_1 and k_2, are given by

$$k_1 = (17.320 \pm 7.337) D_{f1}^{1.9171 \pm 0.9958} \tag{12.5a}$$

$$k_2 = (0.000253 \pm 0.000026) D_{f2}^{14.1666 \pm 2.871} \tag{12.5b}$$

The fit is reasonable. Only three data points are available. The availability of more data points would lead to a more reliable fit. The binding rate coefficient k_1 is sensitive to the degree of heterogeneity on the biosensor surface or the fractal dimension D_{f1} as noted by the close to second (=1.9171) order of dependence exhibited.

The binding rate coefficient k_2 is extremely sensitive to the degree of heterogeneity on the biosensor surface or the fractal dimension D_{f2} as noted by the higher than 14th (=14.1666) order of dependence exhibited.

Table 12.3 and Figure 12.11c show for a dual-fractal analysis the increase in the ratio of the binding rate coefficients k_2/k_1 with an increase in the fractal dimension ratio D_{f2}/D_{f1}. For the data presented in Table 12.3 and Figure 12.11c, the ratio of the binding rate coefficients k_2/k_1 is given by

$$k_2/k_1 = (5.9673 \pm 0.4984)(D_{f2}/D_{f1})^{0.8223 \pm 0.2817} \quad (12.5c)$$

The fit is good. Only three data points are available. The availability of more data points would lead to a more reliable fit. The ratio of the binding rate coefficients k_2/k_1 is sensitive to the degree of heterogeneity on the biosensor surface or the ratio of fractal dimensions D_{f2}/D_{f1} as noted by the 0.8223 order of dependence exhibited.

For the data presented in Table 12.3 and in Figures 12.12a–c for the dissociation of ATP in the absence and in the presence of ascorbate in solution to the ATP biosensor, the dissociation rate coefficient k_d, k_{d1}, and k_{d2} are given by

$$k_d = (1.0874 \pm 0.3)D_{fd}^{6.021 \pm 1.923} \quad (12.6a)$$

$$k_{d1} = (3.8771 \pm 0.6141)D_{fd1}^{5.5645 \pm 0.8088} \quad (12.6b)$$

$$k_{d2} = (12.871 \pm 0.746)D_{fd2}^{3.0767 \pm 1.7552} \quad (12.6c)$$

In the first case, the fit is reasonable and in the other two cases the fit is good. Only three data points are available. The availability of more data points would lead to a more reliable fit. The dissociation rate coefficient k_d is very sensitive to the fractal dimension in the dissociation phase D_{fd} as noted by the slightly higher than sixth (=6.021) order of dependence exhibited.

The dissociation rate coefficient k_{d1} is very sensitive to the fractal dimension in the dissociation phase D_{fd1} as noted by the order between 5 and 6 (=5.5645) exhibited.

The dissociation rate coefficient k_{d2} is very sensitive to the fractal dimension in the dissociation phase D_{fd2} as noted by the slightly higher than third (=3.0767) order exhibited. In this case the fit is not good.

As indicated in the earlier chapters in this book, affinity K values are of interest to chemists, biochemists, and practicing biosensorists. Table 12.3 and Figures 12.13a–c show the increase in the affinity K (= k/k_d), K_1 (= k_1/k_{d1}), and K_2 (= k_2/k_{d2}), with an increase in the fractal dimension ratios, D_f/D_{fd}, D_{f1}/D_{fd1}, and D_{f2}/D_{fd2}, respectively. For the data

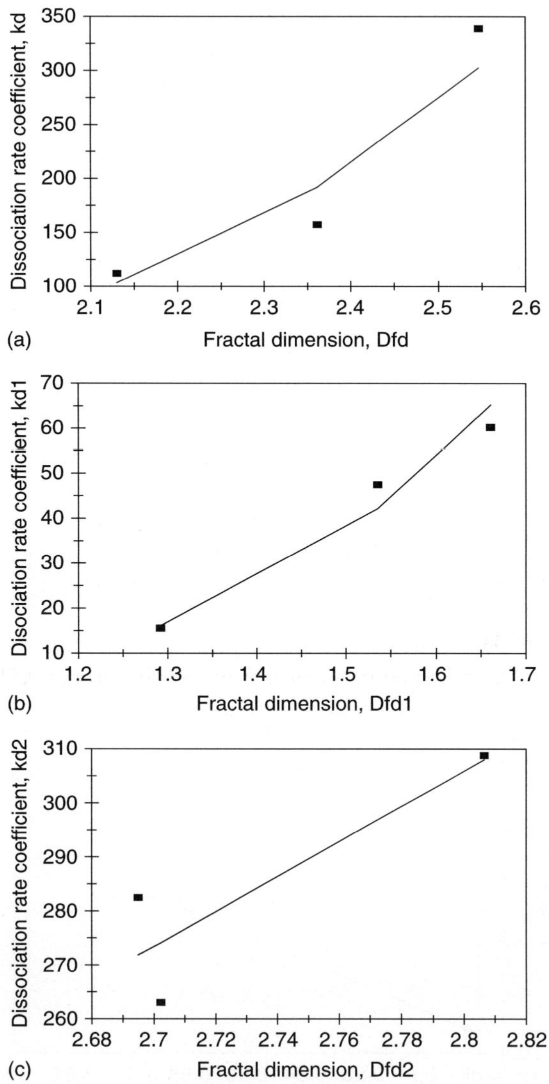

Figure 12.12 (a) Increase in the dissociation rate coefficient k_d with an increase in the fractal dimension D_{fd}. (b) Increase in the dissociation rate coefficient k_{d1} with an increase in the fractal dimension D_{fd1}. (c) Increase in the dissociation rate coefficient k_{d2} with an increase in the fractal dimension D_{fd2}.

presented in Table 12.3 and in Figures 12.13a–c, the affinity K, K_1, K_2, are given by

$$K(=k/k_d) = (0.7227 \pm 0.0675)(D_f/D_{fd})^{3.6949 \pm 0.3975} \tag{12.7a}$$

$$K_1(=k_1/k_{d1}) = (0.8555 \pm 0.2807)(D_{f1}/D_{fd1})^{4.6062 \pm 1.0255} \tag{12.7b}$$

$$K_2(=k_2/k_{d2}) = (0.9774 \pm 0.0123)(D_{f2}/D_{fd2})^{5.8762 \pm 0.1057} \tag{12.7c}$$

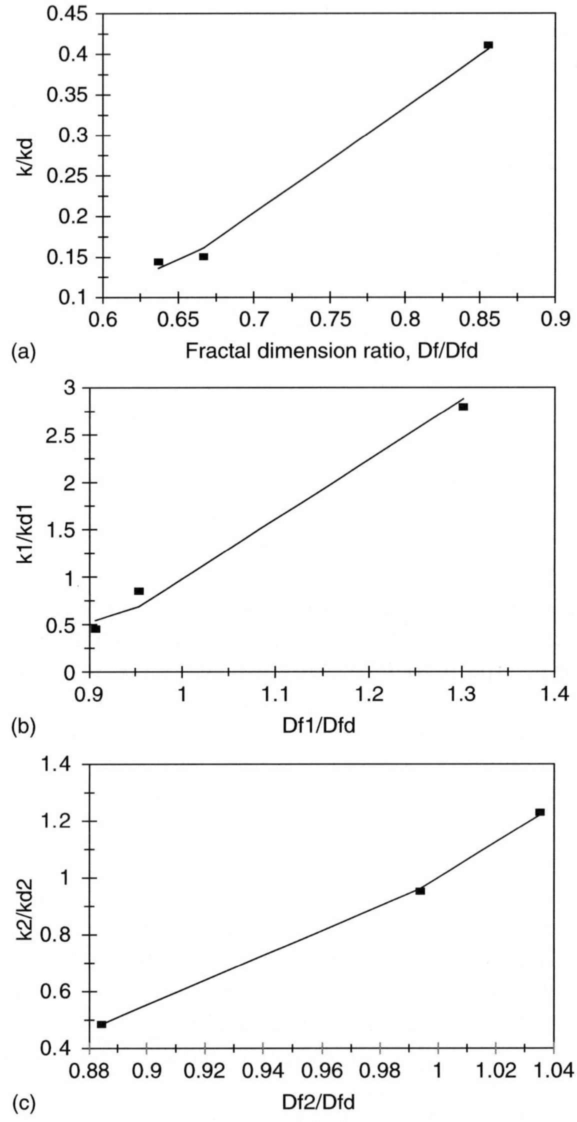

Figure 12.13 (a) Increase in the affinity K (= k/k_d) with an increase in the fractal dimension ratio D_f/D_{fd}. (b) Increase in the affinity K_1 (= k_1/k_d) with an increase in the fractal dimension ratio D_{f1}/D_{fd1}. (c) Increase in the affinity K_2 (= k_2/k_d) with an increase in the fractal dimension ratio D_{f2}/D_{fd2}.

The fit is very good. Only three data points are available. The availability of more data points would lead to a more reliable fit. The affinity K (= k/k_d) is very sensitive to the ratio of the fractal dimensions D_f/D_{fd} as noted by the order of dependence between 3½ and 4 (=3.6949) exhibited.

The affinity K_1 (= k_1/k_{d1}) is very sensitive to the ratio of the fractal dimensions D_{f1}/D_{fd1} as noted by the order of dependence between 4 and 5 (= 4.6062) exhibited.

The affinity K_2 (= k_2/k_{d2}) is very sensitive to the ratio of the fractal dimensions D_{f2}/D_{fd2} as noted by the close to sixth order of dependence (=5.8762) exhibited.

Mistrik *et al.* (2004) have recently used the Biacore biosensor to analyze the binding and dissociation kinetics of leptin–leptin interactions. They presented evidence for a 1:1 (analyte:receptor) stoichiometry. These authors indicate that leptin is produced in quite a few organs in the body, and is involved in a broad range of metabolic processes (Hirschberg, 1988; Haynes *et al.*, 1997). Furthermore, the hormonal protein, leptin is involved in energy homeostasis, and inhibits food intake, stimulates energy expenditure, and influences insulin secretion, lipolysis, and sugar transport (Cunningham *et al.*, 1999; Cohen *et al.*, 1996; Kieffer *et al.*, 1996; Mistrik *et al.*, 2004). Mistrik *et al.* (2004) do indicate that leptin is a member of the cytokine family.

Figures 12.14a–d show the binding and dissociation of (1) 1 (2) 3, (3) 5, and (4) 10 nM human leptin in solution to the leptin receptor (ObR) immobilized on a sensor chip surface (Mistrik *et al.*, 2004). In case (1), a single-fractal analysis is adequate to describe the binding and dissociation kinetics. The values of (a) the binding rate coefficients k and the fractal dimensions D_f for a single-fractal analysis, and (b) the dissociation rate coefficient k_d and the fractal dimension D_{fd} for a single-fractal analysis are given in Table 12.4a and b.

In cases (2), (3), and (4) a dual-fractal analysis is adequate to describe the binding kinetics and a single-fractal analysis is adequate to describe the dissociation kinetics. The values of the binding rate coefficients k and the fractal dimensions D_f for a single-fractal analysis, (b) the binding rate coefficients, k_1 and k_2, and the fractal dimensions, D_{f1} and D_{f2}, for a dual-fractal analysis, and (c) the dissociation rate coefficient k_d and the fractal dimension D_{fd} for a single-fractal analysis are given in Table 12.4a and b. It is of interest to note that as the human leptin concentration in solution increases from 1 to 3 nM in solution, there is a change in the binding mechanism, since a single-fractal analysis is adequate to describe the binding kinetics at 1 nM human leptin concentration in solution, whereas a dual-fractal analysis is required to describe the binding kinetics at the 3 nM human leptin concentration in solution. There is no change in the dissociation mechanism since at both the 1 and 3 nM human leptin concentration in solution a single-fractal analysis is adequate to describe the dissociation kinetics. Also, it is of interest to note that as the fractal dimension increases by 68.7% from a value of D_{f1}=1.5956 to D_{f2}=2.6924, the binding rate coefficient increases by a factor of 15.26 from a value of k_1=0.1797 to k_2=2.7432.

The analysis for Figures 12.4c and d is similar to that of Figure 12.14b and is not presented here, in order to avoid repetition.

For the 3–10 nM human leptin concentration in solution, Figures 12.15a and b and Table 12.4a show for a dual-fractal analysis the increase in the binding rate coefficients, k_1 and k_2, with an increase in the human leptin concentration (in nM) in solution. For the data presented in Table 12.4a and in Figures 12.15a and b, the binding rate coefficients, k_1 and k_2, are given by

$$k_1 = (0.0365 \pm 0.0110)\left[\text{human leptin concentration, in nM}\right]^{1.3574 \pm 0.3153} \tag{12.8a}$$

$$k_2 = (0.6949 \pm 0.0773)\left[\text{human leptin concentration, in nM}\right]^{1.2049 \pm 0.1233} \tag{12.8b}$$

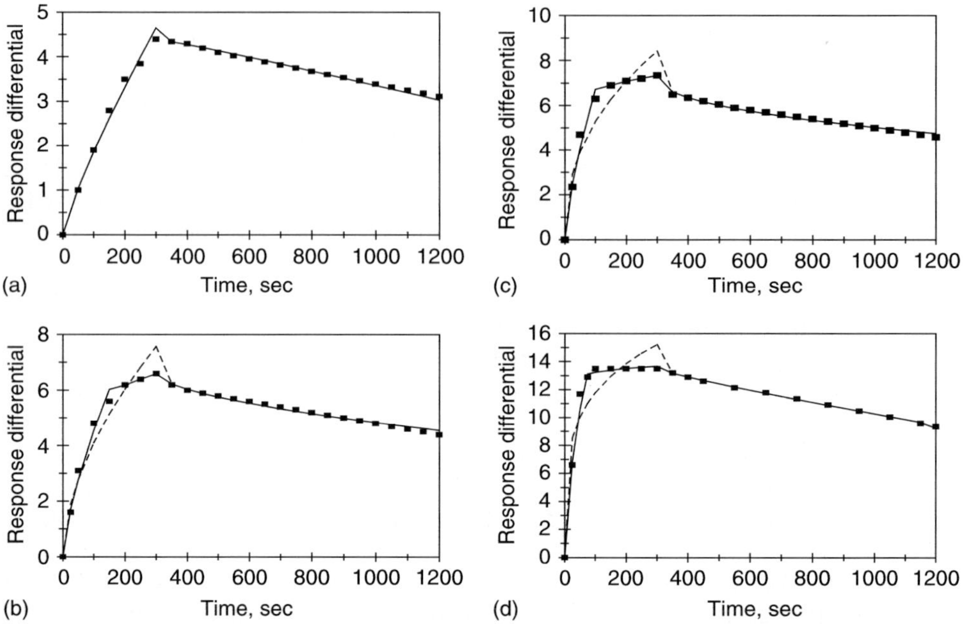

Figure 12.14 Binding and dissociation of different concentrations (in nM) of human leptin in solution to leptin receptor immobilized on a sensor chip surface (Mistrik *et al.*, 2004): (a) 1, (b) 3, (c) 5, (d) 10. When both a dashed line (-------) and a solid (______) line are used, then the dashed line is for a single-fractal analysis and the solid line is for a dual-fractal analysis. In this case, the solid line provides the better fit.

The fit is good. Only three data points are available. The availability of more data points would lead to a more reliable fit. The binding rate coefficient k_1 exhibits close to a one and one-half (=1.3574) order of dependence on the human leptin concentration (3–10 nM) in solution. The fractional order of dependence exhibited lends support to the fractal nature of the system.

The binding rate coefficient k_2 exhibits an order of dependence between one and one and one-half (=1.2049) on the human leptin concentration (3–10 nM) in solution. Note that the order of dependence on the human leptin concentration in solution exhibited by the binding rate coefficient k_2 (=1.2049) is less than that exhibited by the binding rate coefficient, k_1 (=1.3574). Once again, the fractional order of dependence exhibited by the binding rate coefficient k_2 on the human leptin concentration (in nM) lends support to the fractal nature of the system.

Figures 12.15c and d and Table 12.4a and b show the increase in the binding rate coefficient k_1 with an increase in the fractal dimensions, D_{f1} and D_{f2}. For the data presented in Figures 12.15c and d and in Table 12.4a and b, the binding rate coefficients, k_1 and k_2, are given by

$$k_1 = (0.00167 \pm 0.000076) D_{f1}^{15.536 \pm 5.145} \tag{12.8c}$$

$$k_2 = (2.9\text{E-}07 \pm 0.8\text{E-}07) D_{f2}^{16.128 \pm 3.831} \tag{12.8d}$$

Table 12.4

(a) Binding and dissociation rate coefficients and (b) Fractal dimensions for the binding and dissociation phase for different concentrations (in nM) of human leptin in solution to leptin receptor (ObR) immobilized on a sensor chip surface (Mistrik *et al.*, 2004)

(a)

Analyte in solution/receptor on surface	k	k_1	k_2	k_d	k_{d1}	k_{d2}
1 nM leptin/ObR (leptin receptor)	0.04034 ± 0.00251	NA	NA	0.000717 ± 0.000062	NA	NA
3 nM leptin/ObR (leptin receptor)	0.3144 ± 0.0467	0.1797 ± 0.02100	2.7432 ± 0.0059	0.03671 ± 0.00200	NA	NA
5 nM leptin/ObR (leptin receptor)	0.7496 ± 0.1393	0.2545 ± 0.0451	4.4349 ± 0.0136	0.1331 ± 0.0053	NA	NA
10 nM leptin/ObR (leptin receptor)	4.0391 ± 0.7062	0.8911 ± 0.1045	11.551 ± 0.1901	0.01029 ± 0.00042	NA	NA

(b)

Analyte in solution/receptor on surface	D_f	D_{f1}	D_{f2}	D_{fd}	D_{fd1}	D_{fd2}
1 nM leptin/ObR (leptin receptor)	1.3350 ± 0.0814	NA	NA	0.7794 ± 0.05018	NA	NA
3 nM leptin/ObR (leptin receptor)	1.8814 ± 0.1244	1.5956 ± 0.1622	2.6924 ± 0.01509	1.8183 ± 0.02310	NA	NA
5 nM leptin/ObR (leptin receptor)	2.1512 ± 0.1534	1.5744 ± 0.3332	2.8233 ± 0.01183	2.1274 ± 0.00258	NA	NA
10 nM leptin/ObR (leptin receptor)	2.5344 ± 0.1426	1.7334 ± 0.3076	2.9407 ± 0.03314	1.2452 ± 0.02564	NA	NA

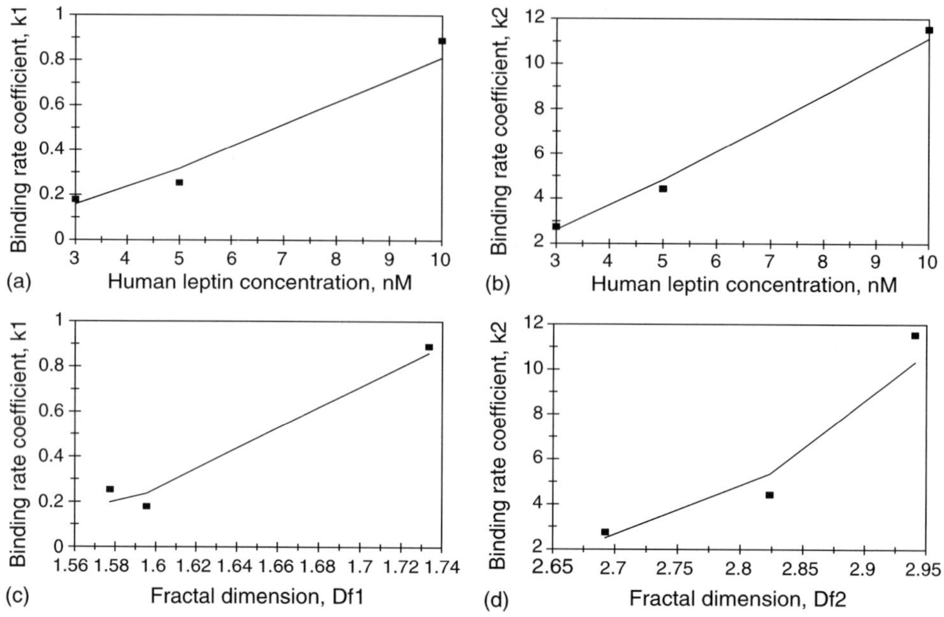

Figure 12.15 (a) Increase in the binding rate coefficient k_1 with an increase in the human leptin concentration (in nM) in solution. (b) Increase in the binding rate coefficient k_2 with an increase in the human leptin concentration (in nM) in solution. (c) Increase in the binding rate coefficient k_1 with an increase in the fractal dimension D_{f1}. (d) Increase in the binding rate coefficient k_2 with an increase in the fractal dimension D_{f2}.

The fit is good. Only three data points are available. The availability of more data points would lead to a more reliable fit. The binding rate coefficient k_1 is extremely sensitive to the fractal dimension in the binding phase D_{f1} as noted by the higher than 15th (=15.536) order exhibited.

The binding rate coefficient k_2 is extremely sensitive to the fractal dimension in the binding phase D_{f2} as noted by the higher than 16th (=16.128) order exhibited.

Figures 12.16a and b and Table 12.4a and b show the increase in the affinity K_1 $(=k_1/k_d)$ and K_2 $(=k_2/k_d)$ with an increase in the fractal dimension ratios D_{f1}/D_{fd} and D_{f2} D_{fd}, respectively. For the data presented in Table 12.4a and b and Figures 12.16a and b, the affinity, K_1 and K_2, are given by

$$K_1 = (10.448 \pm 0.131)(D_{f1}/D_{fd})^{5.702 \pm 0.0269} \tag{12.8e}$$

$$K_2 = (6.496 \pm 0.726)(D_{f2}/D_{fd})^{6.012 \pm 0.2449} \tag{12.8f}$$

The fit is very good. Only three data points are available. The availability of more data points would lead to a more reliable fit. The affinity K_1 is very sensitive to the ratio of fractal dimensions (D_{f1}/D_{fd}) as noted by the order between 5 and 6 (=5.702) exhibited.

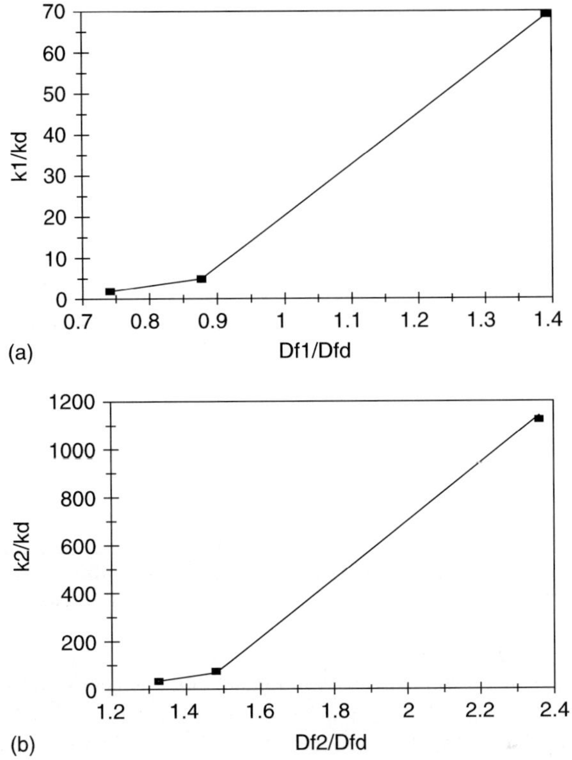

Figure 12.16 (a) Increase in the affinity K_1 (= k_1/k_d) with an increase in the fractal dimension ratio D_{f1}/D_{fd}. (b) Increase in the affinity K_2 (= k_2/k_d) with an increase in the fractal dimension ratio D_{f2}/D_{fd}.

The affinity K_2 is very sensitive to the ratio of fractal dimensions D_{f2}/D_{fd} as noted by the slightly higher than sixth order (=6.012) exhibited.

Figures 12.17a–d show the binding and dissociation of (1) 30, (2) 10, (3) 5, and (4) 3 nM mouse leptin concentration in solution to the leptin receptor (ObR) immobilized on a sensor chip surface (Mistrik *et al.,* 2004). In cases (1) to (3) a dual-fractal analysis is required to describe the binding kinetics and a single-fractal analysis is adequate to describe the dissociation kinetics. The values of (a) the binding rate coefficients k and the fractal dimensions D_f for a single-fractal analysis, (b) the binding rate coefficients, k_1 and k_2, and the fractal dimensions, D_{f1} and D_{f2}, for a dual-fractal analysis, and (c) the dissociation rate coefficient k_d and the fractal dimension D_{fd} for a single-fractal analysis are given in Table 12.5a and b.

It is of interest to note that as the mouse leptin concentration in solution decreases by a factor of 3 from 30 to 10 nM in solution, for a dual-fractal analysis (a) the binding rate coefficient k_1 decreases by a factor of 3.22 from a value of 5.3286 to 1.6528, and (b) the binding rate coefficient k_2 decreases by a factor of 1.5 from a value of 10.826 to 7.216.

The trends are similar as noted for cases (1) and (2) and are not presented here to avoid repetition.

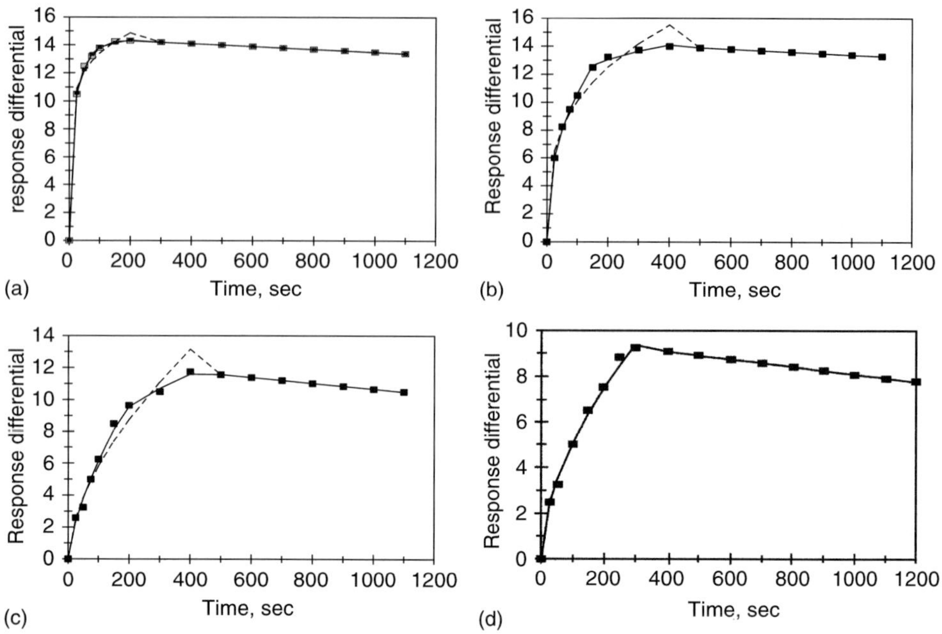

Figure 12.17 Binding and dissociation of different concentrations (in nM) of mouse leptin in solution to the leptin receptor (ObR) immobilized on a sensor chip surface (Mistrik *et al.*, 2004): (a) 30, (b) 10, (c) 5, (d) 3. When both a dashed line (-------) and a solid (______) line are used, then the dashed line is for a single-fractal analysis and the solid line is for a dual-fractal analysis. In this case, the solid line provides the better fit.

In case (4) note that a single-fractal analysis is adequate to describe the binding and dissociation kinetics. The values of the binding and dissociation rate coefficients k and the fractal dimensions D_f for a single-fractal analysis, and (b) the dissociation rate coefficient k_d and the fractal dimension D_{fd} for a single-fractal analysis are given in Table 12.5a and b. Note that there is a change in the binding mechanism as one goes from the 3 nM mouse leptin concentration in solution to the 5–30 nM mouse leptin concentration in solution since a single-fractal analysis is adequate at the 3 nM concentration, whereas a dual-fractal analysis is required to describe the binding kinetics at the 5–30 nM mouse leptin concentration in solution.

An attempt was made to see if there was a correlation between the binding rate coefficients, k_1 and k_2, and the concentrations for the mouse and human leptin (in nM) used. No trend was observed for the numbers available in the Tables 12.4 and 12.5 presented. At 5 nM leptin concentration the mouse and human leptin binding rate coefficient k_1 values were 0.2596 and 0.2545, respectively (mouse is higher than human). Also, the binding rate coefficient k_2 values at 5 nM leptin concentration for the mouse and human leptin were 2.1812 and 4.4349, respectively (mouse is lower than human). For the 10 nM leptin concentration in solution for the mouse and leptin trends in the rate coefficients k_1 and k_2 were reversed.

Table 12.5

(a) Binding and dissociation rate coefficients and (b) Fractal dimensions for the binding and dissociation phase for different concentrations (in nM) of mouse leptin in solution to leptin receptor (ObR) immobilized on a sensor chip surface (Mistrik *et al.*, 2004)

(a)

Analyte in solution/receptor on surface	k	k_1	k_2	k_d	k_{d1}	k_{d2}
30 nM leptin/ObR (leptin receptor)	6.7890 ± 0.2587	5.3286 ± 0.1305	10.826 ± 0.099	0.001 ± 1.78E-14	NA	NA
10 nM leptin/ObR leptin receptor)	2.4101 ± 0.1863	1.6528 ± 0.0374	7.2157 ± 0.1086	0.001 ± 0.0	NA	NA
5 nM leptin/ObR (leptin receptor)	0.3841 ± 0.0474	0.2596 ± 0.0303	2.1812 ± 0.0552	0.001820 ± `0.000005	NA	NA
3 nM leptin/ObR (leptin receptor)	0.3997 ± 0.0181	NA	NA	0.001746 ± 0.000016	NA	NA

(b)

Analyte in solution/receptor on surface	D_f	D_{f1}	D_{f2}	D_{fd}	D_{fd1}	D_{fd2}
30 nM leptin/ObR (leptin receptor)	2.7036 ± 0.0442	2.5742 ± 0.0614	2.8937 ± 0.03718	2.0 ± 1.78E-14	NA	NA
10 nM leptin/ObR (leptin receptor)	2.3780 ± 0.0604	2.1908 ± 0.0430	2.7760 ± 0.03980	2.0 ± 1.78E-14	NA	NA
5 nM leptin/ObR (leptin receptor)	1.8202 ± 0.09442	1.6270 ± 0.161	2.4418 ± 0.1016	1.0042 ± 0.003474	NA	NA
3 nM leptin/ObR (leptin receptor)	1.9934 ± 0.01988	NA	NA	1.0100 ± 0.008490	NA	NA

Figure 12.18a,b and Table 12.5a show for a dual-fractal analysis the increase in the binding rate coefficients k_1 and k_2 with an increase in the mouse leptin concentration (5–30 nm) in solution. For the data presented in Figure 12.18a,b and in Table 12.5a the binding rate coefficient k_1 and k_2 is given by

$$k_1 = (0.02478 \pm 0.01826)[\text{mouse leptin concentration, nM}]^{1.6300 \pm 0.4322} \quad (12.9a)$$

$$k_2 = (0.7035 \pm 0.4183)[\text{mouse leptin concentration, nM}]^{0.8469 \pm 0.3651} \quad (12.9b)$$

The fit is reasonable. Only three data points are available. The availability of more data points would lead to a more reliable fit. The binding rate coefficient k_1 exhibits an order of dependence between one and one-half and two (=1.63) on the mouse leptin concentration (5–30 nm) in solution. The non-integer order of dependence once again lends support to the fractal nature of the system. It is of interest to note that the order of dependence exhibited by the binding rate coefficient k_1 on the mouse leptin concentration in

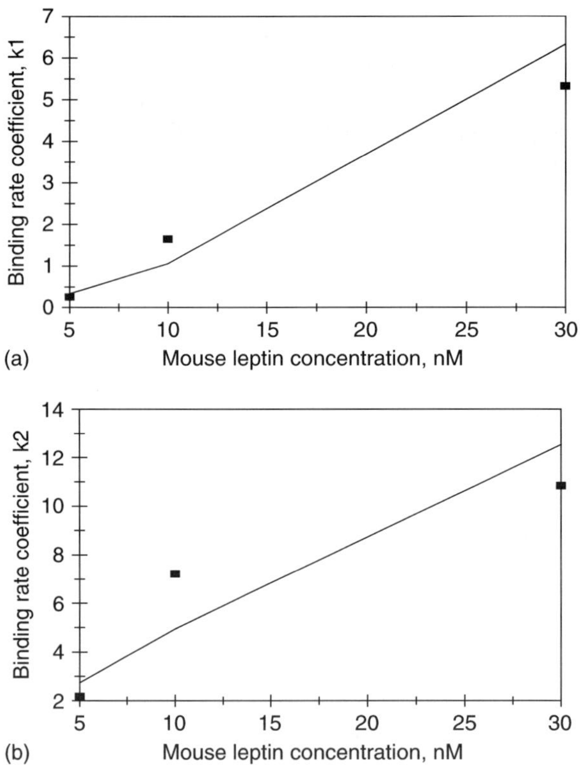

Figure 12.18 (a) Increase in the binding rate coefficient k_1 with an increase in the mouse leptin concentration (5–30 nM) in solution. (b) Increase in the binding rate coefficient k_2 with an increase in the mouse leptin concentration (5–30 nM) in solution.

solution (=1.630) is almost twice that exhibited by k_2 on the mouse leptin concentration in solution (=0.8469).

Figures 12.19a and b show the increase in the affinities (1) K_1 (= k_1/k_d) and (2) K_2 (= k_2/k_d) with an increase in the mouse leptin concentration (5–30 nM) in solution. For the data presented in Figures 12.19a and b and in Table 12.5a, the affinities K_1 and K_2 are given by

$$K_1(= k_1/k_d) = (9.673 + 12.95)\left[\text{mouse leptin concentration, nM}\right]^{1.934 \pm 0.6650} \quad (12.10a)$$

$$K_2(= k_2/k_d) = (275.035 + 315.49)\left[\text{mouse leptin concentration, nM}\right]^{1.150 \pm 0.598} \quad (12.10b)$$

In case (1) there is scatter in the fit of the data. This is reflected in the error in the estimated value of the affinity K_1 presented. Only the positive error is provided since the estimated affinity K_1 value cannot be negative. Only three data points are available. The availability of more data points would lead to a more reliable fit. The affinity K_1 exhibits close to a second (=1.934) order of dependence on the mouse leptin concentration (5–30 nm) in solution.

As observed for the affinity K_1 there is scatter in the fit of the data for the affinity K_2 when plotted against the mouse leptin concentration (5–30 nm) in solution. The scatter in the fit of the data is reflected in the error in the estimated value of the affinity K_2 presented. Once again, only the positive error is presented since the estimated affinity K_2 value cannot be negative. Only three data points are available. The availability of more data points would lead to a more reliable fit. The affinity K_2 exhibits slightly higher than first- (=1.150) order of dependence on the mouse leptin concentration in solution. Note that the affinity K_2 exhibits a lower order of dependence (=1.1502) than the affinity K_1 (=1.934) on the mouse leptin concentration (5–30 nM) in solution.

Figures 12.20a and b and Table 12.5a and b show for a dual-fractal analysis the increase in the binding rate coefficients k_1 and k_2, with an increase in the fractal dimensions, D_{f1}.and D_{f2}, respectively. For the data presented in Figures 12.20a and b and in Table 12.5a and b, the binding rate coefficients k_1 and k_2 are given by

$$k_1 = (0.0615 + 0.1545)\, D_{f1}^{4.752 \pm 3.411} \quad (12.11a)$$

$$k_2 = (0.000489 \pm 0.000006) D_{f2}^{9.4094 \pm 0.08697} \quad (12.11b)$$

In the first case, there is scatter in the data. This is reflected in the error in the estimated value of the binding rate coefficient k_1. Only the positive error is presented since the binding rate coefficient k_1 cannot have a negative value. Only three data points are available. The availability of more data points would lead to a more reliable fit. The binding rate coefficient k_1 is very sensitive to the degree of heterogeneity or the fractal dimension, D_{f1} that exists on the biosensor surface as noted by the close to fifth (=4.752) order of dependence exhibited.

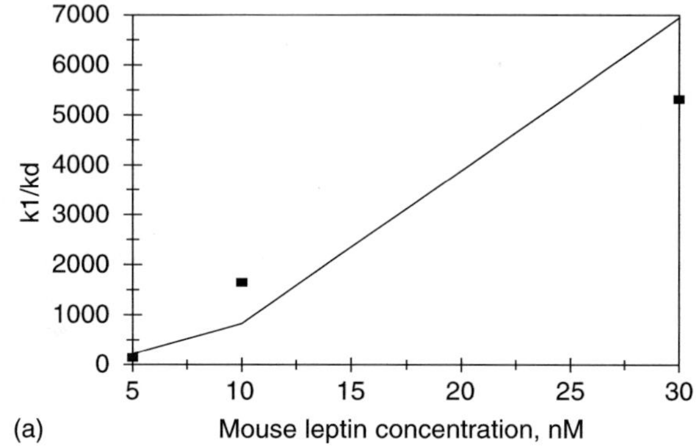

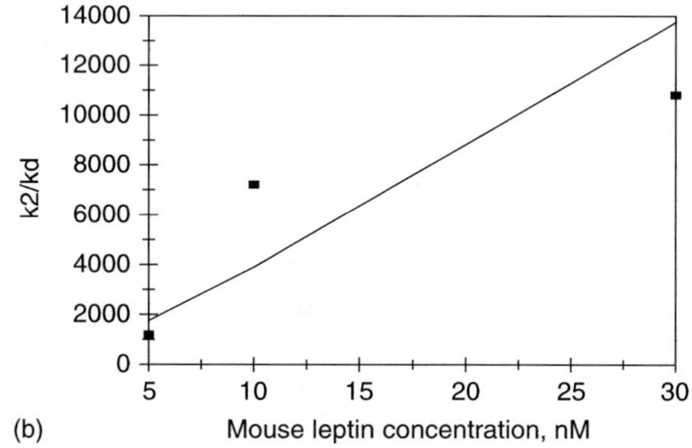

Figure 12.19 (a) Increase in the affinity K_1 with an increase in the mouse leptin concentration (5–30 nM) in solution. (b) Increase in the affinity K_2 with an increase in the mouse leptin concentration (5–30 nM) in solution.

In the second case, the fit is very good. The R^2 (sum of least squares) value is 0.9999. Only three data points are available. The availability of more data points would lead to a more reliable fit. Once again, the binding rate coefficient k_2 is very sensitive (extremely sensitive in this case) to the degree of heterogeneity or the fractal dimension D_{f2} that exists on the biosensor surface as noted by the higher than ninth (=9.409) order of dependence exhibited.

It would be of interest to see the influence of mouse and human leptin concentration in solution on the fractal dimension on the biosensor surface, and its subsequent influence on the binding rate coefficient k_1 for a dual-fractal analysis. Figure 12.21a, Table 12.4a and b, and Table 12.5a and b show the increase in the binding rate coefficient $k_1$with an increase in the fractal dimension D_{f1}. Once again, the data for human and mouse leptin are plotted on Figure 12.21a. For the data presented in Figure 12.21a, the binding rate coefficient k_1 is given by

$$k_1 = (0.01438 \pm 0.00734)\, D_{f1}^{6.258 \pm 0.9076} \qquad (12.12a)$$

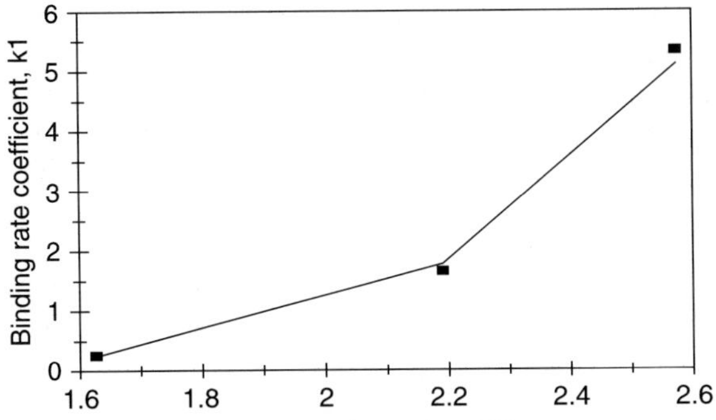

(a)

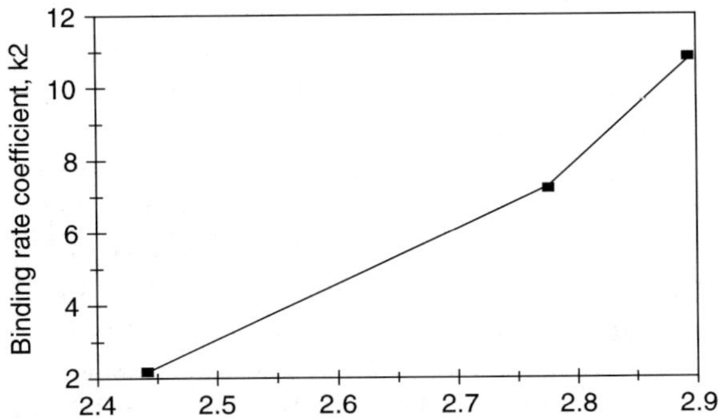

(b)

Figure 12.20 (a) Increase in the binding rate coefficient k_1 with an increase in the fractal dimension D_{f1}. (b) Increase in the binding rate coefficient k_2 with an increase in the fractal dimension D_{f2}.

The fit is very good in spite of the fact that data for human and mouse leptin are plotted together. Six data points are available. The fact that the fit is very good indicates that both the human and the mouse leptin concentrations affect the degree of heterogeneity on the biosensor surface in a similar fashion, and subsequently influence the binding rate coefficient k_1 too in a similar fashion. The binding rate coefficient k_1 is very sensitive to the degree of heterogeneity or the fractal dimension D_{f1} on the biosensor surface as noted by the higher than sixth (=6.258) order of dependence exhibited.

Figure 12.21b, Table 12.4a and b, and Table 12.5a and b show the increase in the binding rate coefficient k_2 with an increase in the fractal dimension D_{f2}. Once again, the data for human and mouse leptin are plotted in Figure 12.21b. For the data presented in Figure 12.21b, the binding rate coefficient k_2 is given by

$$k_2 = (0.00048 \pm 0.000219) D_{f2}^{9.1954 \pm 2.5204} \tag{12.12b}$$

The fit is reasonable. Six data points are available. The binding rate coefficient k_2 is extremely sensitive to the degree of heterogeneity or the fractal dimension D_{f2} on the

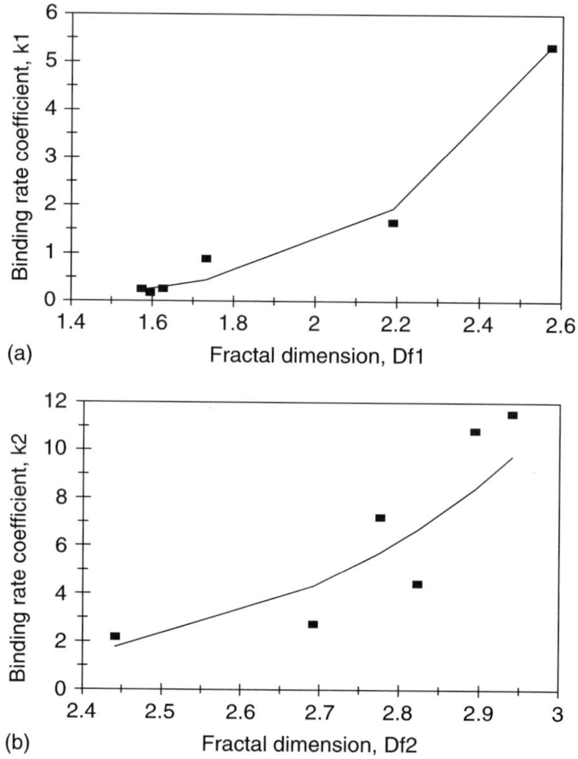

Figure 12.21 (a) Increase in the binding rate coefficient k_1 with an increase in the fractal dimension D_{f1}. (b) Increase in the binding rate coefficient k_2 with an increase in the fractal dimension D_{f2}.

biosensor surface as noted by the higher than ninth (= 9.1954) order of dependence exhibited.

Grote *et al.* (2005) indicate that though the SPR biosensor has been used frequently to analyze the binding and dissociation kinetics of different analyte–receptor systems, it has one major disadvantage, that is lack of structural information provided by the instrument. These authors indicate that matrix-assisted laser/desorption ionization (MALDI)-MS seems to be a perfect complement to the SPR biosensor. Grote *et al.* (2005) indicate that this technique not only does permit the determination of molecular masses, but it is also tolerant to hydrogel surfaces. These authors indicate that attempts have been made to integrate the SPR and the mass spectrometry (MS) technique synergistically (Williams and Addona, 2000; Lofas, 2003; Nedelkov and Nelson, 2003; Seymour, 2003; Zhukov *et al.*, 2002a, b, 2004).

Grote *et al.* (2005) have used the SPR technique in combination with MS to analyze the binding of human myeloid-related protein 14 (MRP14) in solution to antibody against human myeloid-related protein 14 immobilized on an IBIS instrument (IBIS Technologies, BV, Hengelo, The Netherlands). A TofSpec-2E (Micromass, Manchester, United Kingdom) was used for the MS. Figure 12.22a shows the binding of 0.1 mg/ml human myeloid-related protein 14 in 5 nM sodium acetate (NaAc) in solution to the antibody against human myeloid-related protein immobilized on a sensor chip surface. A single-fractal analysis is adequate to describe the binding kinetics. The values of the binding rate coefficient k and the fractal dimension D_f for a single-fractal analysis are given in Table 12.6.

Figure 12.22b shows the binding of 0.035 mg/ml human brain FABP to the rabbit polyclonal antibody against human brain FABP (anti-hB-FABP) immobilized on a sensor chip surface. Once again, a single-fractal analysis is adequate to describe the binding kinetics. The values of the binding rate coefficient k and the fractal dimension D_f for a single-fractal analysis are given in Table 12.6.

Serra *et al.* (2005) have recently indicated that the determination of ALP activity is important in the milk industry (Claeys *et al.*, 2002). ALP is an important parameter for pasteurization control, and it is also used for disease diagnostic purposes. Serra *et al.*

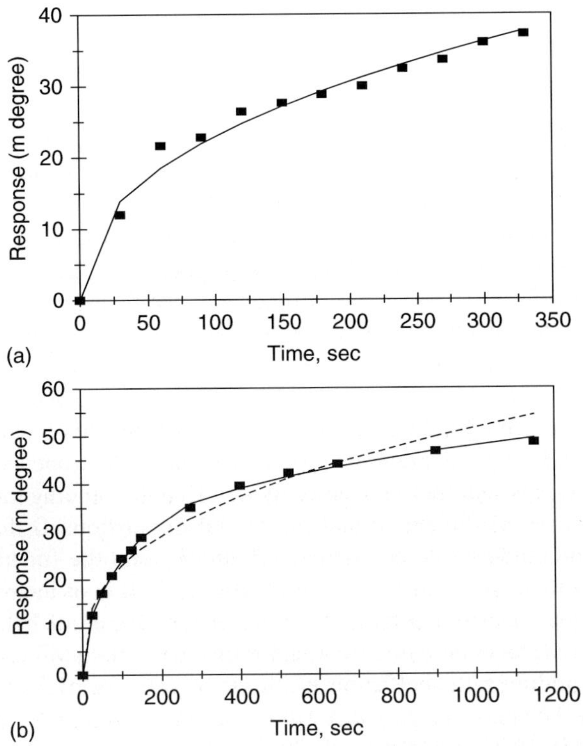

Figure 12.22 (a) Binding of 0.1 mg/ml human myeloid-related protein 14 in 5 nM sodium acetate (NaAc) in solution to the antibody against human myeloid-related protein immobilized on a sensor chip surface (Grote *et al.*, 2005). (b) Binding of 0.035 mg/ml human brain FABP to the rabbit polyclonal antibody against human brain FABP (anti-hB-FABP) immobilized on a sensor chip surface (Grote *et al.*, 2005). When both a dashed line (-------) and a solid (____) line are used, then the dashed line is for a single-fractal analysis and the solid line is for a dual-fractal analysis. In this case, the solid line provides the better fit.

Table 12.6

Binding rate coefficients and fractal dimensions for (a) human myeloid-related protein 14 (MRP14) in solution and (b) human brain FABP in solution to a SPR biosensor chip surface (Grote *et al.*, 2005)

Analyte in solution/ receptor on surface	k	k_1	k_2	D_f	D_{f1}	D_{f2}
0.1 mg/mL MRP14/ rabbit polyclonal antibody against human myeloid-related protein 14	3.3627 ± 0.0614	NA	NA	2.1664 ± 0.06586	NA	NA
0.035 mg/mL human brain FABP/rabbit polyclonal antibody against human brain fatty acid binding protein	4.5366 ± 0.2414	NA	NA	2.2954 ± 0.03626	NA	NA

(2005) further indicate that ALP has been used as a tracer in immunoassays (Che *et al.*, 2000; Ruan *et al.*, 2002; Abad-Villar *et al.*, 2000; Pemberton *et al.*, 2001).

Serra *et al.* (2005) emphasize that their electrode produces a sensitive detection of ALP. Phenyl phosphate (PP) is used as a substrate. ALP generates phenol from PP. Phenol is oxidized by the tyrosinase. These authors further indicate that this quinone (the oxidized product) is reduced to catechol at the electrode surface. This action results in a cycle between the tyrosinase substrate and the electroactive product. Serra *et al.* (2005) emphasize that the cycle amplifies the response and leads to a sensitive detection of ALP.

Figure 12.23a shows the binding of 25 μL of 0.01 M PP plus 5 μL of 2.5×10^{-6} M ALP in solution at pH 10 to the graphite–teflon–tyrosinase electrode (Serra *et al.*, 2005). A dual-fractal analysis is required to adequately describe the binding kinetics. The values of (a) the binding rate coefficients k and the fractal dimensions D_f for a single-fractal analysis, and (b) the binding rate coefficients, k_1 and k_2, and the fractal dimensions, D_{f1} and D_{f2}, for a dual-fractal analysis are given in Table 12.7. It is of interest to note that for a dual-fractal analysis, as the fractal dimension, or the degree of heterogeneity on the biosensor electrode surface increases by a factor of 1.61 from a value of D_{f1}=1.6750 to D_{f2} =2.6930, the binding rate coefficient increases by a factor of 3.516 from a value of k_1 =0.1499 to k_2 =0.5271.

Serra *et al.* (2005) indicate that phenyl phosphate can undergo a spontaneous hydrolysis, especially in an alkaline medium (Bauer *et al.*, 1996). Serra *et al.* (2005) analyzed the binding of 5.0×10^{-5} M phenyl phosphate plus 2.5×10^{-11} M ALP in solution to the graphite–teflon–tyrosinase electrode in the pH range 6.5–10.

Figure 12.23b shows the binding of 5.0×10^{-5} M PP plus 2.5×10^{-11} M ALP in solution at pH 6.5 to the graphite–teflon–tyrosinase electrode (Serra *et al.*, 2005). A single-fractal analysis is required to adequately describe the binding kinetics. The values of the

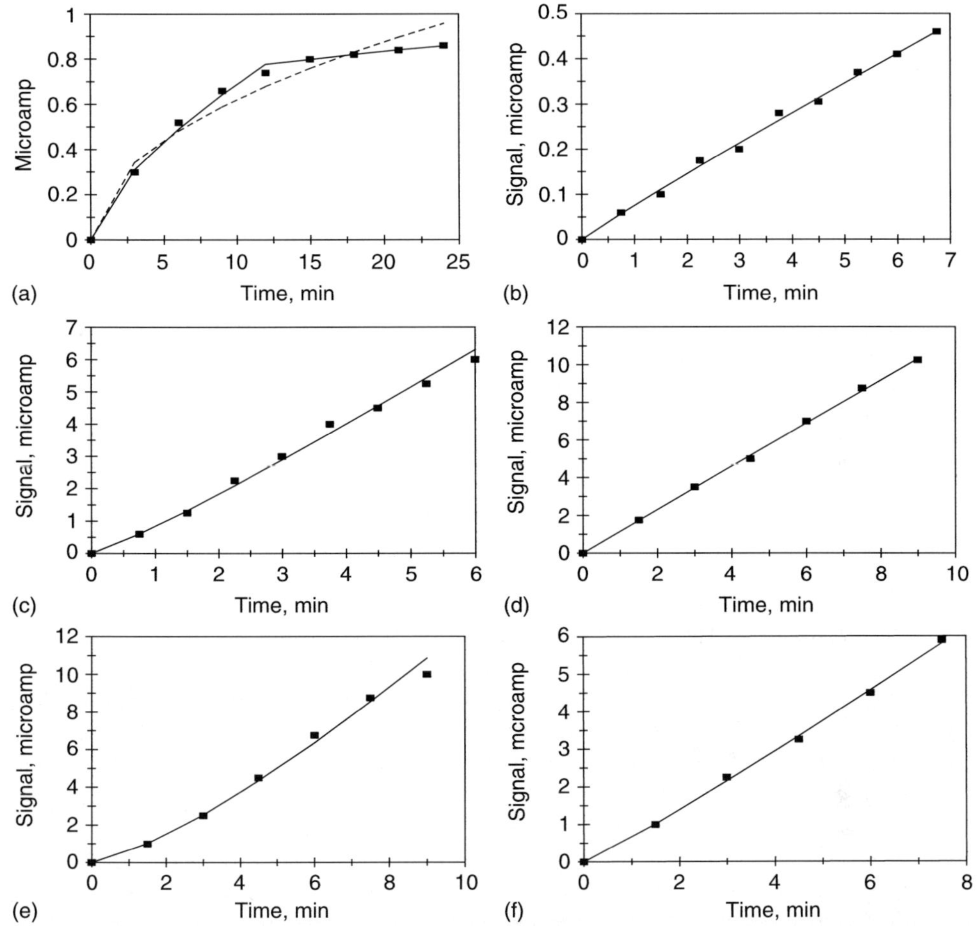

Figure 12.23 (a) Binding of 25 μl of 0.01 M PP plus 5 μl of 2.5×10^{-6} M ALP in solution at pH 10 to the graphite–Teflon–tyrosinase electrode (Serra *et al.*, 2005). Binding of 5.0×10^{-5} M PP plus 2.5×10^{-11} M ALP in solution at different pH to the graphite–Teflon–tyrosinase electrode (Serra *et al.*, 2005): (b) 6.5, (c) 7.5, (d) 8.5, (e) 9.5, and (f) 10

binding rate coefficient k and the fractal dimension D_f for a single-fractal analysis are given in Table 12.7.

Figure 12.23c shows the binding of 5.0×10^{-5} M PP plus 2.5×10^{-11} M ALP in solution at pH 7.5 to the graphite–Teflon–tyrosinase electrode (Serra *et al.*, 2005). Once again, a single-fractal analysis is required to adequately describe the binding kinetics. The values of the binding rate coefficient k and the fractal dimension D_f for a single-fractal analysis are given in Table 12.7.

Figures 12.23d–f show the binding of 5.0×10^{-5} M ALP at pH 8.5, 9.5, and 10.0, respectively, in solution to the graphite–Teflon–tyrosinase electrode. A single-fractal analysis is adequate to describe the binding kinetics in each of these cases. The values of

Table 12.7

Influence of pH on the binding rate coefficient and the fractal dimension for ALP in solution using phenyl phosphate as a substrate to a graphite–Teflon composite tyrosinase biosensor (Serra *et al.*, 2005)

Analyte in solution/graphite–teflon–tyrosinase biosensor electrode	k	k_1	k_2	D_f	D_{f1}	D_{f2}
(a) 25 μL of 0.01 M phenyl phosphate + 5 μL of 2.5×10^{-6} M ALP; pH 10/graphite–teflon–tyrosinase electrode	0.1988 ± 0.0213	0.1499 ± 0.0095	0.5271 ± 0.0012	2.0138 ± 0.1091	1.6750 ± 0.1180	2.6930 ± 0.10283
(b) 5.0×10^{-5} M phenyl phosphate + 2.5×10^{-11} M ALP; pH 6.5/graphite–Teflon–tyrosinase electrode	0.07542 ± 0.00481	NA	NA	1.1062 ± 0.0607	NA	NA
(c) 5.0×10^{-5} M phenyl phosphate + 2.5×10^{-11} M ALP; pH 7.5/graphite–Teflon–tyrosinase electrode	0.8426 ± 0.0493	NA	NA	0.7522 ± 0.6114	NA	NA
(d) 5.0×10^{-5} M phenyl phosphate + 2.5×10^{-11} M ALP; pH 8.5/graphite–Teflon–tyrosinase electrode	1.1659 ± 0.0260	NA	NA	1.1054 ± 0.02976	NA	NA
(e) 5.0×10^{-5} M phenyl phosphate + 2.5×10^{-11} M ALP; pH 9.5/graphite–Teflon–tyrosinase electrode	0.5962 ± 0.0348	NA	NA	0.3582 ± 0.07656	NA	NA
(f) 5.0×10^{-5} M phenyl phosphate + 2.5×10^{-11} M ALP; pH 10/graphite–Teflon–tyrosinase electrode	0.5630 ± 0.0223	NA	NA	0.8256 ± 0.0527	NA	NA

the binding rate coefficient k and the fractal dimension D_f for a single-fractal analysis in each of these cases is given in Table 12.7.

Figure 12.24a and Table 12.7 show for a single-fractal analysis the increase in the binding rate coefficient k with an increase in the fractal dimension D_f. For the data plotted in Figure 12.24a the binding rate coefficient k is given by

$$k = (0.9346 \pm 0.2391)\, D_f^{0.4924 \pm 0.2742} \tag{12.13}$$

There is scatter in the fit of the data. Only four data points are available. The binding rate coefficient k is only mildly sensitive to the fractal dimension D_f or the degree of heterogeneity on the electrode surface as noted by the slightly less than one-half (=0.4924) order of dependence exhibited. The estimated fractal dimension D_f values and the corresponding binding rate coefficient k values were obtained for the different pH values analyzed.

Figure 12.24b and Table 12.7 show the dependence of the binding rate coefficient k on the pH in the range 6.5–10. Very few data points are available. For the data points available, the highest value of the binding rate coefficient k is obtained at pH 8.5. It is perhaps instructive to model the increasing and decreasing segments of the binding rate coefficient k as a function of pH even though very few points are available.

In the pH range 6.5–8.5, Figure 12.24b shows that the binding rate coefficient, k for a single-fractal analysis exhibits an increase with pH and is given by

$$k = (3.7 + 4.4)\text{E-}10(\text{pH})^{10.376 \pm 4.101} \tag{12.14a}$$

There is scatter in the fit of the data. This is reflected in the error of the binding rate coefficient k. Only the positive error is presented since the binding rate coefficient cannot have a negative value. Only three data points are available. The availability of more data points would lead to a more reliable and better fit. In the pH range 6.5–8.5, the binding rate coefficient k is extremely sensitive to the pH as noted by the higher than tenth (=10.376) order of dependence exhibited. The non-integer order of dependence exhibited by the binding rate coefficient k on the pH lends support to the fractal nature of the system.

In the pH range 8.5–10, Figure 12.24b shows that the binding rate coefficient k for a single-fractal analysis exhibits a decrease with pH and is given by

$$k = (27969 \pm 4130)(\text{pH})^{-4.729 \pm 1.172} \tag{12.14b}$$

The fit is good. Once again, only three data points are available. The availability of more data points would lead to a more reliable and better fit. Note that the pH 8.5 point (the maximum value of k is exhibited here) is used for both the increasing and the decreasing sections of the binding rate coefficient k with pH. The binding rate coefficient k is very sensitive to the pH in the range 8.5–10 as noted by the negative order of dependence between 4½ and 5 (= −4.729) exhibited. The non-integer order of dependence exhibited by the binding rate coefficient k on the pH once again lends support to the fractal nature of the system.

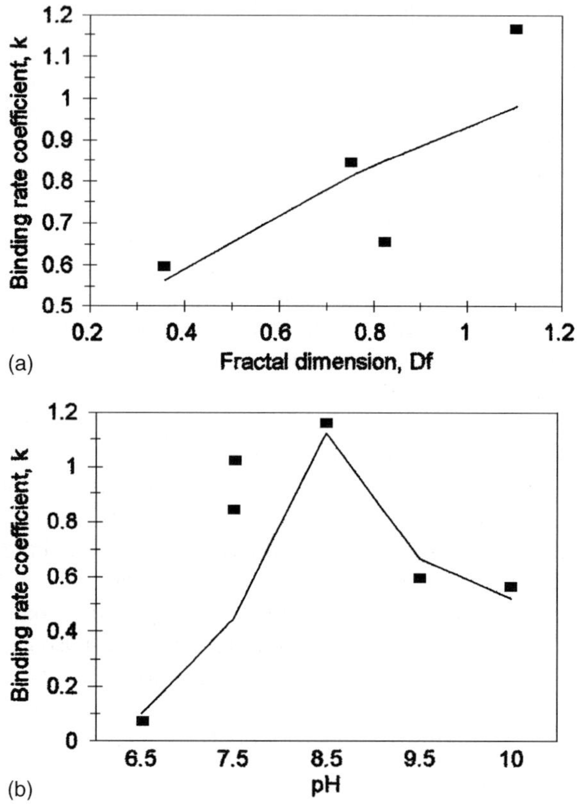

Figure 12.24 (a) Increase in the binding rate coefficient k with an increase in the fractal dimension D_f. (b) Change in the binding rate coefficient k with a change in the pH.

12.4 CONCLUSIONS

A fractal analysis is used to analyze the binding and dissociation (if applicable) of different compounds on biosensor surfaces. The systems analyzed include (a) ATP in solution (with and without the interferent, ascorbate) to a Pt microelectrode coated with a layer containing glycerol kinase plus glycol-3-phosphatase (Laiudet *et al.,* 2005), (b) human and mouse leptin (Ob) in solution to a leptin receptor (ObR) immobilized on a sensor chip surface (Mistrik *et al.,* 2004), (c) MRP 14 in solution to rabbit polyclonal antibody against MRP 14 immobilized on a sensor chip surface, and human FABP in solution to rabbit polyclonal antibody against human FABP immobilized on a sensor chip surface (Grote *et al.,* 2005), and (d) the influence of pH on ALP in solution to a graphite–Teflon composite biosensor using phenyl phosphatase as a substrate (Serra *et al.,* 2005).

The binding kinetics is described by either a single- or a dual-fractal analysis. A dual-fractal analysis is used only when a single-fractal analysis did not provide an adequate fit. This was done using Corel Quattro Pro 8.0 (1997). The fractal dimension provides a quantitative measure of the degree of heterogeneity present on the biosensor chip surface. Note that, as indicated in the different chapters in this book, the fractal dimension for the binding and dissociation phase, D_f and D_{fd}, respectively, is not a typical independent variable, such as analyte concentration, that may be directly manipulated. It is estimated from eqs. 12.1a–c, and one may consider it as a derived variable.

An increase in the fractal dimension value or the degree of heterogeneity on the surface leads, in general, to an increase in the binding and dissociation rate coefficient(s). For example, for the binding of ATP in solution in the pH range 6.5–8.0, and for a dual-fractal analysis, the binding rate coefficient k_2 exhibits greater than sixth (=6.582) order of dependence on the fractal dimension D_{f2} that exists on the ATP biosensor surface. This indicates that the binding rate coefficient k_2, in this case, is very sensitive to the degree of heterogeneity on the ATP biosensor surface.

Predictive relations are also developed, for example, for (a) the affinity K (= ratio of rate coefficients present in the binding and dissociation phases) as a function of the corresponding ratio of fractal dimensions present in the binding and in the dissociation phases, respectively, (b) the dissociation rate coefficient k_d as a function of the fractal dimension D_{fd} present in the dissociation phase, (c) the dissociation rate coefficient k_d as a function of pH in solution, (d) the ratio of the binding rate coefficients k_2/k_1 as a function of the ratio of the corresponding fractal dimensions D_{f2}/D_{f1} (Liaudet *et al.*, 2005), (e) the binding rate coefficient k_1 and k_2 as a function of the human leptin concentration in solution, (f) the binding rate coefficient k_2 as a function of the mouse leptin concentration in solution, and (g) the affinities K_1 (= k_1/k_d) and K_2 (= k_2/k_d) as a function of mouse leptin concentration in solution (Mistrik *et al.*, 2004). For the data analyzed for mouse and human leptin concentration in solution, the binding rate coefficients, k_1 and k_2, exhibit higher than sixth (=6.258) and ninth (9.1954) order of dependence on the fractal dimensions obtained when these concentrations are used, respectively. This indicates that, at least for this case, the binding rate coefficients are extremely sensitive to the degree of heterogeneity that exists on the biosensor surface.

The predictive relationships developed for the binding rate coefficients k_1 and k_2, and the dissociation rate coefficient k_d as a function of the fractal dimension or the degree of heterogeneity that exists on the surface, and for the affinities as a function of the ratio of the fractal dimensions in the binding and in the dissociation phase that exist on the biosensor surface are of considerable value because they directly link the binding and dissociation rate coefficients and affinity values to the degree of heterogeneity that exists on the biosensor chip surface. This provides a means by which these rate coefficients or affinity values may be manipulated by changing the degree of heterogeneity on the sensor chip surface. As far as affinity values are concerned this may require a little ingenuity since by changing the heterogeneity on the surface one may be affecting the heterogeneities for the binding and dissociation phases, simultaneously. Furthermore, note that, and as indicated above, in the different chapters in the book, these rate coefficients are, in general, sensitive or very sensitive to the degree of heterogeneity on the biosensor surface.

Quite a few different examples are presented in this chapter, emphasizing that the degree of heterogeneity that exists on the biosensor surface does significantly affect, in general, the rate coefficient and affinity values, and subsequently the kinetics in general. More such studies are required to determine whether the binding and dissociation rate coefficient(s), and subsequently the affinity values are sensitive to their respective fractal dimensions on the biosensor chip surface.

A better understanding of all the possible parameters that influence the kinetics of binding and dissociation of different analyte–receptor systems on biosensor surfaces is critical. This will be of considerable assistance, for example, to help select the correct drug of choice from a list of possible candidates. More often than not, the influence of diffusion and heterogeneity on the biosensor surface on the kinetics is neglected. As indicated in this chapter and in the book, the degree of heterogeneity significantly influences the binding as well as the dissociation kinetics occurring on biosensor surfaces. It would behoove the practicing biosensorists to start beginning pay more attention to this aspect of kinetics on biosensor surfaces. One may perhaps argue that the influence of diffusional limitations may be minimized or perhaps even be eliminated if the biosensor is run properly. But, the influence of heterogeneity on the biosensor surface should be included in the kinetic analysis of analyte–receptor reactions occurring on biosensor surfaces. Ideally, one would like to separate the influence of diffusional limitations and heterogeneities, on the biosensor surfaces, so that they may be controlled and their influence better analyzed. Note that the present analysis presented in this chapter and throughout the book does not do that.

REFERENCES

Abad-Villar, E.M., Fernandez-Abedul, M.T., and Costa-Garcia, A. (2000). Flow injection enzyme immunoassay based on the use of gold bands. *Analytica Chimica Acta*, **409**, 149–158.

Adriaensen, D., and Timmerhaus, J.P. (2004). *Current Opinion in Pharmacology*, **4**, 207–214.

Amal, R., and Raper, J.A. (1993). *Particle Particle Systems Character*, **10**, 239–245.

Bauer, C.G., Eremenko, A.V., Ehrentreich-Fortser, E., Bier, F.F., Makower, A., Halskall, H.B., Heinemann, W.R., and Scheller, F.W. (1996). Zeptomole-detecting biosensor for alkaline phosphates in an electrochemical immunoassay for 2,4-dichlorophenoxyacetic acid. *Analytical Chemistry*, **68**(15), 2453–2458.

Bohr, H., Kuhle, K., Sorenson, A.H., and Bohr, J. (1997). *Z. Physik* D, **40**, 513–515.

Bowser, D.N., and Khakh, B.S. (2004). *Journal of Neuroscience*, **24**, 8606–8620.

Che, Y.H., Li, Y.B., Slavik, M., and Paul, D. (2000). Rapid detection of *Salmonella typhimuriumin* in chicken carcass wash-water using an immuno electrochemical method. *Journal of Food Protection*, **63**, 1043–1048.

Claeys, W.L., Van Loey, A.M., and Hendrix, M.E. (2002). Intrinsic time temperature integrators for heat treatment of milk. *Trends in Food Science and Technology*, **13**, 293–311.

Cohen, B., Novick, D., and Rubinstein, M. (1996). Modulation of insulin activities by leptin. *Science*, **274**, 1185–1188.

Cook, S.P., and McClesky, E.W. (2002). *Pain*, **95**, 41–47.

Cook, S.P., Vulchanova, L., Hargreaves, K.M., Elde, R., and McClesky, E.W. (1997). *Nature*, **387**, 505–508.

Corel Quattro Pro, 8.0. (1997). Corel Corporation, Ottawa, Canada.

Cunningham, M.J., Clifton, D.K., and Steiner, R.D. (1999). Leptin's action on the reproductive axis: perspective and mechanisms. *Biology of Reproduction*, **60**, 216–222.

Gourley, P.L. (1988). Nanolasers, *Scientific American*, (March), 56–61.

Gourley, P.L., and Sasaki, D.Y. (2001). Biocavity lasers. *Am. Sci.*, **89**, 152–159.

Gourley, P.L. (2003). Biocavity laser for high-speed cell and tumor biology. *Journal of Physics D: Applied Physics*, **21**, R228–R239.

Gourley, P.L. (2005). Brief overview of BioMicroNano Technologies. *Biotechnology Progress*, **21**, 2–10.

Gourley, P.L., McDonald A.E., Hendricks, J.K., Copeland, R.G., Hunter, J., Akhil, O., Dunne, J.L., Skirboll, S.L., Nihlen, L., and Capps, H.H. (2000). Nanolaser/microfluidic biochip for real time tumor pathology. *Biomedical Microdevices*, **2**, 111–122.

Grosskinsky, S., Timme, M., and Naudorf, B. (2002). Universal attractors of reversible aggregate-reorganization processes. *Physics Reviews Letters*, **88**, 24501–24504.

Grote, G., Dankbar, N., Gedig, E., and Koening, S. (2005). Surface plasmon resonance/mass spectrometry interface. *Analytical Chemistry*, **77**, 1157–1162.

Havlin, S. (1989). Molecular diffusion and reaction. In: *The Fractal Approach to Heterogeneous Chemistry: Surfaces, Colloids, Polymers*, (ed. D. Avnir), Wiley, New York, pp. 251–269.

Haynes, W.G., Morgan, D.A., Walsh, S.A., Mark, A.L., and Sivitz, W.I. (1997). Receptor-mediated regional sympathetic nerve activation by leptin. *Journal of Clinical Investigation*, **100**, 270–278.

Hirschberg, A.L. (1988). Hormonal regulation of apetite and food uptake. *Annals of Medicine*, **30**, 7–20.

Katsu, T., Yang, X.R., and Rechnitz, G.A. (1994). *Analytical Letters*, **27**, 1215–1224.

Kieffer, T.J., Heller, R.S., and Habener, J.F. (1996). Leptin receptors expressed on pancreatic β-cells. *Biochemical and Biophysical Research Communications*, **224**, 522–527.

Kitakaze, M., and Hori, M. (2000). *Expert Opinion and Investment in Drugs*, **9**, 2519–2535.

Kunapuli, S.P., Ding, Z., Dorsam, R.T., Murugappan, S., and Quinton, T.M. (2003). *Current Pharmaceutical Design*, **9**, 2303–2316.

Lee, C.K., and Lee, S.L. (1995). Multi-fractal scaling analysis of reactions over fractal surfaces. *Surface Science*, **325**, 294.

Liaudet, E., Sonja, H., Droniou, M., and Dale, N. (2005). Microelectrode biosensor for real-time measurement of ATP in biological tissue. *Analytical Chemistry*, **77**, 3267–3273.

Lofas, S. (2003). *American Biotechnology Laboratory*, **21**, 16F–16G.

Magazu, S., Maisano, G., and Mallamace, F. (1989). *Physics Review* A, **39**, 4195–4200.

Miele, M., and Fillenz, M.J. (1996). *In vivo* determination of extracellular brain ascorbate. *NeuroscienceMethods*, **70**, 15–19.

Mistrik, P., Moreau, F., and Allen, J.M. (2004). BiaCore analysis of leptin–leptin receptor interaction: evidence of 1:1 stoichiometry. *Analytical Biochemistry*, **327**, 271–277.

Molina-Bolivar, J.A., Galisteo-Gonzalez, F., and Hidalgo-Alvarez, F. (1989). Cluster morphology of protein-coated colloids. *Journal of Colloid & Interface Science*, **298**, 445–454.

Murphy, L.J., and Galley, P.T. (1994). Measurement *in vitro* of human plasma glycerol with a hydrogen peroxide detecting microdialysis enzyme electrode. *Analytical Chemistry*, **66**, 4345–4353.

National Science Foundation Program Solicitation. (2005). NSF 05-526, Sensors and Sensor Networks (Sensors), National Science Foundation Funding Opportunity Document, National Science Foundation, Arlington, Virginia, USA, Proposals due March 3.

Nedelkov, D., and Nelson, R.W. (2003). *Trends in Biotechnology*, **21**, 301–305.

Pemberton, R.M., Hart, J.P., and Mottram, T.T. (2001). An electrochemical immunosensor for milk progesterone using a continuous flow system. *Biosensors & Bioelectronics*, **16**, 715–723.

Ramakrishnan, A., and Sadana, A. (2001). A single-fractal analysis of cellular analyte-receptor binding kinetics utilizing biosensors. *Biosystems*, **59**(1), 35–45.

Rice, M.E. (2000). *Trends in Neuroscience*, **23**, 209–216.

Ruan, C., Wang, H., and Li, Y. (2002). A bienzyme electrochemical biosensor coupled with immunomagnetic separation for the rapid detection of *Escherichia coli* 015:H7 in food samples. *Transactions of ASHRAE*, **45**, 249–255.

Sadana, A. (2001). A fractal analysis approach for the evaluation of hybridization kinetics in biosensors. *Journal of Colloid and Interface Science*, **234**, 9–18.

Sadana, A., and Chen, Z. (1996). A fractal analysis of the influence of non-specific binding on antigen-antibody binding kinetics for biosensor applications. *Biosensors and Bioelectronics*, **11**, 769–781.

Sarkar, S., Lee, L.K., Hart, S.L., Hailes, H.C., Levy, S.M., Tabor, A., and Shamlou, P.A. (2005). The fractal structure of polycation–DNA complexes. *Biotechnology and Applied Biochemistry*, **41**, 127–136.

Schlenk, J.O., Miller, E., Gaddis, R., and Adams, R.N. (1982). Homeostatic control of ascorbate concentration in CNS extracellular fluid. *Brain Research*, **253**, 353–356.

Schweibert, E.M., and Kishore, B.K. (2001). *American Journal of Physiology and Renal Physiology*, **280**, 943–965.

Serra, B., Marales, M.D., Reviejo, A.J., Hall, E.A.H., and Pingarron, J.M. (2005). Rapid and highly sensitive electrochemical determination of alkaline phosphatase. *Analytical Biochemistry*, **336**, 289–294.

Seymour, C. (2003). *Proteomics*, **3**, 809–810.

Shamlou, A.P., Stavrinides, S., and Titchener-Hooker, N.J. (1996). *Bioprocess Engg.*, **14**, 237–243.

Sim, J.A., Young, M.T., Sung, H.Y., North, R.A., and Surprenant, A. (2004). *Journal of Neuroscience*, **24**, 6307–6314.

Stamford, J.A., Kruk, Z.L., and Millar, J. (1984). Regional differences in extracellular ascorbic acid levels in the rat brain determined by high speed cyclic voltametry. *Brain Research*, **299**, 289–295

Williams, C., and Addona, T.A. (2000). *TIBTECH*, **18**, 45–48.

Zhukov, A., Buijs, J., and Suckau, D. (2002a). *Biochemist*, **24**, 21–23.

Zhukov. A., Suckau, D., and Buijs, J. (2002b). *American Biotechnology Laboratory*, **20**, 10, 12.

Zhukov, A., Schurenberg, M., Jannson, O., Malmqvist, M., and Roepstorff, P. (2004). *Journal of Biomolecular Technology*, **15**, 112–119.

– 13 –

Market Size and Economics for Biosensors

13.1 INTRODUCTION

The application of biosensors in different areas recently has experienced a substantial growth resulting in an increasing demand and projections of market size and emphasis on economics. There is continual pressure to decrease the biosensor cost for the different applications and make them competitive with current best available technology (BAT) detection systems. A recent book on the commercial aspects of biosensors is available from the Institute of Bioscience and Technology at Cranfield University, Silsoe, United Kingdom (Newmann *et al.*, 2004). This is a seminal book and Professor A. Turner and his colleagues were one of the first to place and introduce biosensors as a discipline to a wider audience. These authors carefully analyze the glucose biosensor market as a discipline and compare this with other emerging applications for biosensors. This is an expensive book and costs $1200 or €1000. This (expense) is a common trend in these types of books or reports as indicated later on in the text.

Quite a few reports are available that project the biosensor market in the different areas. The projections are generally from 5 to 10 years since the inception of these reports. More often than not, these reports are expensive (in the range of thousands of U.S. dollars), and understandably so since considerable effort is taken by the respective authors to put these types of financial and marketing reports together. These reports are perhaps affordable only by organizations and not by single individuals. For example, the Freedonia report #1547 (2002) on Chemical Sensors provides forecasts from the year 2006 to 2011. This report indicates that the United States demand for gas sensors is estimated to be $2.7 billion in the year 2006. Although this report is relatively old, it costs $3700. The report predicts that the biosensor prices will continue to fall due to (a) strong competition, (b) increased economies of scale, and (c) improved manufacturing technology. Furthermore, the report indicates that optical sensors will grow at the fastest rate amongst the different sensor types. The detection of glucose for the effective management of diabetes is still the primary driving force for biosensor sales.

Other studies by the Freedonia group are also available. For example, the U.S. market for *in vitro* diagnostics (IVD) (Freedonia Report # 1424, 2001) stood at $9 billion in the year 2001, when the report was written. This report also costs $3700. Both these reports by the Freedonia group provide historical data for the years 1992, 1996, and 2001, and projections for the years 2006 and 2011. Other reports from the Freedonia group are also available (2005a,b).

Frost and Sullivan (2005a) in a very recent report (cost $5500) have highlighted the advances in point-of-care (POC) technologies. This report emphasizes that the POC industry providers need to indicate that POC testing is not less accurate, not more expensive, and not more difficult to manage than testing in central laboratories. The report emphasizes that this POC industry needs to demonstrate these aspects to the end users (physicians and patients) of their products besides their cost-effectiveness.

In another report by Frost and Sullivan (2005b) the authors analyze the world biosensors market and indicate that companies are investing in biosensors R&D in anticipation of emerging applications for these biosensors. This report does emphasize that the high costs of the development of biosensors does present a challenge, especially since the market, besides that for glucose, is very small. Furthermore, quite a few 'big' and established players are already in the glucose detection market, and inroads into this area would be difficult unless one comes out with a novel strategy, or there is a 'quantum jump' as far as the 'patient' is concerned, such as non-invasiveness or substantially less pain and inconvenience are involved. The above-mentioned report does emphasize that in order to speed up commercialization of biosensors, better development strategies are essential. Finally, the report indicates that applications of biosensors to non-biological applications are beginning to appear. This is a step in the right direction.

A very recent report by Takeda Pacific (2005), provides estimates of the global market for medical biosensors, and emphasizes diabetes management and POC-related applications. This report indicates that the biosensor market for diabetes was $7.1 billion in the year 2004. This market was estimated to grow by 9.7% till the year 2008. This report emphasizes that the increase in obesity and the subsequent disturbing increase in diabetics is the driving force for better glucose control and diabetes management devices. According to this report the medical biosensor market was estimated to be in $billion 7.79, 8.55, 9.38, and 10.3 for the years 2005, 2006, 2007, and 2008, respectively. This report was careful to analyze the commercial and research development activities on biosensors at academic institutions and at public and private organizations. This is an 80-page report and costs $1995.

A $4995 report by Kalorama Information (2004) on the United States market for over-the-counter (OTC) diagnostics analyzes the promising OTC market. The testing kits analyzed include glucose, pregnancy, blood pressure, ovulation prediction, urine, and lancets. Market size and growth estimates are provided till the year 2008. Information was also provided for the testing of (a) drugs of abuse, (b) HIV testing at home, (c) cancer screening at home, and (d) others. Factors such as market trends, regulatory trends, and market share movements were also considered in the forecast projections.

Kalorama Information (2005a) have also come out with a recent report on the United States markets on Analytical Chip technology. This 480-page, $3500 report analyzes the market for cell, tissue, probe, and microbiological microarrays. This is the second edition of this report by this company. Kalorama Information (2005b) in a recent report have also estimated the worldwide market for cancer diagnostics. These authors state that according to the World Health Organization (WHO) in the year 2003, there were estimated to be 25 million people living with cancer. This number according to WHO is increasing by 11 million new cases each year. Thus, in the years 2006, 2007, and 2008, the estimated people living with cancer are $25+3\times11 = 58$, 69, and 80 million, respectively. According to the Kalorama Information report (2005b), IVD should play a significant role in matching the drug to cancer and subsequently to the patient. Furthermore, the detection device (or IVD) could play a critical role in monitoring the action of the drug on the disease.

The above-mentioned report mentions that there are a couple of drawbacks or handicaps in *in vitro* testing: (a) high cost of $350–3500 and (b) competition to penetrate the traditional clinical market which includes tissue biopsies and serum assays. The report anticipates a continually changing dynamics between the traditional array methods and the present or subsequently to-be-developed IVD devices. This 328-page report, like the ones mentioned above is expensive and costs $3500.

Frost and Sullivan (2005c) in a recent strategic analysis of the European Molecular Diagnostics market indicate that molecular diagnostics is the fastest growing market in the IVD industry in Europe. These authors define molecular diagnostics to include the detection of infectious diseases, cancer, and pharmacogenomics. Finally, this is a very expensive report that costs $5450.

Rajan (2002) in an old report estimates the United States (U.S.) market for biosensors at $1.82 billion for the year 2006. This author estimated the market at $1.32 billion for the year 2001, with an average annual growth rate of 6.6%. Thus, in the years 2002, 2003, 2004, 2005, and 2006 the estimated U.S. market for biosensors is/was in $billion 1.41, 1.50, 1.60, 1.70, and 1.82, respectively. The numbers have been rounded off to three decimal places. Let us compare the U.S. market in the year 2003 ($1.5 billion) with the worldwide market size for biosensors estimated to be $7.3 billion (Goliath Industry Information, 2004). This indicates that in the year 2003, the U.S. fraction of the worldwide market for biosensors was $1.5/7.3 = 20.5\%$. Incidentally, a Fuji-Keizai, USA, Inc. report (2004) also estimated the worldwide biosensor market at $7.3 billion for the year 2003. This report also projected the biosensor market to grow at an annualized rate of 10.4% to (in $billion) 8.06, 8.90, 9.82, and 10.8 in the years 2004, 2005, 2006, and 2007, respectively. Both the Fuji-Keizai (2004) and the Goliath (2004) reports have a publication date of April 2004.

Zarkoff (2002) in an old report had estimated the average annual growth rate (AAGR) of 7.5% for the biosensor market. This author had analyzed the following market segments for biosensor applications: agriculture, medical analysis, food monitoring, high-throughput screening, and nanobiotechnology. In the year 2001, the highest to lowest market estimated by this author was in the following sequence: High-throughput screening, medical analysis, agriculture, food monitoring, and nanobiotechnology. For the year 2006, Zarkoff (2002) predicted the highest to lowest market for biosensors in the following sequence: High-throughput screening, medical analysis, agriculture, nanobiotechnology, and food monitoring. Considering the effort and resources being spent worldwide it is not surprising that nanobiotechnology has climbed from the last place to the fourth place, displacing food monitoring to the last place. Once again, considering the recent realization of the importance of nanobiotechnology one may reasonably anticipate nanobiotechnology to play a bigger role in biosensors and nanobiosensors. This is consistent with the miniaturization required to facilitate the commercialization of biosensors.

The-Infoshop (2002) in a report on Biosensors and Bioelectronics with a publication time (or application) of the years 2002–2009 mentions the thought leaders in the United

States in this area that include the Oak Ridge National Laboratory, Ames Laboratory, The Cornell Center for Nanobiotechnology, and the Sandia National Laboratory, Albuquerque. The report outlines the reasons for doing this report, its methodology and information sources, along with the analyst credentials. The last point is of interest since market forecasts and projections as indicated elsewhere in this chapter is not a 'sure thing.' The analyst's credentials should help in increasing the reliability of the report. The report also briefly mentions the technological obstacles for biosensor development that include production of materials, nonspecific binding, longevity of materials, and bioelectronics. It also describes the effect of current events on the markets, and the effect of different parameters such as manufacturing, microfluidics, non-invasive devices, and automation on the future market potential of biosensors. Emphasis is also placed on the miniaturization of components. The report analyzes the development of biosensors in the following areas: agriculture, high-throughput screening, medical analysis, glucose monitoring devices, diagnostics, and food monitoring.

The detection of nerve agents in real-time is a very possible threat, and there is a global market demand for first responders. Governments around the world are concerned about this aspect of detection. Louie (2002) in an excellent lecture entitled "Commercial Development of Biosensors—The Business of Science," presented at the *Conference on Research, Technologies, and Applications in Biodefense* emphasizes that the Commercial development implies transforming a technological approach to a product offering. This author indicates that the global market for nerve agent detection in real time is $10 million in the year 2001. Assuming a conservative annual increase of 10% in the market for this type of biosensor, the market projections for the years 2006, 2007, 2008, 2009, and 2010 are $million 16.1, 17.7, 19.5, 21.4, and 23.6, respectively. Needless to say even a single actual nerve agent attack anywhere in the world would skyrocket the annual increase for real-time detection devices for nerve agents. Perhaps, a biosensor platform needs to be designed that could be modified to detect the different nerve agents. A successful and modifiable biosensor platform would considerably assist in reducing the costs of these types of biosensors.

It is of interest to get an order of magnitude estimate of the surveillance cost (using biosensors) to protect a city in the United States or even the entire land area in the United States. Dasey (2005) in (Finding the Bugs, http://www.ndu.edu/ctnsp/biosensing_02.htm, reference 22) indicates that to protect the city of Washington, DC (estimated area equal to 180 square kilometers) approximately 1000 sensors are required in a surveillance mode. JASON (2005) in (Finding the Bugs, http://www.ndu.edu/ctnsp/biosensing_02.htm, reference 21) indicates that to protect the entire land area of the United States with approximately one sensor per kilometer an approximate annual cost of $10–15 billion is estimated. Clearly, the above-mentioned report indicates that at the present state of technology, biosensors may be economically applied only at high-valued areas. Apparently, a cheaper method is required to economically permit the surveillance mode in wider range of areas. Perhaps, a combination of biosensors and other detection devices need to be employed for better, more reliable, and economic surveillance over a wider range of areas.

Kharif (2003) in Finding the Bugs, http://www.ndu.edu/biosensing 02.htm, indicates that with the gradual increase in the threat of a possible attack on U.S. cities (as evidenced by the threat color code issued by the Department of Homeland Security on a daily basis), the number of biosensors that may be used in the surveillance mode is also gradually increasing. Officials at the Department of Homeland Security (DHS) estimate that the present cost of sensors in a surveillance mode in cities is around $2 million per year per city. The major cost involved here is labor, wherein the filter papers from the air samplers are to be taken to the laboratories for the detection of the possible presence of pathogens. A wireless transmission of this type of filter paper data to testing laboratories or a centralized location would very significantly assist in the decrease of the biosensor surveillance cost. This would also help eliminate human error involved.

In case the filter paper comes out positive for the possible presence of pathogens, then this centralized location or testing laboratory could alert the local authorities. Obviously, there is a need for better communication between the emergency services involved at the local, state, and national level, and a chain-of-command that comes into play when a possible threat situation is detected. The lack of this for a different type of situation, emergency, however (e.g. The Katrina, a Category 5 hurricane, episode in, New Orleans, Louisiana, U.S.) was clearly demonstrated. In this case, the emergency situation was not immediately life threatening, but did involve the movement of a large number of individuals from a serious situation to a safer place. As expected, the time factor is going to be much smaller if, and one hopes that does not happen, a possible biological threat is sensed or created. Ideally, one would prefer to have a very sensitive surveillance biosensor system installed that permits the most possible time available to move individuals from 'harm's way.' However, this may involve some compromise such as the increase of false positives. Moving populations from cities, and then discovering that this was a false alarm, is going to be not only very expensive but also exceedingly frustrating.

The Venture Planning Group (2005) have come out with a new worldwide strategic business outlook book entitled "Biosensors 2006: U.S., Europe, and Japan." In that the authors forecast the biosensor market for the years 2005, 2010, and 2015 for the different geographic locations that include the United States, Europe, Japan, and other countries. They also discuss new product development opportunities with significant market potential. These authors also carefully discuss the handicaps and risks involved in the different biosensor application areas that include agriculture, biotechnology, diagnostics, defense, environmental monitoring, and food and beverage. Lists of companies that market biosensor-based products are also included. Finally, these authors present the names of private, governmental, and university-based research groups where significant biosensor research is presently being conducted. These authors emphasize that one of the challenges facing biosensor manufacturers in the next decade is integrating new technology planning with business and corporate strategies to help minimize the cost of manufacturing these biosensors. This cost-conscious environment is bound to produce increasing pressure on individuals involved in the manufacture of biosensors, especially if these biosensors are to have a chance to compete with existing and well-established detection technologies used in laboratories and elsewhere.

In this chapter, the different aspects and the steps that may be involved in the market size and economics of biosensors are reviewed. Such type of information is not available free of charge in the open literature. As expected, the different biosensor companies will continue to guard this type of economic information very carefully. Furthermore, market projections and forecast by their very 'definition' provide estimates only. The accuracy of these estimates is based on who is making these estimates, what are the credentials of the individual or group writing the report, what assumptions have been used to make these estimates, and how much faith and trust may be placed on these estimates. With this caveats in mind this chapter examines (a) biosensor trends and collaborations between companies, universities, and state and governmental organizations; (b) the commercial development of biosensors including the handicaps and hindrances involved; (c) the biosensor technologies and techniques that are being developed and which exhibits promise; (d) examples of biosensor companies (small, medium, and large) that are successful, the number of employees and sales; (e) factors that could help increase biosensor markets, and (f) examples of size of grants from different sources required to help develop a biosensor for a particular application. Details gathered from the open published literature and from internet sources (there may be a reliability factor here) will be examined, analyzed, and placed in the above-mentioned categories.

It is hoped that this chapter will provide the readers with an initial estimate and rationale of market size and economics of biosensors for their specific application. If one may be allowed to draw an analogy between the solution of a non-linear problem by an iterative solution technique, which requires an initial solution or estimate, then similarly the information presented in this chapter should hopefully provide the means for an initial estimate of the market size for the different (or one's specific) biosensor applications. A healthy dose of an individual's experience if applied appropriately should greatly enhance in providing for more accurate estimates of the biosensor market size and economics for biosensors for one's particular case.

13.2 BIOSENSOR TRENDS AND COLLABORATIONS BETWEEN COMPANIES AND STATE AND GOVERNMENTAL AGENCIES

In the state of Indiana, United States, major private and academic institutions got together on April 6, 2005 for the second Indiana Biosensor Symposium to raise awareness in the state with regard to this emerging technology and foster collaborations. (Roche Diagnostics, Press Release, 2005). The event was attended by researchers, venture capitalists, and entrepreneurs. Research being performed at the different companies and at academic institutions was highlighted. The participants included Roche Diagnostics, the University of Notre Dame, Purdue University, Purdue's Discovery Park, Indiana University School of Medicine, Indiana University, and the Rose-Hulman Institute of Technology. A program specialist in the division of SBIR/STIR from the National Institutes of Health (NIH) was also present. This program provides technology start-ups in the nation with the much-needed financial support during the early stages of development.

Indiana's Life Science Initiative, Biocrossroads, had as early as the year 2002 identified biosensors as one of the eight key science areas that exhibited the most potential for business formation activities. This report (Roche Diagnostics, Press Release, 2005) emphasizes that since biosensors can accommodate a wide range of therapeutic and diagnostic tasks to enhance human health, biosensors are poised to be one of the rapidly growing areas in the life sciences.

The above-mentioned report also indicates that a Biosensor Maturity Assessment made by IBM in the year 2004 indicated that the worldwide biosensors market is estimated to be \$9.4 billion. In the year 2004, the Diagnostics Division of Roche had sales of 7.8 billion Swiss Francs. Biosensors is just one section of the total diagnostic sales. One Swiss Franc = \$0.7570 U.S. (Exchange rate December 1, 2005). Thus, 7.8 billion Swiss Francs is equal to \$5.90 billion U.S.

A similar collaborative arrangement between public and private institutions called BEACON was formed in the State of Connecticut (CASE reports, 1999). BEACON stands for Biomedical Engineering Alliance for Central Connecticut. The participants included Trinity College, the University of Hartford, the University of Connecticut, and the University of Connecticut Health Center. One of the main goals of BEACON is to increase university and industry collaboration in the fields of biomechanics and biosensors/bioinstrumentation. The basic idea behind BEACON is to increase educational opportunities for biomedical engineering students. These authors note that a similar arrangement based on the BEACON model is set up in Southern Germany. This includes the University of Freiberg Medical School, the University of Applied Sciences at Furtwangen and Sigmaringen. Finally, though a biomedical program/discipline/department is cost intensive for universities, it should help alleviate the decrease in student enrolment faced by quite a few universities in the United States, especially those that are not in top third. Thus, there may be a double-benefit here: a much-needed increase in engineering enrolment at U.S. universities, and 'tapping into' an emerging area of business opportunities.

Finally, at the first BEACON-sponsored symposium on biosensors the reasons that were hindering the commercialization of biosensors were discussed. These included cost, stability, and sensitivity. Development cost and ease of manufacture were also examined. The primary determinant was small market size and justification. Development cycles were also considered to be too long, and more interdisciplinary interactions were required to aid in the successful development of biosensors.

Roth (2005) also very recently indicates that high development costs and other challenges are limiting the use of biosensors. However, as indicated above, miniaturization, improved transducer technology, and increased sensitivity should assist in the commercialization of biosensors.

The German microfluidics company THXXS (DrugResearcher.com, 2003) has developed a novel modular microfluidic construction kit. This is a lab-on-a-chip (LOC) device that permits researchers to configure required chip functions. The company (THXXS) indicates that their injection molding process inexpensively produces disposable chips, cards, or discs. These form a large part of the commercial manufacturing process. Their LOC devices are easier to manufacture, and permit the development of lower priced and more flexible biosensors. The company, THXXS emphasizes that their microfluidics kit is targeted at the biosensor industry and at drug discovery detection devices, and is particularly suited to meet the ever-increasing market pressure to reduce high development costs.

Israeli biotechnology companies and companies with similar interest in the State of Virginia, United States met at a VITAL (Virgina-Israel Technology Alliance) summit held in Roanoke, Virginia, United Sates (Diamond, 2002). Israel's progress in science and in converting this science to real-life products was acknowledged. These biotechnology products could be developed, manufactured, and marketed. For example, Israel's leadership in biotechnology and physics has led to the biosensor system design (BSD). This novel system is based on electrostatic interaction, and may be used to detect pathogens in air, food, and water as well as in medical testing.

The above are only three examples that indicate the rapid evolution of collaborations or clusters where work on biosensors may be rapidly enhanced to help develop faster, better, more reliable, cheaper, and more efficient biosensors as a whole. In the United States, focused areas such as the Tennessee Technology Corridor in the State of Tennessee, The Research Triangle Park in North Carolina, and clusters in the East and West Coast of the United States are rapidly developing on biosensors and in the Life Sciences to foster the advancement of emerging technologies. Surely, such clusters for biosensors and other emerging technologies of interest are forming or have already formed in Europe, Japan, Far East, and elsewhere in the world.

13.3 COMMERCIAL DEVELOPMENT OF BIOSENSORS

Louie (2002) defines commercial development as "the process of reducing a technological approach to product offering." He refers to the following points that are involved in being successful in the market place:

(a) *Technology*: This involves novel technology, a platform, and adequate patent protection. Platform technologies are helpful in the sense that they are modifiable to detect different pathogenic threats. Patent protection should assist in the prevention of theft of one's ideas and market opportunities.

(b) *Technological Hurdles*: Louie asks the following relevant questions: Have technical data being gathered to make the product realizable? Have regulatory requirements being met? The product should also be stable and manufacturable, and have adequate performance characteristics. Performance characteristics such as stability, reproducibility, reliability, sensitivity, easy to use, robust, rapidity, and incorporation into sensor networks are a few of the necessary requirements for efficient biosensors.

(c) *Market*: Define the market, and estimate the market size. Examine the competition and see if one can identify unmet needs in the market place. A niche market would significantly affect profit and the 'bottom line'. Louie (2002) indicates that the present emphasis is on the detection of Anthrax and small pox. One should also be prepared to detect other biological threats such as Ebola, hantavirus, ricin, and plague. Real-time biosensors are required for the military, and real-time inexpensive biosensors are required for first-time responders. Note that the detection of very dilute analytes such as pathogens in air would lead to a significant degree of variability in the identification, and in making quantitative the amount of the pathogen in the air to be detected. This would contribute quite significantly to the cost of a reliable and an efficient biosensor.

(d) *Business*: Anticipate business hurdles (Louie, 2002), define a road map to reach one's objectives, examine the manufacturability aspects, and finally one should have the finances to manufacture the prototype. Different avenues by which one may raise finances should be carefully examined and their pros and cons should be considered. For example, one may consider government grants, private financing, personal finances/mortgages, and venture capitalists. A well-defined commercial plan is exceedingly helpful since it helps one to organize and accelerate the time to the market. Eventually this will show up in a positive sense in the bottom line. One may also consider different options with regard to manufacturability, which includes outsourcing and partnering (which would help in conserving cash).

Louie (2002) concludes by saying that government markets for biosensors are driving the process, and as is the case for any profession that is personnel intensive where many man-hours must be spent to develop a process (in our case a successful biosensor), one should be able to enjoy the process.

Rogers and Mascini (2005) have recently analyzed the use of biosensors for analytical monitoring and pointed out the advantages and disadvantages. They have specifically indicated the reasons for the lack of or slow progression toward commercial markets. These authors emphasize that biosensors do exhibit the potential to complement laboratory-based and field analytical methods for environmental monitoring. However, in order that these biosensors become commercially viable, the authors suggest that the following general requirement should be met: (a) diversity of compounds and complexity of matrices in environmental samples and (b) variability of data quality requirements among different environmental programs. More specifically, with regard to biosensor applications, the difficulties include: (a) high development costs for single analyte systems, (b) limited shelf life, and (c) limited operational lifetimes for pre-manufactured biorecognition elements. These authors emphasize that the initial commercially viable environmental biosensors will be successful only in rather narrow areas of application.

13.4 BIOSENSOR TECHNOLOGIES AND TECHNIQUES

In this section, we examine some promising biosensor detection techniques for different applications. Costs are provided wherever they are available.

ISIS Pharmaceuticals, Carlsbad, California has very recently indicated that its TIGER biosensor system has characterized an epidemic in a military camp (Globeinvestor, 2005). TIGER is an acronym for Triangular Identification for Genetic Evaluation of Risks. The Ibis division of ISIS used its expertise in RNA to develop a biosensor for the identification of bacteria and viral organisms.

The results of the TIGER biosensor system for the detection of respiratory pathogens has recently been published (Ecker *et al.*, 2005). A broad range of infectious organisms could be detected by the TIGER biosensor including those that are newly emerging, genetically altered, and unculturable. The particular advantage of the TIGER biosensor is that it rapidly identifies and makes quantitative the infectious organism without actually knowing what one is looking for. The particular strain of the organism may also be detected. For example, these authors indicate that at the Marine Corps Recruit Depot (MCRD) in San Diego, United States, the TIGER biosensor system rapidly identified a virulent form of *Streptococcus pyogenes*, and prevented the epidemic of a severe respiratory disease.

The costs for developing this TIGER biosensor system was around $38.5 million which was obtained from U.S. Governmental agencies such as the Defense Advanced Research Projects Agency (DARPA), the National Institute of Allergy and Infectious Diseases (NIAID), the Center for Disease Control (CDC), the Federal Bureau of Investigation (FBI), and the Department of Homeland Security (DHS). An additional $9.0 million has been approved, and there is a potential for additional funding.

Weibring (2003) discusses the use of LIDAR (light detection and ranging) for cloud detection. The recognition is based on the same principles as RADAR except that instead of bouncing longer wavelength radio waves of a target, higher energy light waves are used. In the LIDAR technique, light waves in the infrared, the ultraviolet, and the electromagnetic spectrum are generated. The multiple energy wavelength of LIDAR furnishes three-dimensional imaging. The authors emphasize that under controlled conditions detection of aerosolized clouds at long distances is possible. The limitations of this technique are cost ($4000 for a simple LIDAR used for speed monitoring) and limited distance capability. The U.S. Army uses LIDAR-based technology on its UH-60 Blackhawk helicopters to detect aerosol clouds from a long distance. The technique is capable of distinguishing biological agents from non-biological agents. This is based on the excitation of intracellular fluorescent compounds. False positives are also possible by this technique. The false positive may be obtained for pollen, mold, and organic excreta.

Coronado (2005) indicates that the researchers at the Univeristat Autonoma de Barcelona have developed miniaturized biosensors to detect DNA. These researchers have developed biosensors as thin as a fingernail. The advantage of this very small size of biosensors is that they reduce the time required to detect and identify the DNA chains from several hours to minutes, depending on the DNA chain. The authors (Salvador Alegret, Manuel del Valle, and Maria Isabel Pividori of the Sensors and Biosensors group at the Universitat Autonoma de Barcelona (UAB)) indicate that these DNA sensors may be applied for paternity tests, to detect genetically modified food or genetically modified organisms (GMOs) (this is a big item in Europe), help in identifying bacterial strains in food-borne illnesses, and in testing for toxicity in new drugs. Coronado (2005) indicates that once these biosensors are mass produced, the costs of these tests should be similar to that of pregnancy test kits available at pharmacies.

The UAB researchers indicate that their developed biosensor is able to decrease the time required from 2 days to just 30 min for the detection of infection from *Legionella* (causes Legionnaire's disease). Also, the authors indicate that their new sensor can detect *Salmonella* in about 4–5h compared to 3 or 5 days required by the traditional microbiological methods. This is a reduction in the detection time by more than an order of magnitude. The authors indicate that the sensor could also be used to detect other pathogens such as *Campylobacter* and *Lysteria*. The intention of these authors is to mass produce these types of biosensors so that they may be readily available at local pharmacies. Finally, they state that these miniaturized electrochemical biosensors meet the need to develop DNA sensors that may be used for analysis, they cost less, and may also be used by unsupervised personnel.

Professor Lekkala's laboratory at the Tampere University of Technology in Finland has very recently started developing an implantable biosensor (MedGadget, 2005). This laboratory is developing a wireless implantable biosensor that can work reliably for long periods of time under the skin. Other requirements include (a) efficient operation in a moist and corrosive environment even when for some reason or the other the protective coating is damaged or rendered useless. The authors indicate that this sensor's circuit boards are flexible, and thin enough (of the order of 0.1 mm) to bend with the patient's movements. Furthermore, this implantable biosensor can be customized by attaching a layer that contains antibodies to the implantable coating. These antibodies may be released as and when required (controlled release). It should, however, be pointed out that this research is in its early stages.

Different researchers have attempted to attach proteins to gold surfaces using self-assembled monolayers of thiols and sulfides. Kalaugher (2004) indicates that in the presence of oxidizing agents such as ozone and hydrogen peroxide these attached proteins exhibit poor stability. This author indicates that investigators at the Technical University of Munich and r-Bets coating in Austria have developed a means of attaching proteins to nanocrystalline diamond. These researchers attached catalase to a diamond film and made a biosensor that could detect hydrogen peroxide. Jose Garrido of this group emphasized the important advantages of using diamond: chemical stability, biocompatibility, surface modification is simple, and a large electrochemical window. The researchers indicate that the strength of the carbon–carbon bond at the diamond/biolayer interface is a distinct advantage over attaching proteins to metal surfaces such as gold. Attempts are being made to develop biosensors using other enzymes such as glucose oxidase (for the detection of glucose in diabetes patients). This is a promising development, especially with regard to the stability offered by diamond surfaces. Needless to say, for commercialization purposes the cost of the diamond needs to be taken into account. Presumably, extremely thin layers of diamond would have to be eventually used if the process is moved toward the commercialization phase.

Alt (2005) in a write-up for Hoovers™ on Advanced Biosensor, Inc., located in Columbia, South Carolina, U.S. indicates that this company makes Holter monitors (that record a person's ECG (electrocardiogram)) and blood pressure monitors. There are 20 employees in the company, and the annual sales are \$4.0 million (U.S.). This author emphasizes that there are six competitors for Advanced Biosensor, Inc., and in order to analyze the competitive landscape one needs to compare the following items with its top three competitors: profitability, growth, operations, and valuation. A column-to-column of these items as suggested by Alt (2005) is helpful.

Flurry (2003) in Georgia Faces indicates that the researchers at the University of Georgia, Athens, Georgia, U.S. have been awarded a 4-year grant of \$1 million (U.S.) from the National Science Foundation to develop and analyze three-dimensional nanoscale structures in biosensing. Implantable glucose biosensors is a promising development in the management and treatment of diabetes. The researchers at the University of Georgia plan to examine key issues such as long-term calibration and aging effects on biosensors. The intent is to make these implantable biosensors more sensitive and stable. These researchers have developed a novel nanofabrication technique called glancing angle deposition that fabricates nanoscale 3-D pillars.

Jun Li, a research scientist at NASA (National Aeronautical and Space Administration) in the United States (InfoZine, 2005) has developed a microscopic (carbon-nanotube biosensor) sensor, which can monitor spaceship water quality as well as detect biohazards.

This biosensor, the size of a stamp is an electronic chip which is able to detect multiple pathogens, such as bacteria and viruses.

Chu (2005) has very recently indicated that although medical applications continue to dominate the biosensor market, there is increasing pressure from the pharmaceutical industry for rapid assay biosensors which would speed up the drug discovery process. Silver (2005) too has very recently indicated that biochips and bioarrays are an integral part of the drug discovery and development process. Furthermore, different types of biosensors are at the core of the POC diagnostic medical devices. At the recent International Symposium on Emerging Rapid Assay Technologies held in Taipei, China in November 2005, presentations on newer biosensor technologies were made that included new DNA sensors for environmental monitoring, use of adhesive tape in biosensor design, and autoimmune protein arrays. The detection of autoimmune diseases using biosensors and microarray devices is a rapidly advancing area, which does exhibit the potential to help alleviate the suffering, quality of life, and cost associated with these insidious diseases.

Aragon (2004) indicates that Khosla, a well-respected technology investor in the industry has criticized the recent IPO (initial pubic offering) by Nanosys. This Palo Alto, California company declared its intent to the SEC (Security Exchange Commission) to raise \$115 million (U.S.) through its public offering. This was to be the first publicly traded Nanotech company. Khosla advises that, and we quote, "When you invest, invest in an application, than in a technology. Invest in an application that makes an impact." Khosla emphasized that Nanosys does not have a predictable business model. There were too many open questions unanswered for it to be a publicly traded company.

Supporters of Nanosys, however, indicate that Nanosys has locked up over 200 patents from major nanotechnology centers at universities such as Harvard, Cornell, MIT, and UCLA, and has development agreements with Intel Corporation and Matsushita Electric. It also has a grant from the National Institutes of Health to develop nanotechnology-based biosensors. Finally, only time will tell whether this company will be successful or not.

The Frost and Sullivan Report entitled "World Biosensors Markets" (2005b) emphasizes that better development strategies are essential to speed up the commercialization of biosensors. The report indicates that support from financial agencies is difficult to obtain due to long development periods and low successful rates. Other problems with the commercialization of biosensors include sterilization, reproducibility, reliability, selectivity, and cost of manufacture. The report points out the advantages of miniaturization on the commercialization of biosensors: it will hasten commercialization as well as permit the widening of application of biosensors into different, hitherto untouched markets.

Nanotechnology and nanobiotechnology are expected to impact quite a few areas in biotechnology, and biosensors is expected to be one of them. This is further evidenced by the significant amount of resources being allocated by government sources in the United States, Europe, Japan, and elsewhere in the world. This Frost and Sullivan report (2005b) further indicates that leading edge integrated circuit and wireless technology is being successfully integrated into the high-end biosensor applications. Finally, this report analyzes and forecasts the biosensor markets for (a) research laboratories, (b) home diagnostics, (c) point-of-care, (d) process industries, (e) biodefense, and (f) environmental monitoring.

In a recent program solicitation for research proposals, the National Science Foundation (National Science Foundation Program Solicitation, 2005, NSF 05-526) was seeking proposals on biosensors that would impact national security, health care, the environment, energy, food safety, and manufacturing. This program entitled "Sensors and Sensor Networks (Sensors)" had a proposal deadline of March 03, 2005. One of the goals of the program solicitation is to improve performance and lifetime, and decrease life cycle costs. The program solicitation emphasizes that there is a need for sensor fabrication, and newer methods of sensor fabrication, manufacture, and encapsulation. Furthermore, the NSF (National Science Foundation) program solicitation emphasizes the need to do research on robustness during manufacture. Other aspects of biosensor research emphasized in the program solicitation include: miniaturization, reduced cost, and lower power consumption.

MarketResearch.com (2005) in a recent report on the Indian Biotechnology Industry indicates that India with its 280 biotechnology companies is contributing to the total biotechnology market of $100 billion. This is a $1600 report that analyzes the Indian Biotechnology Industry and emphasizes the partnering with American and Chinese companies. India has an abundant supply of well-qualified academics, a good educational infrastructure of academic institutions, and the labor is relatively cheap. Many companies such as big pharmaceuticals and software companies are investing in India. Indian companies are making a concerted effort to develop better business models and improve product commercialization. Intel has recently announced that it would invest $1 billion in India (CNBC Business (2005), TV Channel, 6:00 a.m., December 05). It is not entirely unexpected that in the future more research and development activities, including biosensor development would move overseas to India as well as to China and to other countries (which offer similar infrastructure, an excellent pool of educational talent, and cheap labor). This viewpoint is consistent with the current popular book entitled, " The World is Flat: A Brief History of the Twenty-First Century" (Friedman, 2005). Publishers Weekly (2005) in a review of this book indicates that cheap and ubiquitous communication has eliminated the obstacles to international competition. The review emphasizes that economic stability is not guaranteed, and flexibility and ruggedness in entrepreneurship will gain predominance. If one may take the liberty to expound on this a little more using an analogy from biochemistry and enzymes, then one may state that 'an enzyme is most active at or near the point of instability.'

13.5 EXAMPLES OF SMALL AND MEDIUM BIOSENSOR COMPANIES THAT ARE EITHER SUCCESSFUL OR ARE ON THE WAY TO BE SUCCESSFUL

Examples of small and medium biosensor companies will be presented to primarily indicate the time, effort, and resources required to start, and successfully run a biosensor company, which makes a profit. Companies that are presently making a loss are also presented. Hopefully, these companies will start making a profit in the near future. Examples have been taken from the open literature in a random fashion, and in no way reflect on the way the company is run or should be run. The examples presented are for illustrative purposes. Any conclusions drawn or statements made, are once again, for educational purposes only, and there is a possibility that some conclusions or statements made may be incorrect. No recommendations for investing in any of these companies are to be inferred in any way by the analysis to be presented in this section.

Let us arbitrarily select a small company size as less than or equal to 75 employees. A medium size company is considered to arbitrarily employ between 76 and 300 employees. A company that has more than 300 employees may be considered as a large company.

13.5.1 Examples of small biosensor companies

Ambri Limited located in Sydney, Australia is a pioneer in integrating biotechnology, nanotechnology, and electronics (Ambri, 2005). It has patented the ICS™ (Ion Channel Switch) biosensor, which acts as a biological switch, and is capable of detecting a range of biological and chemical substances such as drugs, hormones, bacteria, and DNA in blood, urine, and saliva. The ICS™ technology is a patented, self-assembling synthetic bio-membrane. The company indicates that its technology permits it to make sensors the size of 1/5th of a millimeter—the size of a pinprick. Essentially, each chip has a gold base coated with biosensor properties. The Ambri literature (Ambri, 2005) indicates that each chip has multiple biosensors. Thus, replicates of the same test can be performed on a single disposable chip. This is a rapid response, single-step method.

Ambri posted a loss of $11 million (Australian) in 2004. This is equivalent to U.S.$8.17 million (Exchange rate: $1 (Australian) = U.S.$0.7429, Exchange rate, December 16, 2005). For the year 2005 (ending June 30), the company posted a loss of $8.95 million (Australian) (FED, 2005). This is equivalent to U.S.$6.65 million. The loss has decreased by 18.6% from the previous year, but the company is still a few years away from turning a profit. No dividend was paid to share holders. In an effort to cut down employee expenses, Ambri cut down its number of employees from 76 to 32. This decreased the employee expenses from $6.3 to $4.6 million (Australian).

FED (2005) indicates that AMBRI has put out feelers for a possible merger with a compatible enterprise. Ambri's Chairman indicates that the ICS technology is still a few years away before it is a commercial product available for sale. A possible suitor would be a private company that is looking to improve its range of products. Ambri emphasizes that market feedback indicates that there is still a need (as yet unmet) for a multi-test medical diagnostic product, which can be produced at a low cost and in volume.

Biophage Pharma Inc. (2005) has recently announced its third quarter financial results. The company was founded in 1995, and is located at the Biotechnology Research Institute in Montreal, Canada. The company employs 15 people, including 13 researchers. This company is building a portfolio of new diagnostic tools based on nanosensor technologies. Its primary focus is in the development of nanosensor platforms to detect pathogens and their toxins. Those agents that may be used as biological warfare agents are also a focus in their research and application.

The company has recorded a loss of $165,973 (Canadian) for the third-quarter period of the year 2005 (June–August 2005). This is equivalent to $143,567 (U.S.) (Exchange rate $1 (Canadian) = U.S.$0.865). For the same period (quarter) last year, the company had recorded a loss of $109,575 (Canadian). This is equivalent to $124,185 (U.S.). In order that the company may focus on the bottom line, their President and CEO indicates that they will concentrate on increasing revenues, decrease their burn rate, and prioritize their ongoing programs. They are pushing forward to the pre-commercialization of their PDS® biosensor. They are looking forward to marketing an end-to-end live pathogen detection and control system.

This company has already been in existence for just over 10 years, and if the last 2 years are any indication, then they still have a few years to go before they turn a profit.

Isis Pharmaceuticals is located in Carlsbad, California. Its Ibis division has developed the TIGER (Triangulation Identification Genetic Evaluation of Risk) biosensor technology (Isis Pharmaceuticals, 2005). Isis Pharmaceuticals uses its expertise in RNA to discover and develop novel drugs for its product pipeline. It develops antisense drugs to treat metabolic, cardiovascular, inflammatory diseases, and cancer. It is appropriate to focus on the Ibis division, which has developed the TIGER biosensor technology. No numbers of employees for either the Ibis division or for Isis Pharmaceuticals as a whole are available (Isis Pharmaceuticals, 2005). It has been arbitrarily placed in the small company category. The TIGER biosensor system being developed and commercialized by the Ibis division is a revolutionary system that helps identify unknown and emerging infectious organisms.

Government sources are the primary sources for grants to help develop the TIGER biosensor system to rapidly detect and identify infectious organisms. Different governmental agencies, such as the Department of Homeland Security (DHS), the FBI (Federal Bureau of Investigation, the CDC (Center of Disease Control in Atlanta, Georgia, USA), and the National Institutes of Health (NIH) are among the agencies that perceive an urgent need for this type of biosensor systems, and awarded grants to Ibis/Isis to help develop the TIGER biosensor system.

The Isis Pharmaceuticals (2005) Financial Report document indicates that they received U.S.$3.4 million and U.S.$8.7 million from governmental sources for the periods January 2005–March 2005 and January 2005–August 2005, respectively. These numbers may be compared with U.S.$2.2 million and U.S.$8.9 million for the same periods in the year 2004. Table 13.1 shows the losses in operations by the Ibis Division incurred for the different time periods in the years 2004 and 2005.

Though only data for a couple of years (partial) are presented in Table 13.1, the Ibis division of Isis Pharmaceuticals is still apparently 3–5 years from turning a profit. Isis as well as presumably its Ibis division have taken steps to cut down costs by reorganization and by focusing its resources on key programs.

Table 13.1

Losses from operations incurred by the Ibis division of Isis pharmaceuticals (Isis Pharmaceuticals, 2005)

Year	January 2005–March 2005	January 2005–September 2005
2005	$97,000	$1,600,000
2004	$803,000	$2,300,000

The Ibis division of Isis is very progressive and it is anticipated that they will expand into other areas where their TIGER biosensor will be effectively used to detect different types of analytes. Finally, it is appropriate to indicate that the numbers presented in Table 13.1 are taken from Isis Pharmaceuticals (2005). Any type of errors in the data or data interpretation are unintentional.

Although only three examples of small biosensor companies are given above, it seems reasonable to say that it takes at least between 10–15 years, if not more, for a company to turn a profit. This author presumes that Venture Capitalists are aware of this since they generally have had years and years of experience dealing with different sorts of start-ups, besides biosensor companies. Needless to say quite a few ingredients are involved in making a start-up successful, besides a good and viable idea, financing, hard work, etc. Patience and fortitude, including other attributes or ingredients, are also high up on the list. Perhaps, the timing is now right to give an example of a successful biosensor company.

13.5.2 Example of a medium-sized biosensor company

Biacore located in Uppsala, Sweden markets its surface plasmon resonance (SPR) biosensor. The company was started in 1984, and as of September 2005 it had 278 permanent employees (Biacore, Life Sciences, 2005). The SPR biosensor provides unique kinetic data on protein interactions. The software that comes along with the SPR biosensor provides values of the binding and dissociation rate coefficients, and the affinity values (if applicable). The SPR biosensor may be used (a) for antibody characterization, (b) proteomics, (c) for drug screening, and (d) in the food analysis market to name a few of its applications. The company continuously brings out better and better models that are forever expanding the applications of these SPR biosensors. Biacore's customer market includes life science research centers, leading global companies, and a large number of biotechnology companies.

Biacore has sub-divided its worldwide sales market into three areas: Americas, Europe, and Asia-Pacific regions. Table 13.2a and b show the sales figures by the regions mentioned above for the years 2004 and 2005. Table 13.2b shows the net income for the years 2004 and 2005 (Biacore, Life Sciences, 2005). The numbers are given in SEK (Swedish Kroner). The numbers given in brackets are the corresponding amounts in U.S. dollars. (Exchange rate one SEK = U.S.$0.1270, December 16, 2005).

The total sales has increased by 11.29% from the year 2004 (337 million SEK) to the year 2005 (375.5 million SEK). Note that one is comparing sales figures for the months January to September 2005 for each year. The sales for the Americas has increased by 6.95% from 132.3 SEK million for the year 2004 to 141.5 SEK million for the year 2005. In total amounts the sales in the Americas is the highest for the year 2005. The sales for the Asia-Pacific region has increased by 15.68% (the largest percentage increase by region) from 69.7 to 109.3 SEK million for the year 2005. For the Europe region, there is actually a decrease in total sales from 135.4 SEK million by 7.9% in the year 2004 to 124.7 in the year 2005. Net income for each region was not available from the report (Biacore, Life Sciences, 2005). Thus, it is not presented. That would, however, have been instructive.

Biacore did report a net income of 141.8 million SEK between the months of January–September 2005 (see Table 13.2b). This is equivalent to U.S.$18 million.

Table 13.2

Biacore worldwide (a) sales by region and (b) net income for all regions in SEK (Swedish Kroner) million for the years 2004 and 2005. U.S. dollars are given in brackets (Biacore, Life Sciences, 2005)

	Year	
	2004	2005
(a) Region		
Americas	132.3 (16.80)	141.5 (17.97)
Asia-Pacific	69.7 (8.852)	109.3 (13.88)
Europe	135.4 (17.196)	124.7 (15.84)
Total	337.4 (4.285)	375.5 (47.69)
(b) Net Income		
January–March	−1.0 (−0.127)	12.0 (1.524)
April–June	0.6 (0.0762)	113.3 (14.39)
July–September	−24.5 (−3.035)	16.5 (2.029)
October–December	24.6 (3.124)	Not available
Total	−0.3 (−0.0381)	141.8 (18.00)

Hopefully, this net income will keep on increasing in the years to follow. Biacore is a very progressive company and it is anticipated that they will expand into other areas where their SPR biosensor will be effectively used to detect different types of analytes. Finally, it is appropriate to indicate that the numbers presented in Table 13.2a and b are taken from Biacore, Life Sciences (2005). Any type of errors in the data or data interpretation are unintentional.

13.6 FACTORS THAT COULD HELP INCREASE BIOSENSOR MARKETS

Vogel and Zinkl (1997) have indicated the general conditions required for the development of a biosensor industry in Switzerland. These authors emphasize that the well-developed chemical and pharmaceutical industry is conducive to the development and production of biosensors. There is good infrastructure in place in that there is an excellent university system, start-up companies, and industry that could support the development of a biosensor industry. Furthermore, these authors emphasize the following ingredients should be present or are prerequisites for successful biosensor start-up companies: (a) presence of excellent technology, (b) access to export markets, (c) transfer of technologies between universities and the private sector, (d) availability of low-cost resources, (e) climate that fosters entrepreneurship, and (f) a political climate that is receptive to start-ups with a minimum of regulatory requirements. The authors emphasize that the government and the public sector has put in the necessary resources to help build the biosensor industry. Another aspect that needs to be emphasized is manufacturing and distribution. Both these factors will play a significant role in biosensor economics.

One of the problems associated with the biosensor industry, and as indicated elsewhere in this chapter, is the sum of start-up money required. The SBIR (Small Business Innovation Research)/STTR (Small Business Technology Transfer) programs run by the National Science Foundation (NSF) and the National Institutes of Health (NIH) provide such start-up monies for successful proposals which are carefully reviewed. Presumably, other countries too have similar or better programs that provide start-up monies for biosensors and other industries.

In late 2001, there were public hearings on the current and future workforce needs of California's biotechnology industry of which biosensors are a part (Work Force Survey Results—CA (California) Needs, 2001). Both short- and long-term needs were addressed. At that point in time it was noted that there was a critical shortage of skill sets associated with the transition of research ideas from the bench scale to the industry, followed by subsequent FDA (Federal Drug Administration) approval, commercialization and manufacture for biotechnology as a whole.

The report and the hearings suggested more involvement in translational research, and applied research and development. All of these factors should help in bringing down the development costs of biotechnological products as a whole, and biosensors in particular. The report emphasized the growth in a company as it grows from a typical R&D boutique to process and development, and finally the manufacturing stage. For example, in a small medical therapeutics company, the number of personnel involved in research may be from 1 to 49. The company will typically grow from 50 to 149 employees as the processes and products are developed. Finally, as the company goes into manufacturing and production, the number of employees in the company will be more than 150. The report (Work Force Survey Results—CA (California) Needs, 2001) emphasizes that though research and discovery is involved throughout the process, the major growth, as the company matures, is in the areas of operation, development, clinical, regulatory, quality control, marketing, sales, finance, and business development.

13.7 EXAMPLES OF SIZE OF GRANTS AND FINANCING FROM DIFFERENT SOURCES

Cambridge Evening News (2005) and ResearchResearch (2005) indicate that the Biotechnology Sciences Research Council has awarded £1 million or U.S.$1.763 million (£1 = U.S.$1.76297, Exchange rate, November 15, 2005) to Professor Piero Migliorato in the Department of Engineering at Cambridge University and Paul Ko Ferrigno of the Medical Research Council's Cancer Cell Unit in Cambridge to develop low-cost and disposable biosensors to detect cancer and infectious diseases. This emphasis on low cost and portability has also been emphasized in the recent National Science Foundation Program Solicitation, 2005, NSF 05-526, Sensors and Sensor Networks (Sensors). Professor Migliorato indicates that the biosensor they have in mind is flexible, and may be used in different areas of application including infectious disease and bioterrorism.

Detection of infectious diseases is a hot market, and Chief Executive David Creed of ITI (Intermediary Technology Institute) of Glasgow, Scotland indicates that the market for infectious disease testing is estimated at around £2.6 billion or U.S.$4.58 billion

(£1 = U.S.$1.76297, Exchange rate, November 15, 2005) (Scotsman.com, 2005). The research and development program at ITI Techmedia will be spending around £10 million (or U.S.$17.63 million) to build a biosensor platform in about 4 1/2 years. It should be pointed out that ITI Techmedia develops intellectual property for the Scottish economy. Their function should serve as a good role model, wherein they assist in commercializing projects that the market is, in general, reluctant to pursue and take risks on. Other countries should also have these types of institutions or organizations, if they already do not have them.

Adnavance Technologies (TechFinance.ca, 2005) has very recently announced that it has completed a round of financing and raised $3.85 million to advance its m (messenger)-DNA technology to help develop ultra-high sensitivity biosensors for medical diagnosis and disease prevention. This is the third round of financing that has been completed by Adnavance. Venture capitalists, banks, and others were involved in this round of financing. A director of Adnavance, Yad Garcha indicates that Adnavance's innovative technology platform has important consequences for the molecular diagnostics and biosensor markets.

DARPA (Defense Advanced Research Project Agency) has awarded the School of Engineering at Arizona State University (ASU), Tempe, Arizona, USA with a $11.7 million multicollaborative research project to develop novel hybrid biomolecular nanodevices to assist in disease detection and for pharmaceutical drug testing (Nano techwire.com, 2005). A consortium of universities in U.S., United Kingdom, and Germany are involved. These hybrid biomolecular nanodevices interface biological matter with electronic circuitry. This creates a silicon chip, which can interpret molecular changes, and can convert these molecular changes into real-time digital information. The Principal Investigators (PIs), Professors Stephen Goodnick and Trevor Thornton, of the Electrical Engineering Department at ASU indicate that their ion-channel biosensor may be used to detect, for example, biological agents. More specifically, ion-channel proteins will bind to specific target biological agents, and thereby change the electrical behavior of these channels. The PIs indicate that another possible application for this ion-channel biosensor is in drug testing.

Yeates (2005) has recently indicated that the U.S. Army has begun testing a hand-held detector that costs less than $2000, and can identify bioterrorism agents in trains, buses, and in buildings within 60 s. The device weighs less than 2 pounds. This was developed by AnzenBio, a company located in Utah. No costs for development were provided. Also, the time taken to develop the biosensor by AnzenBio was not given. The CEO, of AnzenBio, Gary Crocker indicates that the biosensor is generic in the sense that it can be loaded with different antibodies to target different pathogens. Furthermore, not only is the pathogen identified, but also its concentration is made available. Their biosensor can also be easily modified to another biosensor chip for home use to detect allergens and fungus.

Biophage Pharma, Inc. (2005) has very recently indicated that it has developed a biosensor to detect antibiotic resistant strains of bacteria within 7 h. Once again, no costs were given to develop this biosensor, and the time taken to develop it. This biosensor method does take less than one-half the time required by the disk diffusion or the tube dilution technique, which take between 18 and 24 h. Antibiotic resistance is a major public health hazard around the world. Patients affected by drug-resistant organisms have longer hospital stays, and are treated by less-effective and more toxic and more expensive drugs. The use of this biosensor to detect viable bacteria that are also resistant to antibiotics will lead the way for the diagnosis of nosocomial infections.

Nosocomial infections are those that originate in a hospital (Abedon, 1998). This author indicates that about 20,000 deaths occur in the U.S. every year due to nosocomial infections. Abedon (1998) further adds that about 20 million American hospital patents acquire a clinically significant nosocomial infection. Some of the organisms that cause these nosocomial infections are *Enterococcus spp.*, *Escherchia coli*, *Pseudomonas spp*., and *Staphylococcus aureus*. This Canadian biopharmaceutical company is building its portfolio on the rapid detection and elimination of biological contaminants including those that may be used in biological warfare.

13.8 CONCLUSIONS

This is the last chapter of the book and attempts to present the different facets that may be involved in helping to set up a biosensor industry. Perhaps, not all of the aspects have been presented. However, an attempt has been made, like in the previous two books by this same author on Biosensors (Sadana, 2003, 2005) to provide a perspective, albeit from an individual who has spent most, if not all of his working life in academic institutions. That said, this type of information is not easily available in the open literature for free. More often than not, one has to spend thousands of dollars to obtain this type of financial report from more prestigious commercial institutions. Thus, the need for this chapter. Hopefully, it will provide interested individuals with an 'initial solution', and just like an iterative solution to a non-linear problem they may be able to use some of the material presented here along with some assumptions which are specific to their own project or problem to arrive at a more palatable solution.

Biosensors, at present, is a narrow and focused area, especially if one is viewing it through 'commercial eyes.' The major driving force for the commercialization of biosensors has traditionally been the detection of glucose for the effective management of diabetes. This need will grow as the number if diabetics in the United States and worldwide grow. However, it is hoped that the use of biosensors will expand commercially into other areas, especially for the detection of auto-immune diseases, which are by nature insidious and intractable. Biosensors, are in this author's opinion, and presumably of others too, perfectly poised to assist in the detection of precursors for auto-immune diseases such as arthritis, cancer, systemic lupus erythematosus (SLE), and others. For some of these diseases more than one analyte needs to be detected simultaneously. Biosensors, especially microarrays have the capability of detecting these analytes simultaneously.

This chapter presents only three examples of small biosensor companies that are still showing net income losses. This is a difficult situation, wherein there is continued pressure to show positive results to investors and employees. The time period typically, is apparently more than 10 years before a company begins to show a net profit. The medium-size company, Biacore, in Uppsala, Sweden is beginning to show a profit. Its SPR biosensor is apparently very popular throughout the world. This is a forward-looking and aggressive company, and hopefully, its net profits will continue to grow. The market size for biosensor applications is albeit small, and it takes a few years to turn a net profit, nevertheless it will behoove individuals and companies (at least the 'deep-pocketed' ones) to get into the development of biosensors to other areas of application, such as the detection of different diseases and pathogens, which may be used as weapons of mass destruction. The raising of money (finances) to run these types of operations will, at least in the near future and it is the opinion of this author, be difficult and not so easy to justify. Nevertheless, it is a worthy cause, especially if it helps eliminate or minimize the suffering that ensues with, for example, intractable and insidious auto-immune diseases.

REFERENCES

Abedon, S.T. (1998). Nosocomial Infections, Supplemental lecture, abedon.1@osu.edu, http://www.mansfield, ohio_state.edu/~sabedon/biol2053.htm

Alt, S. (2005). Advanced Biosensor, Inc. Fact Sheet-Hoovers, http://www.hoovers.com/free/co/factsheet.xhtml?COID=100915&cm_ven=PAID&cm_c

Ambri (2005). Biosensor Technology—Critical Care, Company Information, http://www.ambri.com/Content/dispaly.asp?name=Company%20Information

Aragon, L. (2004). KP's Khosla Pulls no Punches, Slams Plans for Nanosys IPO, June 7, http://www.privateequityweek.com/pew/freearticles/1070549978426.html\

Biacore, Life Sciences (2005). Interim Report January–September 2005, October 20, http://www.biacore.com/lifesciences/press/pressreleases/116338/index.html?backurl=%2

Biophage Pharma, Inc. (2005). Biophage Announces Biosensor's Ability to Detect Antibiotic Strains of Bacteria: Biosensor Unit Can Now Detect Antibiotic Resistant Bacteria within 7 Hours, July 14, http://www.pressalert.info/news/SinglePosting.php?ArticleID=2005071467

Cambridge Evening News (2005). One million pounds (British) research boost for scientists, November 30, http://www.cambridge_news.co.uk/news/city/2005/11/30/c33/c9c2-9605-480a-86bf-2f3b5

CASE Reports (1999). BEACON: Public, Private Institutions Unite in Biomedical Alliance, **14**(1), February, http://www.ctcase.org/14_1/beacon.html

Chu, W.L. (2005). Trend Sees Boom in Pharma Analytical Tools, November 15, DrugResearcher.com, http://www.drugresearcher.com/news/ng.asp?n=63923-separation-technology-chromatography

CNBC Business. (2005). Intel to Invest $ One Billion in India, U.S. TV Channel, 6:00 am, December 05.

Coronado, O.L. (2005). New Miniaturized Chip Dramatically Reduces the Time Taken for DNA Analysis, http://www.eurekalert.org/pub_release/2005–04/uadb-nmc040105.php

Dasey, T. (2005). MIT Lincoln Laboratory, Please see Reference 22 in Finding the Bugs, accessed December 06, http://www.ndu.edu/biosensing 02.htm

Diamond, F. (2002). Life science/biotech start-ups command close look. *Enewsbuilder*, **1**(9), November 14, 2002, http://www.enewsbuilder.net/viable_article000102908.cfm

DrugResearcher.com (2003), Breaking News on Drug Discovery, http://drugresearcher.com/productnews/.asp?id=56316

Ecker, D.J., Sampath, R., Blyn, L.B., Eshoo, M.W., Ivy, C., Ecker, J.A., Libby, B., Samant, V., Sannes-Lowrey, K.A., Melton, R.E., Russell, K., Freed, N., Barrozo, C., Wu, J., Rudnick, K., Desai, A., Moradi, E., Kaize, D.J., Robbins, D.W., Hannis, J.C., Harrell, P.M., Massire, C., Hall, T.A., Jiang, Y., Ranken, R., Drader, J.J., White, Neill, McNeill, J.A., Crooke, S.T., and Hofstadler, S.A. (2005). Rapid Identification and Strain-Typing of Respiratory Pathogens for Epidemic Surveillance. *Proceedings of the National Academy of Sciences*, **102**(22), 8012—8017.

FED. (2005). Ambri Puts Feelers Out for Merger, November 29, http://www.theage.com.au/news/Business?FED-Ambri-puts-feelers-out-for-merger/2005/1

Finding the Bugs. (2005). accessed December 06, http://www.ndu.edu/biosensing 02.htm

Flurry, A. (2003). Researchers to Use $ 1 million Grant for Biosensing Study, *Georgia Faces*, http://georgiafaces.caes.uga.edu/viewtext.cfm?id=1937.

Friedman, T.L. (2005). *The World is Flat: A Brief History of the Twenty-First Century*, Farrar, Straus, and Giroux, Publishers, New York, NY.

Freedonia Report #1424, In vitro Diagnostics," June 2001. The Freedonia Group, Inc., 767 Beta Drive. Cleveland, Ohio, 44143-2326, USA, www.freedoniagroup.com

Freedonia Report #1547, Chemical Sensors, April 2002. The Freedonia Group, Inc., 767 Beta Drive. Cleveland, Ohio, 44143-2326, USA, www.freedoniagroup.com

Freedonia Report (2005a). Nanotechnology Biosensor Products—Overview, May 01, 2005, The Freedonia Group, Inc., 767 Beta Drive. Cleveland, Ohio, 44143-2326, USA, www.freedonia group.com

Freedonia Report, Study #1899 (2005b) Nanotechnology in Health Care to 2009, May 01, The Freedonia Group, Inc., 767 Beta Drive. Cleveland, Ohio, 44143-2326, USA, www.freedonia group.com

Frost and Sullivan (2005a). R1-3641, Advances in Point-of-Care Technologies (Technical Insights), June 2005, http://www.mindbranch.cm/products/R1-3641.html

Frost and Sullivan (2005b). Report 1100062, World Biosensors Markets, April 18, 2005. http:www.market.com/land/product.asp?productid=1100062&progid=3602

Frost and Sullivan (2005c). R1-3618, Strategic Analysis of the European Molecular Diagnostics Market, June 2005, http://www.mindbranch.com/products/R1-3618.html

Fuji-Keizai USA, Inc. (2004). U.S. Market and World-wide Biosensor Market, R & D and Commercial Implications, Pub ID FJ 974421, http://marketresearch.com/product/display.asp?productID=974421&xs=r

Globeinvestor (2005). Isis Pharmaceuticals, TIGER Biosensor System Characterizes Epidemic in Military Camp. Carlsbad, California May 26, http://www.globeinvestor.com/servlet/WireFeedRedirect?cf=GlobeInvestor/config&vg=

Goliath Industry Information (2004). Market Size for Worldwide Biosensors Estimated at $7.3 Billion in 2003, April 20, http://goliath.ecnext.com/comsite5/bin/pdinventory.pl?pdlanding=1&referid=2750&item

InfoZine (2005). NASA Scientist Honored for Spaceship Biosensor, November 12, http://infozine.com/news/stories/op/stories/sid/11376/

Isis Pharmaceuticals (2005). Isis Pharmaceuticals Reports Financial Results and Highlights for the Third Quarter 2005: Company Strengthens Financial Position While Advancing Pipeline and TIGER Biosensor System, http://www.prnewswire.com/cgi-bin/stories.pl?ACCT=104 /Story=/www/story/11-08-2

JASON, Biodetection Architectures, The MITRE Corporation, McLean. Virginia, February 2000, http://www.fas.org/irp/agency/dod/jason/biodet.pdf>, accessed December 2003, JASON is a group of distinguished defense contractors; please also see reference 20 in Finding the Bugs, http://www.ndu.edu/boisensing 02.htm, accessed December 06, 2005.

Kalaugher, L. (2004). Editor, Nanotech.web, org, Diamonds could be a Biosensor's Best Friend, News, nanotechweb.org, http://nanotechweb.org/articles/news/3/9/13/1

Kalorama Information (2004). Pub ID KLI 1023975, The U.S. Market for Over-the-Counter Diagnostics: Product Sales and Consumer Survey, November 15, http://www.marketresearch.com/product/display.asp?productid= 1023974&SID=&partner

Kalorama Information (2005a). Pub ID R566-310, The Worldwide Market for Cancer Diagnostics, October 2005, http://www.mindbranch.com/products/R566-310.html

Kalorama Information (2005b). *U.S. Markets in Analytical Chip Technology: Gene, Probe, Tissue, Cell and Microbiological Microarrays*, 2nd edn., http://www.mindbranch/products/R566-304.html

Kharif, O. (2003). A Sharper Node for Danger, *Business Week Online*, May 23, in Finding the Bugs, accessed December 06, http://www.ndu.edu/boisensing 02.htm

Louie, A.S. (2002). Commercial Development of Biosensors—The Business of Science, *Conference on Research, Technologies, and Applications in Biodefense*, November 4, http://www.tiacllc.com/aboutus/pdfs/biosensors_1104-02.pdf#search='biosensor%20market%20size'

MarketResearch.com (2005). Indian Biotech Industry 2005, October 1, 2005, http://marketresearch.com/product//display.asp?productid=1173210&sID=40

MedGadget (2005). Finnish Start Implantable Biosensor Development, November 16, http://www.medgadget.com/archives/2005/11/finnish_start_i.html

Nanotechwire.com (2005). Researchers Target Silicon Chips for Biomolecular Devices, November 2, http://nanotechwire.com/news.asp?nid=2529

National Science Foundation Program Solicitation (2005), NSF 05-526, Sensors and Sensor Networks (Sensors), National Science Foundation Opportunity Document, National Science Foundation, Proposals due March 03, Arlington, Virginia, USA.

Newmann, J.D., Tigwell, L.J., Turner, A.P.F. and Warner, P.J. (2004). Biosensors: A Clearer View, Institute of Bioscience and Technology, Cranfield University, Silsoe, http://www.cranfield.ac.uk/ibst/commercial/biosensor_report2004.htm

Publishers Weekly (2005). The World is Flat: A Brief History of the Twenty-First Century, A Review, http://www.thomaslfriedman.com/worldisflat.htm

Rajan, M. (2002). RB-159, Biosensors and Bioelectronics, U.S. Market for Biosensors & Bioelectronics to Cross $1.81 Billion by 2006, Business Communications Company, Inc., 25 Van Zant Street, Norwalk, CT, 06855, http://www.bccresearch.com/editors/RB-159.html

ResearchResearch (2005). BBSRC Makes £1m grant to Biosensor Consortium, http://www.researchresearch.cm/news.cfm?pagename=newsStory&type=default&element

Rogers, K.R., and Mascini, M. (2005). Biosensors for Analytical Monitoring, U.S. Environmental Protection Agency, Human Exposure & Atmospheric Sciences, http://www.epa.gov.heasd.edrb/biochem/intro.htm

Roth, K. (2005). National Science Foundation grant helps MSU (Michigan State University) Develop Biosensors, Michigan SmallTech News, http://www.michigansmalltech.com/News/detail.asp? ContentId=66C428EC-B77C-4703-8

Sadana, A. (2003). *Biosensors: Kinetics of Binding and Dissociation Using Fractals*, Elsevier, Amsterdam, The Netherlands.

Sadana, A. (2005). *Fractal Binding and Dissociation Kinetics for Different Biosensor Applications*, Elsevier, Amsterdam, The Netherlands.

Scotsman.com (2005). Business—Technology—10 Million Pounds (British) Platform The Perfect Medicine For Scots Firm, http://business.scotsman.com/technology.cfm?id=2327942005

Silver, D., Biotech East (2005). Latest in Assay Technologies Presented at International Forum in Taipei, November 9, http://www.biotecheast.com/modules.php?op=modload&name=News&file=article&sid=

Takeda Pacific Report #R183-013 (2005). Medical Biosensor Applications and Market for 2008-Diabetes Management, Point-of-Care and Related Applications, May, http://www.mindbranch.com/products/R183-013.html

TechFinance.ca (2005). Adnavance Technologies Closes $3.85 Million Series A Financing, November 29, http://www.techfinance.ca/m-topnews+news+tnid-20051129.html

The-Infoshop (2002). Report on "Biosensors and Bioelectronics," http://the-infoshop.com/study/bc11416_biosensors_toc.html

The Venture Planning Group (2005). Biosensors 2006: U.S., Europe, and Japan, 350 Fifth Avenue, Suite 3304, New York, NY, 10118, U.S.A.

Vogel, H., and Zinkl, W. (1997). Biosensors in Switzerland: General Conditions for the Development of a Biosensor Industry in Switzerland, http://www.ta-swiss.ch/www-remain/reports_archive/publications/1997/ta_26_97_e.htm

Weibring, P., Edner, H., and Svanberg, S. (2003). Versatile Mobile LIDAR system for environmental monitoring. *Applied Optics*, **42**(18), 3583–3594.

Work Force Survey Results-CA (California) Needs (2001). The Current and Future Worforce needs of California's Biotechnology Industry, California State University Program for Education and Research in Biotechnology, November, http://www.csuchico.edu/superb/WkFcSurvResults_CA_Needs.html

Yeates, E. (2005). Device Could be Used to Detect Bio-Terrorism Pathogens, KSL News, http://tv.ksl.com/index.php?nid=5&sis=217245

Zarkoff, J. (2002). Promising Growth in Biosensor and Antibody Markets. Medicaldevicelink, November 2002, http://www.devicelink.com/ivdt/archive/02/11/008.html

Index